Functional and Ecological Xylem Anatomy

Uwe Hacke
Editor

Functional and Ecological Xylem Anatomy

Springer

Editor
Uwe Hacke
Department of Renewable Resources
University of Alberta
Edmonton, AB, Canada

ISBN 978-3-319-15782-5 ISBN 978-3-319-15783-2 (eBook)
DOI 10.1007/978-3-319-15783-2

Library of Congress Control Number: 2015934905

Springer Cham Heidelberg New York Dordrecht London

Printed on acid-free paper

Springer International Publishing AG Switzerland is part of Springer Science+Business Media (www.springer.com)

Preface

Trees and wood have fascinated people for centuries, and this book was born out of this fascination. Trees provide a habitat to countless birds and other animals that live in them. Trees provide us with oxygen, timber, shade, and shelter, and they absorb large amounts of carbon from the atmosphere. Similar statements can be made about the other plant groups and growth forms covered in the book.

Vascular plants, whether woody or not, have remarkably complex conducting tissues. The sugars made by photosynthesis as well as many other molecules are distributed in the phloem. Water and nutrients are transported in the xylem. Phloem and xylem, though fundamentally different in the way they function, are in close proximity to each other, and together they connect plant organs that are often quite distant. According to the cohesion-tension theory, water ascends plants in a metastable state under tension (negative hydrostatic pressure). This transport mechanism has profound implications for the structure of the xylem, as explained in several of the book chapters. In woody plants, xylem also provides structural support to the plant body and serves in storage. Water transport and leaf water supply are vitally important for the sustained growth of forests, agricultural crops, and horticultural plants. Globally, the amount of water moving through plant xylem is enormous. Transpiration represents by far the largest water flux from Earth's continents.

Many of my fellow contributors and I have been inspired by Martin Zimmermann's book *Xylem Structure and the Ascent of Sap* and by Melvin Tyree and M. Zimmermann's 2nd edition of the book, which was published in 2002. While the present book contains many ideas and concepts originally described in *Xylem Structure and the Ascent of Sap*, it is different in terms of its organization and content. A large amount of experimental data is synthesized. New and emerging topics are covered.

The overall objective of the book is to explain what is currently known about functional and ecological xylem anatomy. In addition to serving as a source of information to professionals, instructors, and advanced students in plant science and forestry, I hope that the inherent beauty of xylem that is apparent in many images will be a source of inspiration to readers who are not yet familiar with the topic. The book offers a unique combination of scientific insight based on thorough experimental work and beautiful images, which often speak for themselves. The beauty of xylem is evident at different scales; from views of striking tree ring patterns and vascular networks in fern fronds, conifer needles, and angiosperm leaves, to microscopic images of developing pits. As we continue to focus on increasingly small structures that are relevant for water transport in plants, we may eventually arrive at aquaporins, membrane-based water channels that impact water flow in living cells of xylem and phloem, and in other tissues. In short, it is my hope that the book will not only be a source of information, but also of inspiration.

The book begins with a discussion of the xylem of seedless vascular plants by J. Pittermann and co-workers. Ferns and lycophytes have existed for 400 hundred million years, yet we know relatively little about the vascular attributes of these ancestral plants. The following chapters describe the ecological xylem function of conifers, lianas, and drought-adapted chaparral species. The xylem of woody model systems (poplar and grapevine) is discussed in detail.

Each of these plant groups has distinct xylem features, with regard to the size of vessels and tracheids, the way in which these conduits are connected, and in terms of pit structure. Pits connect adjacent conduits to allow water flow while also blocking the spread of gas from embolized conduits. Numerous studies have focused on the function of pits, but less is known about their development. R. Dute synthesizes what we currently know about the development, form, and function of torus-bearing pit membranes in seed plants.

Several chapters explore how wood anatomy and hydraulic traits relate to a plant's ability to cope with drought. As temperature and precipitation regimes are changing, drought-induced tree mortality has been observed in many parts of the Earth in recent decades, and is now a major research topic (see Chap. 9 by W. Anderegg and F. Meinzer).

Plants respond to different and changing environments by producing distinct phenotypes within their life span. Understanding plastic responses of hydraulic traits may be useful in predicting the impacts of climate change on woody plants. In the past, variation in xylem properties has been mostly studied across different species, but we are now learning more about how the xylem of a single species can adjust to different growing conditions.

Many of the anatomical images show that vessels and tracheids do not exist in isolation. Conduits specialized in axial transport of water are intimately connected with living parenchyma cells. As explained by L. Plavcová and S. Jansen, parenchyma cells play important roles in the xylem. They provide opportunities for radial transport between tissues, they store nonstructural carbohydrates, and they facilitate the exchange of water and solutes between the apoplast and symplast.

Our understanding of whole-plant hydraulic architecture would remain incomplete without considering leaves. The leaf represents a major bottleneck in the whole-plant hydraulic system; leaf hydraulic conductance is variable and it influences photosynthetic gas exchange. L. Sack et al. evaluate the contributions of the vein system and the outside-vein system to leaf hydraulic conductance and to its dynamics with leaf water status.

The chapters are written by leading international scholars; I sincerely thank them for joining the project despite many competing demands on their time. I thank Eric Stannard from Springer for inviting me to edit the book, and for his continued support. Finally, thanks to my wife Irene and my parents for their support.

Edmonton, AB
December 2014

Uwe Hacke

Contents

Contributors

William R.L. Anderegg Princeton University, Princeton, NJ, USA

Alex Baer Department of Ecology and Evolutionary Biology, University of California, Santa Cruz, CA, USA

Craig Brodersen Yale University, New Haven, CT, USA

Timothy J. Brodribb School of Plant Science, University of Tasmania, Hobart, TAS, Australia

Thomas N. Buckley IA Watson Grains Research Centre, The University of Sydney, Narrabri, NSW, Australia

Katharine L. Cary Department of Ecology and Evolutionary Biology, University of California, Santa Cruz, CA, USA

Jean-Christophe Domec Bordeaux Sciences, Gradignan, France

Roland R. Dute Auburn University, Auburn, AL, USA

Frank W. Ewers California State Polytechnic University, Pomona, CA, USA

Uwe G. Hacke Department of Renewable Resources, University of Alberta, Edmonton, AB, Canada

Anna L. Jacobsen Department of Biology, California State University, Bakersfield, CA, USA

Steven Jansen Institute of Systematic Botany and Ecology, Ulm, Germany

Daniel M. Johnson Department of Forest, Rangeland, and Fire Sciences, University of Idaho, Moscow, ID, USA

Barbara Lachenbruch Department of Forest Ecosystems & Society, Oregon State University, Corvallis, OR, USA

Tamani F. Lee Department of Biology, California State University, Bakersfield, Bakersfield, CA, USA

Justin A. Martinez Department of Biology, California State University, Bakersfield, CA, USA

Stefan Mayr Department of Botany, University of Innsbruck, Innsbruck, Austria

Frederick C. Meinzer USDA Forest Service, Pacific Northwest Research Station, Corvallis, OR, USA

Mark E. Olson Instituto de Biología, Universidad Nacional Autónoma de México, Mexico DF, Mexico

Marta I. Percolla Department of Biology, California State University, Bakersfield, Bakersfield, CA, USA

Jarmila Pittermann Department of Ecology and Evolutionary Biology, University of California, Santa Cruz, CA, USA

Lenka Plavcová Institute of Systematic Botany and Ecology, Ulm, Germany

R. Brandon Pratt Department of Biology, California State University, Bakersfield, Bakersfield, CA, USA

F. Daniela Rodriguez-Zaccaro California State University, Bakersfield, Bakersfield, CA, USA

Julieta A. Rosell Dept. Ecología de la Biodiversidad, Instituto de Ecología, Universidad Nacional Autónoma de México, Mexico DF, Mexico

Lawren Sack Department of Ecology and Evolutionary Biology, University of California, Los Angeles, CA, USA

Eric Schuettpelz Department of Botany, National Museum of Natural History, Smithsonian Institution, Washington, DC, USA

Paul J. Schulte School of Life Sciences, University of Nevada, Las Vegas, NV, USA

Christine Scoffoni Department of Ecology and Evolutionary Biology, University of California, Los Angeles, CA, USA

Alan R. Smith University of California, Berkeley, CA, USA

Hayden S. Toschi California State University, Bakersfield, Bakersfield, CA, USA

Jessica Valdovinos California State University, Bakersfield, Bakersfield, CA, USA

James E. Watkins Department of Biology, Colgate University, Hamilton, NY, USA

Chapter 1
The Structure and Function of Xylem in Seed-Free Vascular Plants: An Evolutionary Perspective

Jarmila Pittermann, James E. Watkins, Katharine L. Cary, Eric Schuettpelz, Craig Brodersen, Alan R. Smith, and Alex Baer

1 Introduction

The evolution of xylem was critical to the success of embryophytic land plants. According to the fossil record, the earliest terrestrial autotrophs were algae followed by bryophytes, simple organisms that lacked the ability to efficiently transport water (Raven 1984; Sperry 2003; Taylor et al. 2009; Pittermann 2010). Like their ancestors, today's non-vascular plants have essentially two options for survival: inhabit perennially wet substrates such as stream sides or exploit poikilohydry, the capacity to recover from near complete desiccation (Niklas 2000; Proctor and Tuba 2002; Sperry 2003). In the absence of xylem, non-vascular plants are subject to the vagaries of

J. Pittermann, Ph.D. (✉) • K.L. Cary • A. Baer
Department of Ecology and Evolutionary Biology, University of California,
Santa Cruz, CA 95064, USA
e-mail: jpitterm@ucsc.edu; kcary@ucsc.edu; baeranoff@gmail.com

J.E. Watkins
Department of Biology, Colgate University, Hamilton, NY 13346, USA

E. Schuettpelz
Department of Botany National Museum of Natural History, Smithsonian Institution,
P.O. Box 37012, Washington, DC 20013-7012, USA
e-mail: schuettpelze@si.edu

C. Brodersen
School of Forestry and Environmental Studies, Yale University,
195 Prospect Street, New Haven, CT 06511, USA
e-mail: craig.brodersen@yale.edu

A.R. Smith
University Herbarium, University of California,
1001 Valley Life Sciences Building #2465, Berkeley, CA 94720-2465, USA
e-mail: arsmith@berkeley.edu

U. Hacke (ed.), *Functional and Ecological Xylem Anatomy*,
DOI 10.1007/978-3-319-15783-2_1

their habitat and remain small because rehydration is a rapid, metabolically expensive process that supports the rapid recovery of photosynthesis (Proctor and Tuba 2002). Over time, the need for effective spore dispersal combined with increased competition for light selected for progressively taller, homoiohydric vascular plants, peaking with evolution of hydraulically efficient woody xylem across both seed-bearing and extinct seed-free lineages (Niklas 2000; Niklas and Speck 2001; Sperry 2003; Wilson 2013).

Much is known about the structure and function of xylem tissue in trees and shrubs because they are economically important, ecologically dominant, and in the case of angiosperms, highly diverse. However, the xylem in seed-free vascular plants such as ferns and lycopods demands equal scrutiny because (1) ferns and lycophytes have persisted for well over 400 million years—no small feat considering several extinction events, changing climates and the evolution of angiosperm-dominated tropical forests (Rothwell and Stockey 2008; Taylor et al. 2009; Watkins and Cardelus 2012) and (2) many species play critical roles in ecosystem processes and may even control forest growth and regeneration (George and Bazzaz 1999; Watkins and Cardelus 2012). Today, seed-free vascular plants are found on nearly every continent and in a variety of habitats ranging from xeric or salt-water niches up to the tropical forest canopies, which are often dominated by epiphytic ferns (Moran 2008; Watkins and Cardelus 2012). In north temperate regions most ferns are seasonal and terrestrial but in warmer climates, the fern morphospace includes vines, tree ferns, epiphytes, aquatic and floating ferns (Ranker and Haufler 2008; Mehltreter et al. 2010). Boasting an impressive 12,000+ species, the fern lineage is the second most diverse only after the angiosperms (Ranker and Haufler 2008). Despite their long history and ecological breadth, little is known about the vascular attributes of these ancestral plants (Pittermann et al. 2013). The goals of this chapter are to provide an overview of xylem structure and function in modern ferns, and to consider the selection pressures that lead to observed macroevolutionary patterns of xylem structure in extinct and extant seed-free vascular plants. The discussion will be restricted to sporophytic shoots and leaves since these structures are the functional equivalents to stems and leaves in seed plants.

2 Vascular Tissues in Stems and Leaves of Seed-Free Vascular Plants

The fossil record shows that the vascular tissues of seed-free vascular plants appeared prior to those of conifers and angiosperms. Because extant pteridophytes do not produce wood, their primary xylem and phloem tissues are packaged in discrete vascular bundles (meristeles) that are surrounded by an endodermis and often a waxy, circumendodermal band (see stele and vascular patterns in Figs. 1.1 and 1.2). The shape and arrangement of these bundles differs among genera; even within bundles the xylem can adopt a variety of patterns. In contrast to xylem of woody seed plants, which mechanically supports the canopy as well as transports water, the primary xylem of

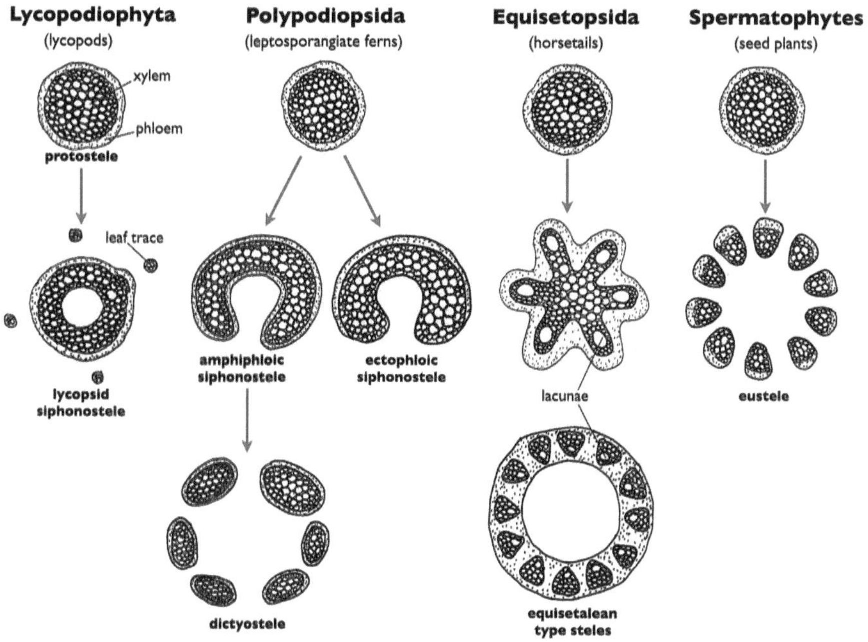

Fig. 1.1 Stem and rhizome stelar patterns in vascular land plants. All fossil and extant vascular arrangements have their origin in the protostele, the simplest of steles. Drawings based on illustrations in Beck (2010) and Taylor et al. (2009)

ferns and lycophytes serves solely for the purpose of water transport (Pittermann et al. 2011, 2013). Despite their developmental canalization, extant pteridophyte shoots exhibit a surprising diversity of vascular arrangements, suggesting that evolution may have acted on fern xylem in response to numerous selection pressures.

2.1 Vascular Arrangement

The primary transport tissues are arranged in central zones of roots or stems known as steles. Stelar theory has traditionally pertained to stems and roots (Beck et al. 1982) but in this discussion, the nomenclature is applied to the arrangement of primary tissues in the main axes of fern leaves (fronds) and less frequently, the stems of lycophytes. One or several vascular traces, each containing xylem and phloem, can constitute a stele and these bundles may be variously situated in ground parenchyma or sclerenchyma tissues (Beck 2010; see Figs. 1.1 and 1.2a–h). Of extant and fossil stelar patterns, the protostelic arrangement characteristic of lycopods is considered the most ancestral (Fig. 1.1; Beck et al. 1982; Taylor et al. 2009). In what is thought to be the very first protostele (also termed a haplostele), the xylem is in the centre of the vascular cylinder and is surrounded by the phloem (Beck et al. 1982). In contrast,

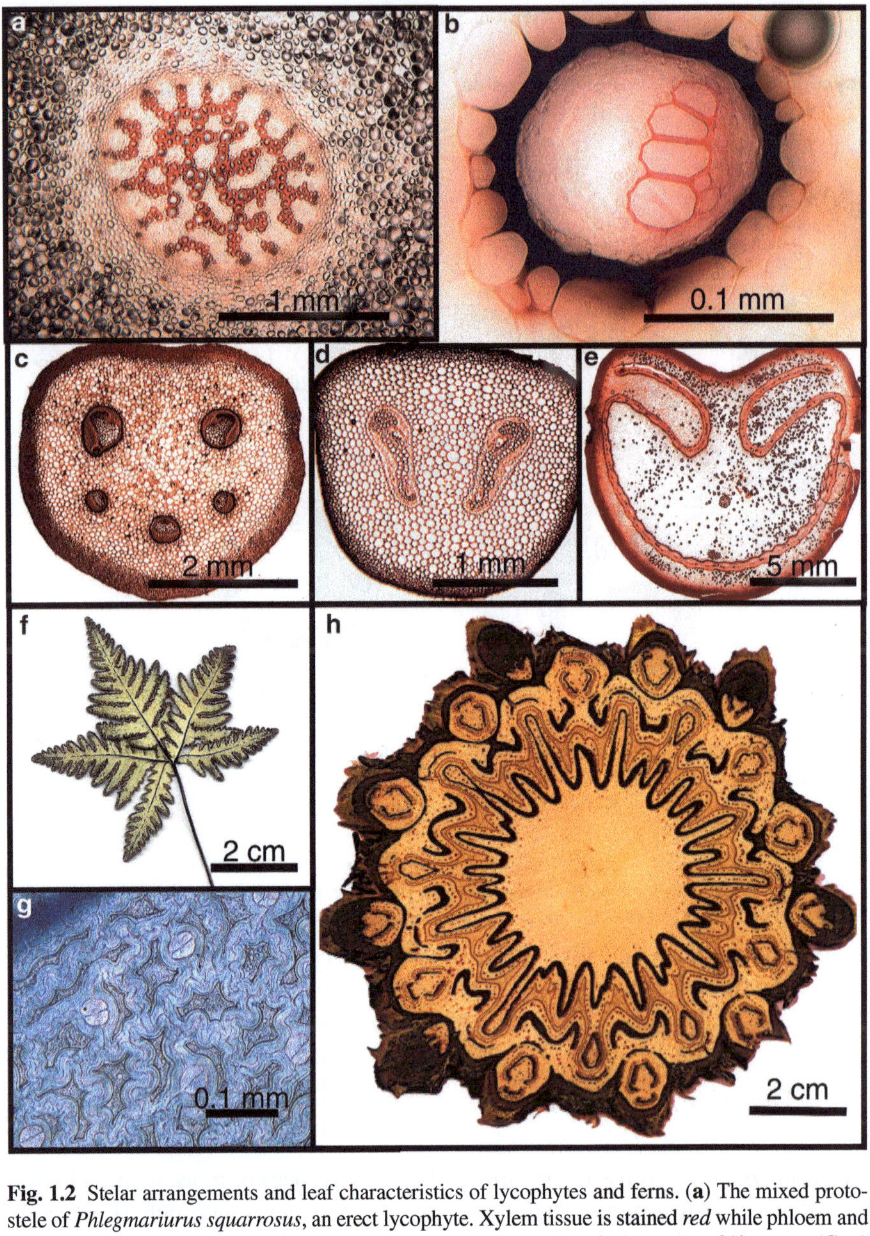

Fig. 1.2 Stelar arrangements and leaf characteristics of lycophytes and ferns. (**a**) The mixed protostele of *Phlegmariurus squarrosus*, an erect lycophyte. Xylem tissue is stained *red* while phloem and parenchyma remain unstained. (**b**) A close-up on a vascular bundle of *Platycerium bifurcatum* (Cav.) C. Chr. showing *red-stained* xylem and clear phloem surrounded by an endodermis with a thick circumendodermal band. (**c**) The dictyostele of *Dryopteris arguta* (Kaulf.) Watt comprising two large vascular bundles and three smaller auxillary ones; the fibrous, peripheral sterome is also stained *red*. (**d**) The dictyostele of *Athyrium filix-femina* (L.) Roth. (**e**) An omega-shaped vascular strand in *Pteris livida* Mett. (**f**) The underside of a frond of *Notholaena standleyi* Maxon, a desiccation-tolerant cheilanthoid fern found in rocky, dry outcrops of the American Southwest; the underside of the leaf is covered with a pale yellow farina that is exposed when the leaf curls while desiccating. (**g**) Stomatal and epidermal patterns on the underside of pinnae in *Pellaea truncata* Goodd., a dry adapted cheilanthoid fern with sclerophyllous leaves. (**h**) Cross section of the trunk of *Dicksonia antarctica* Labill. showing sclerenchyma tissue (*dark brown*) adjacent to a contiguous strand of primary xylem (*dark beige*); leaf traces are evident on the periphery of the trunk. (Panel **h** by Christopher Rico)

the more derived protostele of *Phlegmariurus squarrosus* (G. Forst. Á. Löve and D. Löve; Huperziaceae) exhibits mixed elements of xylem and phloem (see micrograph in Fig. 1.2a; Beck et al. 1982). In the rhizomes and shoots of many lycopods and leptosporangiate ferns, the protostele appears to have given rise to the more derived siphonostele arrangement, which is a hollow or parenchyma-filled cylinder with ectophloic (external) phloem tissue (Fig. 1.1; Beck et al. 1982; Beck 2010). In typical siphonosteles, a diverging leaf trace is evident as an interruption of the cylinder, giving it the resemblance of an arch, but in the absence of leaf traces, the siphonosteles form complete cylinders with phloem that can be on both sides of the xylem (amphiphloic) or simply on the external face (ectophloic) (Fig. 1.1; Beck 2010). The dictyostele is the most dissected and derived of the leptosporangiate fern steles, formed by numerous vascular strands that may or may not come into contact over the length of the rhizome or the frond axis (Figs. 1.1 and 1.2b–d). Unusual among extant steles, Equisetalean stelar arrangements incorporate characteristic air-filled lacunae (Fig. 1.1). Lastly, the primary stem tissues of seed plants form the characteristic eustele, a cylindrical arrangement of numerous vascular bundles (Fig. 1.1).

The three-dimensional arrangement of the vascular network is much less well characterized than the stelar arrangements in transverse sections. The detailed representations of Ogura (1972) and Beck et al. (1982) were probably created from numerous cross sections and careful dissections, but the imaging methods pioneered by Zimmermann and Tomlinson (1974) and Zimmermann and Tyree (2002) improved both efficiency and precision by using sequential photographs of serial stem cross sections to assemble, frame by frame, three-dimensional reconstructions of vessel or vascular bundle arrangements. This cinematographic approach was limited to small pieces of tissue but it provided important insight into the longitudinal and three-dimensional positioning of vascular tissues in both woody and monocot plants (Zimmermann and Tyree 2002). White and Weidlich (1995) used it to document the longitudinal, interdigitated pattern of meristeles in semi-arborescent members of the fern genera *Diplazium* and *Blechnum*, demonstrating that comparable phyllotaxis has evolved convergently in ferns and seed plants. A newer method in which resins and polymers are directly injected into xylem tissue has proven extremely useful for understanding angiosperm vessel networks and has been especially revealing about vessel-to-vessel pit connections (Mauseth and Fujii 1994; Kitin et al. 2004). This approach is also valuable for measurements of conduit length in seed-free vascular plants because tracheids seed-free vascular plants because tracheids are often longer than a few millimetres and frequently resistant to standard solutions of hydrochloric acid and hydrogen peroxide, making tissue macerations problematic (Wheeler et al. 2005; Pittermann et al. 2011).

Newer methods may overcome some of the problems associated with xylem network imaging. Recently, high resolution computed tomography (HRCT) along with straightforward manual frond xylem extractions have revealed interesting differences in the vascular organization of ferns (Brodersen et al. 2012). The HRCT method relies on a synchrotron-generated source of focused X-rays to provide micrometre-range resolution of vascular tissues in three dimensions (Brodersen et al. 2011) and can be used for visualization of water movement and droplet formation within the xylem as well as for reconstruction of tissue arrangements (Brodersen et al. 2010, 2012). Unlike serial sectioning and macerations,

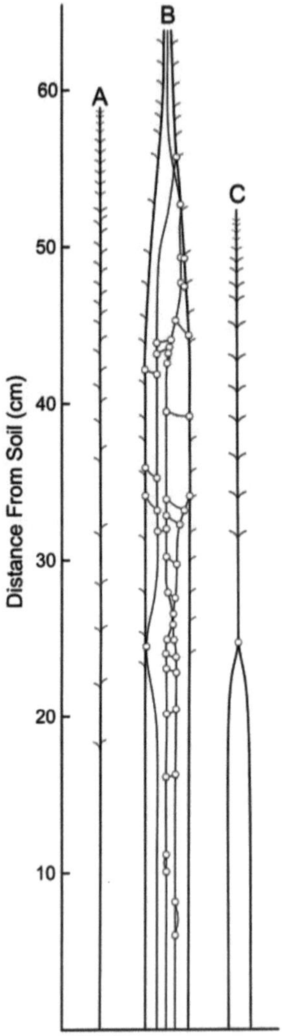

Fig. 1.3 Vascular arrangements along the frond primary axis in *A Osmundastrum cinnamomeum* (L.) C. Presl, *B Polystichum acrostichoides* (Michx.) Schott, *C Onoclea sensibilis* L. Open circles represent the location of connections between vascular bundles

HRCT is non-destructive, but at present, processing time and computing power preclude the imaging of tissues longer than a few centimetres.

Dissections and meristele extractions have revealed that protosteles, such as those in the lycopod *Phlegmariurus squarrosus* and the whisk fern *Psilotum nudum* (L.) P. Beauv. (Psilotaceae) remain invariable along the shoot axis (Pittermann unpublished data), but similar patterns can be seen in *Osmundastrum cinnamomeum* (L.) C. Presl (Osmundaceae) in which a C-shaped stele remains intact over the length of the frond (see xylem map in Fig. 1.3). In more derived pteridophyte clades, the number of vascular bundles as well as their frequency of contact may vary considerably. For example, a 110-cm-long frond of *Pteridium aquilinum* (L.) Kuhn (Dennstaedtiaceae) has anywhere from 23 to 29 vascular bundles at the base of the stipe and between 3 and 6 bundles at the tip of the rachis, with well over 100 bundle connections throughout the main axis of the frond (Brodersen et al. 2012). This is an intricate network and probably an extreme example of stelar dissection. By contrast, only 4–5 basal bundles fuse to a single strand at the tip of the rachis in the simpler network of *Woodwardia fimbriata* Sm. (Blechnaceae) and *Polystichum acrostichoides* (Michx.) Schott (Dryopteridaceae; Brodersen et al. 2012). There is some variation between the base of the frond (petiole) and the rachis region which gives rise to the pinnae (Fig. 1.3). Lateral veins emerge from the large, main vascular strands in the rachis of all three species but network complexity increases in *P. acrostichoides*, whereas it simplifies in the more derived *Onoclea* (Onocleaceae). How selection drives xylem network variation is currently unclear.

2.2 *Xylem and Phloem Ultrastructure*

Phloem function is closely coupled with water transport, and Münch's pressure flow hypothesis posits that mass flow of phloem sap depends in part on water influx from adjacent xylem cells (Pickard and Abraham-Shrauner 2009). Recent models of leaf gas exchange show plant productivity to be coupled to the dynamics of xylem and phloem pressure gradients, and phloem has also been implicated in the recovery of hydraulic function after a water deficit (Salleo et al. 2004; Nikinmaa et al. 2014). The rapid wounding response of phloem makes it a difficult tissue to study but the use of aphid stylets, scanning electron microscopy and models has enabled insight into its functional attributes (Gaupels et al. 2008; Mullendore et al. 2010; Jensen et al. 2012). Unfortunately, studies on phloem tissue in seed-free vascular plants are rare. Early studies of *Equisetum* revealed that the phloem tissue comprises sieve elements and perhaps also parenchyma cells (Agashe 1968). In *Blechnum orientale* L. (Blechnaceae), sieve elements have sieve areas with numerous pores and possibly callose tissue but the presence of companion cells is not clearly established (Fotedar and Shah 1975). Clearly, phloem function in early-derived plants deserves more study in contemporary plant biology.

The xylem structure of seed-free vascular plants has received considerably more attention than phloem. The tracheids of ferns may be substantially longer and wider than those of conifers, reaching widths well over 100 μm and lengths exceeding 1 cm (Veres 1990; Pittermann et al. 2011; see conduit diameter distributions in Fig. 1.4). By comparison, conifer tracheids rarely exceed 40 μm in diameter and 2 mm in length, particularly in stems (Pittermann et al. 2006), with narrower, shorter tracheids found in stiffer wood (Panshin and de Zeeuw 1980). In the absence of a mechanical function, fern tracheids serve solely to transport water, and thus exhibit none of the hydraulic constraints imposed by smaller lumina, relatively thicker walls, and short cells evident in conifer tracheids (Pittermann et al. 2006; Sperry et al. 2006). Several fern genera are believed to have vessel elements in their rhizomes (*Astrolepis* (Pteridaceae), *Marsilea* (Marsileaceae), *Woodsia* (Woodsiaceae) and *Pteridium*; Carlquist and Schneider 2007), with silicon injections confirming vessels in excess of 20 cm in length in the petiole of *Pteridium aquilinum* (Wheeler et al. 2005; Pittermann et al. 2011).

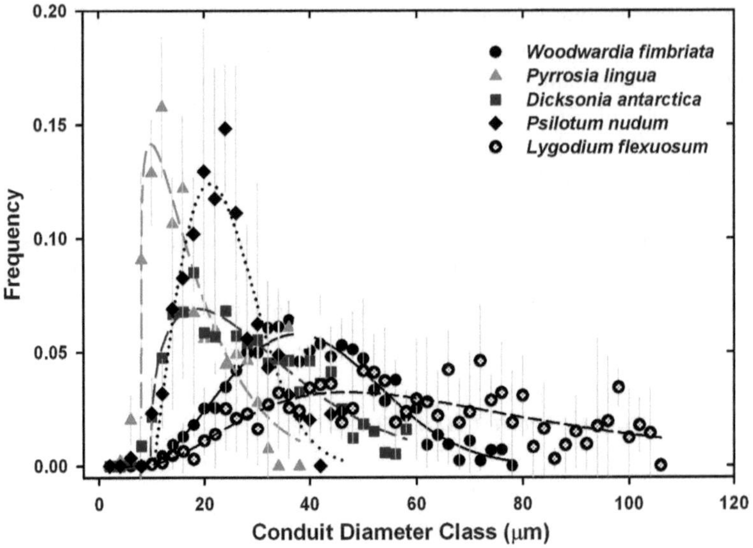

Fig. 1.4 Conduit diameter frequencies in five fern species (reprinted from Pittermann et al. 2011)

Conduits must be perforated by inter-conduit pit membranes to allow water to flow from one conduit to another. As in other vascular attributes, the pit membrane structure in seed-free vascular plants differs substantially from that of conifers and angiosperms. Angiosperms possess what is known as a homogenous pit membrane, which is a partially digested primary wall protected by an overarching, secondary wall border (Choat and Pittermann 2009). These membranes typically occupy radial vessel walls and may be grouped near vessel-to-vessel junctions or located sporadically along the full length of the vessel wall. Conifers, on the other hand, possess a torus–margo pit membrane in which water moves through the porous margo region while the torus, a thickening in the centre of the membrane, serves to protect conduits from the spread of air throughout the xylem (Choat and Pittermann 2009; see Chaps. 2 and 3). The vast majority of ferns and lycopods possess homogenous pit membranes that span the length of the tracheid in a scalariform arrangement and are located on nearly all the tracheid walls (see micrographs in Fig. 1.5a, b). Ranging in thickness from 150 to 350 ηm (Brodersen et al. 2014), pteridophyte pit membranes appear less variable than those of angiosperms, in which range from 70 to 1,892 ηm (Jansen et al. 2009). Similar pit membrane arrangements have been observed in seed-free vascular plants since the Early Devonian (Kenrick and Crane 1997; Taylor et al. 2009). Interestingly, torus–margo pit membranes are present in ferns, but only in the anomalous secondary growth in rhizomes in the genus *Botrychium* (Ophioglossaceae; Morrow and Dute 1998; Rothwell and Karrfalt 2008). The presence of conifer-like pit membranes in *Botrychium* as well as bordered pits in *Psilotum* indicates that ferns have the genetic potential to evolve xylem that is functionally comparable to higher plants, but either selection acted on attributes of fern physiology or life history that override the importance of pit membranes, or the genetic or developmental capacity to capitalize on these traits is absent.

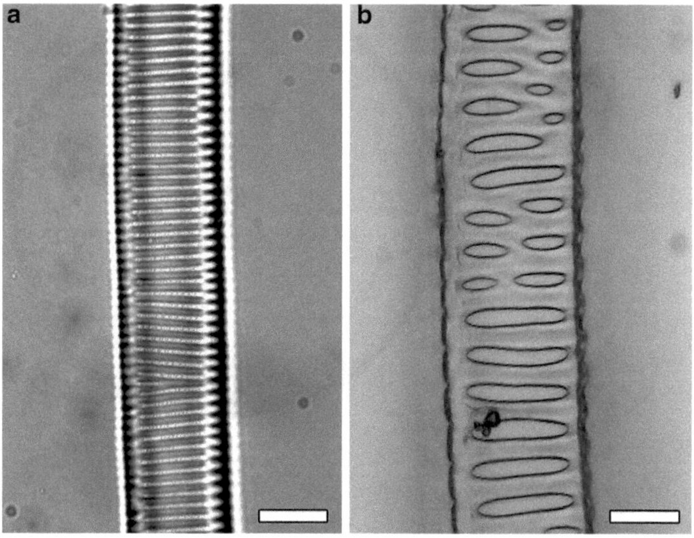

Fig. 1.5 Tracheids of (**a**), *Selaginella pallescens* (C. Presl) Spring (air-seed pressure=3.8 MPa) and (**b**), *Phlebodium aureum* (L.) J. Sm. (air-seed pressure=2.12 MPa). Tracheid walls are perforated by scalariform (*S. pallescens*) and reticulate (*P. aureum*) pit membranes. Scale bar=5 µm

3 The Xylem of Seed-Free Vascular Plants in the Fossil Record

The fossil record opens a window into the fascinating morphological diversity of early land plants, and also into some unusual excursions in plant structure and function. Many of these botanical experiments have failed but others have persisted into the modern flora for reasons that may have little to do with physiology. Aside from physiological attributes, ecological, reproductive and other life history traits certainly have played a role in taxon persistence. In fact, the disappearance of variously woody progymnosperms, sphenopsids and isoëtopsids demonstrates that having secondary xylem is no guarantee against extinction.

3.1 Xylem in Early Land Plants

Fossil spore and possibly micro-fossil records suggest that vascular plants made their appearance in the Silurian (Niklas and Smocovitis 1983; Kenrick and Crane 1997; Taylor et al. 2009), with the Lower Devonian giving rise to a diversity of plants with transport tissues (Niklas 1985). Of these proto-vascular plants, the Rhyniophytes (so-named for the Early Devonian Rhynie Chert locality) are probably the best characterized. Genera such as *Rhynia*, *Aglaophyton* and *Cooksonia* are rhizomatous, with dichotomizing, stomatal-bearing shoots up to 18 cm tall,

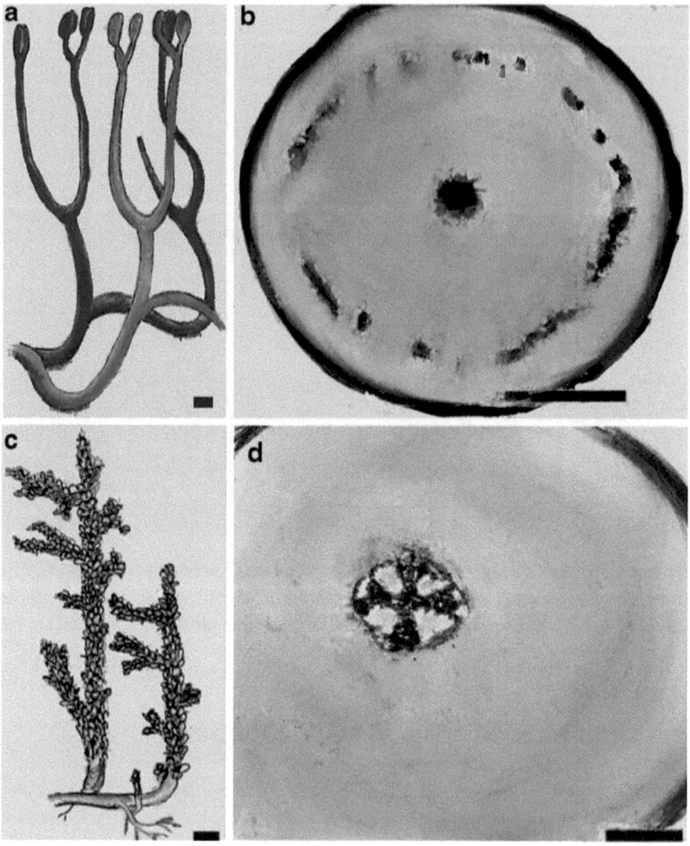

Fig. 1.6 Reconstructions and stem cross sections *Aglaophyton major* (**a** and **b**) and *Asteroxylon mackiei* (**c** and **d**). Scale bars equal 1 cm in the plant reconstructions and 1 mm in the cross sections. Figure adapted from Boyce et al. (2003)

possessing variously developed central conducting strands (Edwards 1986; Taylor et al. 2009; see illustrations of *Aglaophyton* in Fig. 1.6a, b). The vascular tissues of *Rhynia gwynne-vaughanii* and *Aglaophyton major* are described as hybrids between the hydroids of bryophytes and the xylem of more derived plants (Edwards 1986; Kenrick and Crane 1991) but chemical analysis of fossil material suggests that the vascular tissue was not lignified (Boyce et al. 2003). Bearing a close resemblance to bryophytes, *A. major* is categorized as a pro-tracheophyte (see Boyce et al. 2003; Fig. 1.6a, b), with *R. gwynne-vaughanii* falling closest to the next most-derived group, the tracheophytes (Kenrick and Crane 1997; Boyce et al. 2003). *Asteroxylon mackiei*, a more derived eutracheophyte from the Lower Devonian Rhynie Chert, has well-developed lignified primary xylem, the presence of which is consistent with the need to hydrate enations (small flaps of photosynthetic tissue) that cover its stems and increase its evaporative surface area (Boyce et al. 2003; Taylor et al. 2009; see illustrations of *Asteroxylon* in Fig. 1.6c, d).

The conduit structure of early land plants may have been more diverse than that of extant seed-free vascular plants. Moss hydroids are unlignified and, aside from plasmodesmatal-derived pores, they lack additional structural features (Kenrick and Crane 1997; Beck 2010). By comparison, conduits of more derived rhyniophytes show greater complexity and reinforcement against collapse. In *Rhynia*, for example, the transport cells have spongy helical wall thickenings (Kenrick and Crane 1991). In Early Devonian *Gosslingia* and *A. mackiei*, the tracheid walls comprise a two-layered cell wall with a chemically resistant inner layer (Kenrick and Crane 1991; Taylor et al. 2009). The P-type tracheids of *Psilophyton*, a Lower- to Mid-Devonian eutracheophyte, are probably the most structurally robust with what appears to be a higher wall fraction and degradation-resistant cell layer (Kenrick and Crane 1991, 1997). Taken together, the fossil record during the early Devonian trends toward an increase in xylem content and tracheid size, as well as tracheid reinforcement. Wall perforations resembling pit membranes become evident in the eutracheophytes as hydraulic demand must meet increasing plant size.

The Cladoxylopsids, early euphyllous fern-like plants with various degrees of xylem organization, make their appearance in the Middle to Upper Devonian, along with the more derived Rhachophytales, Stauropteridales and Zygopterids (extinct taxa with large, frond-like leaves; Taylor et al. 2009). Collectively, these groups appear to be related to ferns but it is unclear whether they are ancestral or sister taxa. Many of these plants have xylem arrangements that not featured in today's flora (Rothwell and Stockey 2008). For example, the star-shaped protostele of the Cladoxylopsids is defined as an actinostele and is unlike any extant stelar arrangement (Rothwell and Stockey 2008; Taylor et al. 2009; Beck 2010). Some specimens such as *Pseudosporochnus* may have been small and arborescent, but others such as *Eospermatopteris* and *Pietzschia schulleri* reached heights in excess of 8 m (Taylor et al. 2009). In Cladoxylopsids such as the Iriopteridales, the main axes measured over 5 cm in diameter, with cross sections showing dissected cross-shaped or bifurcating star-shaped protosteles that supported leaves with a single vascular trace (Taylor et al. 2009).

3.2 *Secondary Xylem in Seed-Free Vascular Plants*

Increased competition for light and other resources selected for taller plants with more developed canopies in the early evolution of eutracheophytes. Higher rates of transpiration necessitated increased hydraulic demand that was satisfied by greater xylem fraction and wider (and presumably longer) conduits. Fossils show that tracheid diameter increased by an order of magnitude during the Devonian from the 8 μm tracheids of the primitive *Cooksonia* to the 140 μm metaxylem conduits of *Stenomylon primaevum*, an early Carboniferous seed fern (Niklas 1985). Although early vascular plants such as *Rhynia* and *Asteroxylon* were herbaceous, the recent identification of true secondary xylem in putative relatives of *Psilophyton* and *Armoricaphyton* suggests that this trait evolved much earlier than originally thought (Gerrienne et al. 2011; Strullu-Derrien et al. 2014). It appears that *Psilophyton* had a bifacial vascular cambium in which cell initials divide both radially and tangentially.

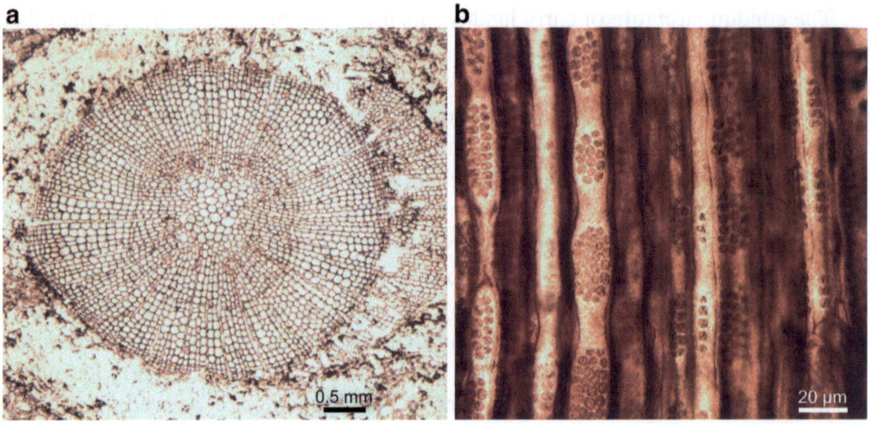

Fig. 1.7 Cross section of a root (**a**) and (**b**) a longitudinal section of stem tissue belonging to *Callixylon-Archaeopteris*, a Devonian progymnosperm. Clustered pit membranes with cruciform apertures are evident in panel **b**. (Material courtesy of Dr. Diane Erwin, UC Berkeley Museum of Natural History)

Secondary xylem, whether unifacial (only tangential cell divisions) or bifacial, is now believed to have evolved on at least five separate occasions, and by the Late Devonian, secondary vascular tissues were present in many dominant seed-free vascular plant lineages including the Lepidodendrales, Sphenopsids, pro-gymnosperms, the Zygopterids and Devonian–Carboniferous seed ferns (Barghoorn 1964; Beck 1970; Cichan 1985a, b; Sperry 2003; Taylor et al. 2009). Unifacial vascular cambia were present in arborescent lycopods such as the Lepidodendrales. Indirect analysis of xylem development suggests that cambial initials divided centrifugally with no anticlinal (radial) division, such that any increase in trunk circumference was accommodated by progressively larger fusiform initials (Cichan 1985a). Similar patterns of development probably occurred in the Carboniferous horsetail relatives including *Sphenophyllum* and *Arthropitys* species (Cichan 1985b).

Prior to the discovery of true secondary xylem in *Psilophyton*, the progymnosperms were thought to represent the earliest known class of plants with true wood derived from a bifacial cambium (Gerrienne et al. 2011). This group of plants possessed the unusual combination of coniferous, pycnoxylic wood with a canopy comprising free-sporing foliage, superficially similar to those of ferns (Beck 1970; Taylor et al. 2009). Growth rings, ray parenchyma and clustered, bordered pits with offset, cruciform apertures were present in both root and trunk xylem of the Archaeopteridales, of which the spore-bearing Devonian/Carboniferous *Archaeopteris-Callixylon* is best studied (Meyer-Berthaud et al. 1999; see micrographs of *Callixylon* xylem in Fig. 1.7a, b). Although it was seed-free, *Archaeopteris* was crowned "the Earth's first tree" because it demonstrates several developmental attributes of more derived plants, including perennial lateral branches (Meyer-Berthaud et al. 1999, 2000).

3.3 Fern Xylem in the Fossil Record

Ferns are functionally defined as megaphyllous plants with spores on the underside of their leaves (Rothwell and Stockey 2008, but see also Smith et al. 2006 for a more thorough treatment), but as *Archaeopteris* and *Psilophyton* demonstrate, specimens from the fossil record can certainly stretch this definition. For clarity, we define ferns as vascular, spore-bearing, megaphyllous plants with only primary growth. The fossil record shows that modern ferns, including the Psilotaceae, Ophioglossaceae, Equisetaceae, Marattiaceae and the leptosporangiate ferns have their origins in the Devonian (Rothwell and Stockey 2008). Although some extinct taxa exhibit stelar patterns that are unlike those of modern ferns today, the vasculature of the true ferns has remained relatively conserved over deep time.

Three major diversification events characterize the evolution of the leptosporangiate ferns, the first of these occurring during the climatically mild Carboniferous period (Rothwell and Stockey 2008). Fossilized petioles (stipes) of Botryopteridaceae and Kaplanopteridaceae fronds show variable tripartite and C-shaped steles surrounded by a thick, fibrous sterome layer (Rothwell and Stockey 2008). Similarly, fossil leaf scars of Marattialean tree ferns (representatives of which are extant in today's tropics) indicate a variety of stelar arrangements including C-shapes as well as more complex, possibly dictyostelic patterns. Stelar arrangements can vary along the length of the frond, but it is difficult to comment on the overall arrangement of the vascular strands with limited fossil material.

The Mesozoic gave rise to the second significant radiation of ferns, and included the Gleicheniales and Schizaeales, descendants of which are found in today's fern flora. Features of other Mesozoic taxa fall within the range of trait variation found in extant plants, as discovered in the Cretaceous remains of *Osmundastrum cinnamomeum*, in which the stipe stelar arrangement (a C-shape) resembles that of modern representatives (Serbet and Rothwell 1999). The steles in the petioles of modern leptosporangiates range from highly dissected, as in the Marattiaceae, to variously C-shaped; given the conserved anatomy of *Osmunda* over deep time, it may be reasonable to suppose that stelar arrangements in the other leptosporangiates with modern representatives are similarly conserved.

The third well-characterized fern radiation probably has origins in the upper Jurassic, with the diversification of the Eupolypod I and II clades (Schneider et al. 2004; Schuettpelz and Pryer 2009), even though most of the fossil diversity is apparent during the Cretaceous. Many taxa are now extinct but the fossil record suggests conserved dictyostelic vascular architecture in the petioles. For example, two large vascular strands resembling those of modern taxa are evident in Eocene specimens of Blechnaceae and Athyriaceae (Rothwell and Stockey 2008; Fig. 1.2c, d). Interestingly, this pattern is prevalent in the most recently derived fern clades, the Eupolypods I and II, a large fraction of which are epiphytic (Schneider et al. 2004; Schuettpelz and Pryer 2009). Characterizing and understanding the biotic and abiotic filters that selected for various vascular attributes over deep time will be a multidisciplinary undertaking requiring expertise in paleobotany, paleoecology, and plant structure and function.

4 Xylem Function in Modern Pteridophytes

Water transport in ferns and other seed-free vascular plants presents an interesting conundrum because in the absence of traits such a bifacial cambium, vessels and torus–margo pit membranes, one might suppose that seed-free vascular plants are at a competitive disadvantage relative to conifers and angiosperms (Pittermann et al. 2011, 2013). However, spore-bearing lineages have thrived since the Devonian, and their physiology certainly warrants a closer look. Indeed, the mechanisms by which ferns and lycophytes compensate for the absence or loss of these features makes for an interesting discussion because it bears not only on the limits of their so-called morphospace but also their evolutionary trajectory over time.

4.1 Hydraulic Efficiency in Ferns and Lycophytes

Of all the "missing" traits, the absence of a cambial layer in modern ferns and lycophytes is perhaps the most costly because it severely limits transport capacity and prohibits lateral branching. Consequently, ferns cannot develop complex canopies or occupy the massive ecological footprints of woody plants (Rowe et al. 2004; Rowe and Speck 2005). Ferns can, however, be dominant elements in tropical and subtropical systems in places such as Hawaii, where the tree fern *Sphaeropteris cooperi* (Hook. ex F. Muell.) R. M. Tryon (Cyatheaceae) has become invasive (Robinson et al. 2010). Native tree ferns are ecologically important in the South American and South Pacific floras where they frequently emerge after disturbances (Coomes et al. 2005; Robinson et al. 2010). In arborescent genera, such as *Dicksonia* and *Cyathea*, fibrous overlapping leaf traces form a several centimetre-thick sheath around the trunk and fibrous roots can also serve a supportive function (Sharpe and Mehltreter 2010; Fig. 1.2h). A second, less conspicuous constraint arising from the absence of wood is an inability to increase canopy size and complexity with age—the tree fern trunk simply elevates the canopy but the number of fronds remains relatively constant. There is little if any room for developmental flexibility when only a limited amount of primary xylem serves the canopy.

The absence of vessels in the vast majority of ferns precludes the leaps in hydraulic efficiency achieved by flowering plants (Brodribb and Feild 2010; Feild et al. 2011). Vessels evolved several times across several lineages including angiosperms, ferns, and possibly in several extinct taxa but for the vast majority of ferns, which are tropical, the xylem remains tracheid based (Calkin et al. 1985; Sperry 2003; Pittermann 2010; Watkins et al. 2010; Watkins and Cardelus 2012). *Pteridium aquilinum*, an aggressive weed, is a conspicuous exception: vessels have been identified in fronds on several occasions, making this species one of the most hydraulically efficient ferns amongst those surveyed (Wheeler et al. 2005; Pittermann et al. 2011; Brodersen et al. 2012). Unlike most pteridophytes, *P. aquilinum* is tolerant of both low light and full sun environments and can support transpiration and photosynthesis rates up to 0.5 mol m^{-2} s^{-1} and 15 µmol CO_2 m^{-2} s^{-1}, respectively, which are on par with those of angiosperms (Page 2002; Marrs and Watt 2006; Pittermann et al. 2011; Brodersen et al. 2012). Over 60 species of pteridophytes have managed to invade both intact and disturbed ecosystems, often outcompeting and even smothering native angiosperms and conifers (Robinson et al. 2010). The reasons behind the ferns' competitive edge are complex but a physiological approach may help explain their rapid rates of spread, as well as the mechanism by which these species push their xylem function and overall physiology beyond the norm. It is possible that vessels may be present in a greater number of fern species than we currently know.

Given that the vast majority of ferns have xylem comprised exclusively of tracheids, the absence of conifer type torus–margo pit membranes seems to add yet another handicap that pteridophytes failed to solve during the course of evolution. Torus–margo pit membranes are found primarily in conifer xylem and are about 60 times more permeable than the average angiosperm homogenous pit membrane (Pittermann et al. 2006; Sperry et al. 2006). This is because water moves from tracheid to tracheid through the highly porous margo region of the pit rather than the dense network of microfibrils and hydrogels characteristic of homogenous pit membranes (Sperry et al. 2006; Domec et al. 2008; Pittermann et al. 2010). Functionally, the increased permeability of torus–margo pit membranes compensates for xylem that consists of otherwise short, single-celled, high-resistance tracheids (Pittermann et al. 2006). If torus–margo pit membranes were replaced with an average angiosperm-type homogenous pit membrane, the tracheid-based vascular system would be 38× less efficient (Pittermann et al. 2005; Sperry et al. 2006).

If woody plants represent paragons of hydraulic optimization, it seems remarkable that ferns manage to transport any water all with what appears to be the least desirable combination of xylem traits. And yet, fronds of *Angiopteris evecta* (G. Forst.) Hoffm. (Marattiaceae) and *Woodwardia fimbriata* can exceed 1.5 m in length with little more than a few strands of primary xylem! The notion that ferns are constrained by primitive, inefficient xylem prevails because the bulk of published data has been generated on tropical taxa, which experience neither high evapotranspirative stress nor a high demand for water. For example, a survey of hydraulic function in tropical terrestrial and epiphytic ferns revealed K_s to be no higher than 8 kg MPa^{-1} m^{-1} s^{-1} (Watkins et al. 2010). These values are corroborated by gas exchange and leaf hydraulic data, which show that ferns barely achieve 20 % of the physiological capacity of angiosperms (Brodribb et al. 2007). However, fern habitats and ecological niches are nearly as varied as those of angiosperms, so when non-tropical taxa are considered, the picture of fern hydraulics changes dramatically. The K_s has consistently been found to be higher than 20 kg MPa^{-1} m^{-1} s^{-1} in the stipes of the seasonally deciduous *P. aquilinum*, the perennial *W. fimbriata* and the stems of climbing *Lygodium japonicum* (Thunb.) Sw. (Lygodiaceae; Pittermann et al. 2011, 2013; Brodersen et al. 2012). Figures 1.8 and 1.9 show numerous species that have K_s in excess of 10 kg MPa^{-1} m^{-1} s^{-1}, exceeding the hydraulic efficiency of temperate ring- and diffuse-porous species by at least a factor of 2 (McCulloh et al. 2010; Lens et al. 2011; see Chaps. 2 and 4). Temperate ferns should be sampled more intensively because these values may be the norm rather than the exception for seasonally deciduous species.

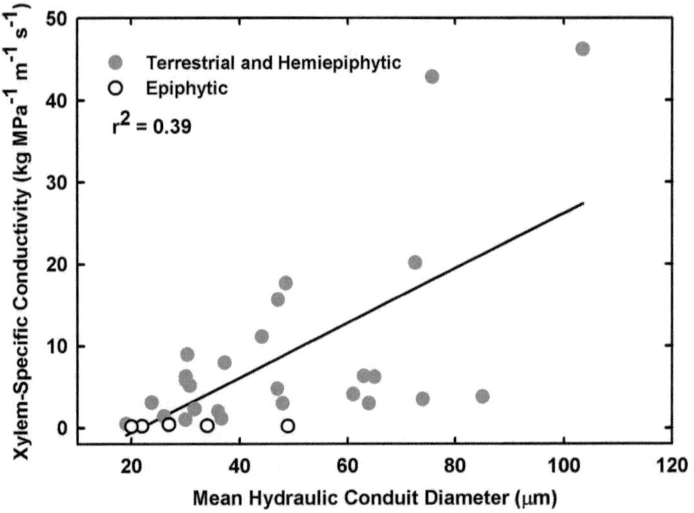

Fig. 1.8 Xylem-specific conductivity as a function of mean hydraulic conduit diameter in terrestrial, hemiepiphytic, and epiphytic ferns from tropical and temperate habitats. Data from Watkins et al. (2010) and Pittermann et al. (2011, 2013)

Three attributes may explain potential high specific conductivities in ferns. First, fern tracheids can explore a surprisingly broad anatomical and functional morphospace because shoot biomechanics are largely dependent on the sterome. Indeed, their central position in the stipe-rachis ensures that they occupy a mechanically neutral zone that is subject to neither the tension nor the compressive stresses imposed on the hypodermal sterome by external loads (Niklas 1992). Hence, species' hydraulic conduit diameters range from 20 to 100 μm, exceeding the tracheid and vessel sizes of several tropical and temperate woody taxa (see fern tracheid dimensions in Figs. 1.4 and 1.10; Veres 1990; McCulloh et al. 2010). Hydraulic efficiency of conduits scales to the fourth power of the lumen diameter (Tyree et al. 1994), so even a modest number of large tracheids can disproportionately contribute to the overall transport capacity to the xylem. A second important consideration is conduit arrangement within the fern vascular bundle. Despite devoting less than 30 % of frond cross-sectional area to xylem tissue, hydraulic efficiency is achieved in part by the close packing of conduits within the vascular bundles, much like tracheids in conifer xylem (see fern xylem structure in Fig. 1.2b–d). In this way, fern xylem combines the best of both worlds: water moves through wide, long, angiosperm-sized conduits that are tightly packed like those of conifers (Pittermann et al. 2013). Lastly, pit membrane attributes also contribute to high K_s in seed-free vascular plants (see pit membranes in Fig. 1.5a). Several studies have shown that the permeability of pteridophyte pit membranes to water is surprisingly similar to

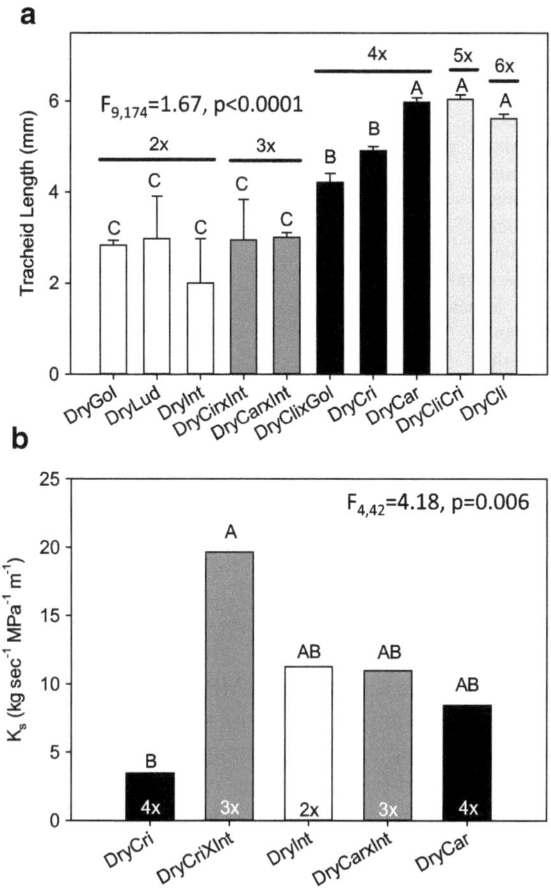

Fig. 1.9 (**a**) Variation in fern stipe tracheid lengths across taxa of hybrid origin and/or ploidy levels. DryGol: fertile diploid *Dryopteris goldiana*; DryLud: fertile diploid *Dryopteris ludoviciana* (Kunze) Small; DryInt: fertile diploid *Dryopteris intermedia* (Muhl. ex Willd.) A. Gray; DryCri × Int (also recognized as × *bootii*): Sterile triploid hybrid between *Dryopteris cristata* (L.) A. Gray and *D. intermedia*; DryCar × Int (also recognized as × *triploidea*): Sterile triploid hybrid between *Dryopteris carthusiana* (Vill.) H.P. Fuchs and *D. intermedia*; DryCli × Gol: fertile tetraploid hybrid between *Dryopteris clintoniana* (D.C. Eaton) Dowell and *D. goldiana*; DryCri: *D. cristata*, DryCar: *D. carthusiana*, both fertile tetraploids; DryCli × Cri: Sterile hexaploid hybrid between *D. clintoniana* and *D. cristata*; DryCli: Fertile heptaploid *D. clintoniana*. Data were taken from a mixture of herbarium specimens maintained at the Cooley Herbarium at Colgate University and from fresh material. In an attempt to control for environment, all taxa were collected or listed on their herbarium label as having been collected in New York state. (**b**) Xylem specific hydraulic conductivity (K_s) of the stipes of DryCri: *D. cristata*; DryCri × Int (also recognized as × *bootii*): Sterile triploid hybrid between *D. cristata* and *D. intermedia*; DryInt: *D. intermedia*, DryCar × Int (also recognized as × *triploidea*): Sterile triploid hybrid between *D. carthusiana* and *D. intermedia*; DryCar: *D. carthusiana*. Ploidy levels are given at the bottom of each bar. Data were taken from species collected in the field at the Huyck Preserve in Upstate New York. Conductivity was determined by the balance method

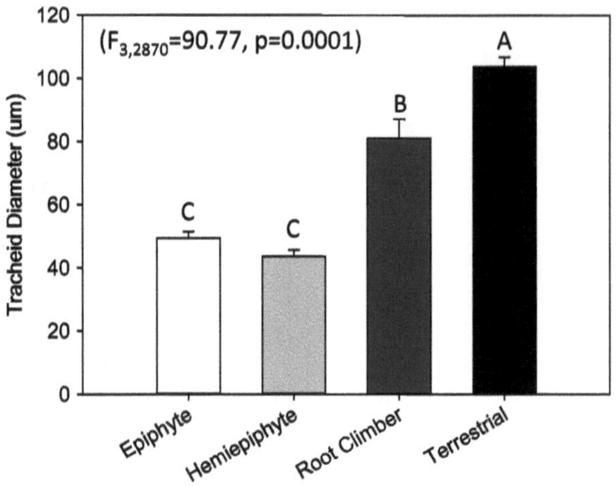

Fig. 1.10 The average diameter of individual tracheids in the stipes of epiphytic, hemiepiphytic (all primary), terrestrial root climbing (see Canestraro et al. 2014), and terrestrial fern species. These data represent multiple species from the tropical lowland forest at La Selva Biological Station in Costa Rica. Stipes were hand-sectioned and a minimum of 30 individual tracheids were measured per stipe section

conifers and vesselless angiosperms. The area-standardized pit resistance (r_{pit}) of ferns and lycophytes ranges from 1.99 to 283 ± 130 MPa s m^{-1}, with an average of 54.6 MPa s m^{-1} (Schulte et al. 1987; Wheeler et al. 2005; Brodersen et al. 2014), whereas r_{pit} in conifers and vesselless angiosperms is estimated at 5.7 and 16 MPa s m^{-1}, respectively (Pittermann et al. 2005; Hacke et al. 2007). By contrast, the average r_{pit} of angiosperm vessels is 336 MPa s m^{-1} (Wheeler et al. 2005; Hacke et al. 2006). These values suggest that, similar to conifers and vesselless angiosperms, highly permeable pit membranes compensate in part for the higher resistance imposed by a tracheid-based vascular system. Neither ferns nor vesselless angiosperms show the predicted 38× loss in hydraulic efficiency associated with non-conifer tracheids because convergent evolution acted to reduce membrane permeability across these distant lineages.

Ferns may be capable of evolving hydraulically efficient xylem, but it is primary xylem nonetheless, and it imposes a fixed limit on physiological performance (Pittermann et al. 2013). Indeed, the vast majority of ferns are found in the understory of moist tropical forests where low light levels combined with high humidity reduce the need for highly efficient xylem. Filmy ferns (Hymenophyllaceae) are an extreme example: their xylem tissue is reduced to less than a dozen conduits in stipe cross section because the pinnae are usually only a single cell thick and lack both a cuticle and stomata (Proctor and Tuba 2002; Proctor 2012). In the understory, there is simply no need for selection to act on increased leaf venation or hydraulic efficiency. Consequently, the data indicate a tight scaling between functional xylem area and the distal frond area (Limm and Dawson 2010; Pittermann et al. 2011). This may have profound implications for leaf gas exchange. Recent studies suggest that stomatal closure in seed-free vascular plants is initiated at higher water potentials than in angiosperms (Brodribb and Holbrook 2004; Brodribb and McAdam 2011), an adaptive response to impending drought stress because ferns lack the hydraulic redundancy that high vein densities provide in angiosperms. High-density reticulate venation offers alternative pathways to water transport should localized embolism occur (Scoffoni et al. 2011). It is easy to argue based on current data that limited productivity in ferns is the penalty imposed by primary xylem, but without seeing a parallel evolutionary trajectory of photosynthetic enzyme function, the reasons behind the physiological canalization of ferns may not be fully understood.

Taken together, the evolution of large tracheids and, in some taxa, vessels, coupled with selection for close conduit packing and permeable pit membranes, appears to offset the hydraulic constraints that are inherent to ancestral, tracheid-based xylem tissue. Yet despite their physiological limitations, it is impressive that seed-free vascular plants managed to not merely persist but even thrive in habitats that are dominated by conifers and angiosperms.

4.2 Hydraulic Failure by Cavitation

All vascular plants contend with low water potentials that can arise from any combination of water deficit or high transpiration rates. Low water potentials render the xylem susceptible to cavitation, that is the entry and subsequent expansion of air within the water column that can lead to an air-vapour embolism. Air enters water-filled conduits through pit membranes because they are mechanically the weakest and most permeable regions of the conduit wall. Embolized conduits cannot transport water, so selection has acted on xylem to keep an appropriate safety margin that protects against cavitation while maintaining efficient rates of water transport. Many studies have examined how xylem structure and function balances cavitation safety with hydraulic efficiency in woody plants (see, for instance the Chap. 7) but only a handful of papers have addressed these issues in seed-free vascular plants.

Despite the general perception of seed-free vascular plants as water-loving understory plants, ferns occupy a wide enough range of habitats and niches to expect that xylem would be subject to adaptive selection. For example, roots are commonly reduced in epiphytic ferns, which helps explain why epiphytes may experience mid-day water potentials lower than −4 MPa as compared to co-occurring terrestrial ferns that stay above −2 MPa (Watkins et al. 2010). Patterns of water use may also vary with life history strategies. A recent study showed that sun-exposed individuals of the perennial *W. fimbriata* retain mid-day water potentials above −0.5 MPa, unlike the seasonally deciduous *P. aquilinum*, in which water potentials frequently dropped below −1.5 MPa (Brodersen et al. 2012). In moist, temperate habitats, the minimum seasonal water potentials varied from −0.2 to −1 MPa in canyon-dwelling *Woodwardia*, *Dryopteris* and *Polystichum* exposed to different light levels (Lo Gullo et al. 2010). Water potentials of lycopods, tree ferns and desiccation-tolerant ferns have yet to be measured, and much more work is needed to characterize fully the water relations of seed-free vascular plants in situ.

Although few studies document fern water potentials in the field, the cavitation response of ferns has received significantly more attention. Evaluating species cavitation resistance involves generating a response curve that shows the percent loss of hydraulic conductivity due to cavitation against progressively decreasing segment water potential. This can be done in any number of ways, including measurement on whole leaves during controlled dehydration (Brodribb and Holbrook 2004), measurement on stipes with the centrifuge method in which water potentials are simulated by spinning at high RPM such that air is pulled in through pit membranes (Alder et al. 1997; Wheeler et al. 2005; Pittermann et al. 2011), and axial air injection whereby air is gradually introduced into one end of a cut stipe, thereby pushing air through pit membranes (Watkins et al. 2010). Several parameters can be derived from this curve, the most widely used being the average water potential at which species experience a 50 % loss of hydraulic conductivity (P_{50}). Early estimates of fern vulnerability to cavitation were performed using whole fronds of tropical species from lowland Costa Rica; these studies generated P_{50} values ranging from about −1.4 to −2 MPa, suggesting that ferns are more sensitive to water deficit than angiosperms (Brodribb and Holbrook 2004). However, in a broad survey of tropical epiphytic and terrestrial fern taxa, Watkins et al. (2010) applied the axial air-injection method to show that P_{50} values can vary from a maximum of −0.4 MPa in terrestrial pteridophytes to a minimum of −4.1 MPa in the cavitation-resistant, epiphytic *Elaphoglossum herminieri* (Bory & Fée) T. Moore. Taken together, P_{50} values in the epiphytes were below −2.5 MPa and significantly lower than what was measured in the terrestrial ferns, an adaptive response consistent with expectations. A broader range of values was reported by Pittermann et al. (2011) who used the centrifuge method to show that several species were unusually cavitation resistant including the perennial *Dryopteris arguta* (Kaulf.) Maxon, *W. fimbriata* and *Pteris cretica* L., which showed P_{50} values ranging from −4 to below −6 MPa, not unlike the P_{50} range experienced by woody plants (Pittermann et al. 2013). The functional significance of such seemingly resistant xylem is unclear because it's not known whether these species experience such low water potentials in situ. One possibility is that low P_{50} values may endow these plants with a significant safety margin to buffer the xylem against excessive cavitation when seasonal water potentials reach their minimum. In these Mediterranean climate-adapted perennials, a several MPa buffer may mean the difference between mere survival versus growth and reproduction. Aside from the Watkins et al. (2010) study, there are few field data with which this question can be explored and certainly no long-term field measures to demonstrate a definitive connection between seasonal water potential and spore production.

The costs of cavitation resistance are well understood in woody plants but much less is known about this in ferns. In cavitation resistant conifer xylem, reduced tracheid diameters and thicker tracheid walls lead to lower specific conductivities and higher wood density because narrower, more reinforced tracheids are necessary to withstand the hoop and bending stresses imposed by the water column during drought (Pittermann et al. 2006). This is quantified by the conduit double-wall thickness (t) to lumen diameter (b) ratio $(t/b)^2$ whereby higher values are associated with denser, more cavitation-resistant xylem (Hacke et al. 2001). Similar costs are evident in angiosperm xylem in which narrower, shorter vessels are mechanically reinforced by thick-walled fibres (Jacobsen et al. 2005). Taken together, xylem structure and function in woody plants is guided by a reasonably predictive safety versus efficiency framework in which cavitation-resistant species with a low P_{50} exhibit reduced transport efficiency and greater xylem construction costs (Hacke et al. 2001; Pittermann 2010). However, no such costs have been identified with cavitation resistance in ferns; neither hydraulic efficiency nor the $(t/b)^2$ metric vary predictably across the broad range of P_{50} values sustained by fern xylem (Watkins et al. 2010; Pittermann et al. 2011). Given that ferns are not equally cavitation resistant, what might drive the observed variation in P_{50}? Brodersen et al. (2012) suggested that vascular organization may control the spread of air but possibly at the cost of hydraulic efficiency. For example, *P. aquilinum* has many vascular bundles that frequently fuse and divide along the length of the frond, a pattern that may reduce hydraulic resistance by creating multiple pathways for water transport, but by the same token, also providing multiple opportunities for network-wide embolism by air-seeding. The P_{50} of *P. aquilinum* ranges from −1.23 to −2.23 MPa, while K_s can exceed 40 kg MPa^{-1} m^{-1} s^{-1} (Pittermann et al. 2011; Brodersen et al. 2012). The opposite pattern was observed in the more cavitation-resistant *W. fimbriata* (P_{50}=−1.11 to −5.21 MPa), in which two large and three smaller vascular bundles were infrequently bridged by small, lateral bundles (Brodersen et al. 2012). The K_s of this species ranges from 10 to 42 kg MPa^{-1} m^{-1} s^{-1} (Pittermann et al. 2011; Brodersen et al. 2012). The anatomical attributes of the vascular bridges were not evaluated, but wider tracheids were consistently observed in the large vascular strands leading Brodersen et al. (2012) to speculate that these perform the majority of water transport, with the smaller bundles acting as a hydraulic reserve should the large bundles fail by embolism. Manipulative experiments indicate that in *W. fimbriata*, the small vascular strands are indeed more resistant to embolism (Brodersen et al. 2012). Whether similar patterns are found in other species remains to be seen.

The safety-efficiency trade-off is also in large part determined by pit membranes, the function of which has been well-studied in both seed-bearing and seed-free vascular plants. The pit membranes of ferns are similar enough to the homogenous pit membranes of angiosperms to allow for some comparisons to be drawn. Typically, thicker pit membranes are associated with higher air-seed pressures, meaning that lower water potentials (or greater air pressures) are required to draw (or push) air across the largest pore present in a pit membrane (Choat et al. 2005; Brodersen et al. 2014). Air-seed pressures are determined experimentally by injecting air into single conduits with glass capillaries and watching for air bubbles at the distal cut end of a stem segment, which is immersed in water (Zwieniecki et al. 2001). In contrast to angiosperms, thicker pit membranes in seed-free vascular plants are only weakly associated with greater air-seed pressures, suggesting that membranes may not be the most direct target of selection (Jansen et al. 2009; Lens et al. 2011; Brodersen et al. 2014). Instead, air-seed pressures were more consistently related to conduit diameter, such that smaller conduits with smaller pit membranes and overall lower pit membrane areas required higher air injection pressures to penetrate the pit membrane (Brodersen et al. 2014; Fig. 1.5b). Smaller pit areas reduce the probability that a large membrane pore will be present acting as the so-called weak link, that allows air to propagate from one conduit to another (Wheeler et al. 2005; Jansen et al. 2009). Functionally, this implies that species with narrow conduits should exhibit low specific conductivities but with the benefit of greater cavitation resistance, while the opposite should be true in taxa with large, hydraulically-efficient conduits (Brodersen et al. 2014). While this was found to be true across tropical terrestrial ferns (Watkins et al. 2010), epiphytic taxa showed no relationship between tracheid diameter, cavitation resistance and K_s despite the clear differences in conduit size (Figs. 1.8 and 1.10; Watkins et al. 2010).

It is not known whether all seed-free vascular plants recover from embolism but desiccation-tolerant plants may hold some clues. Many species of ferns and lycophytes have retained the ability to lose over 95 % of their water yet fully regain their photosynthetic and metabolic activity within hours or days after re-watering (Proctor and Tuba 2002). Poikilohydry is found in small plants such as bryophytes, seed-free vascular plants, and even herbaceous angiosperms, but it is for the most part considered an ancestral trait (Proctor and Tuba 2002; Sperry 2003). However, selection favoured the reversal from homoiohydry to poikilohydry in the filmy fern genus *Hymenophyllum* insofar as members of this genus exhibit highly reduced xylem tissue as well as pinnae that are only one cell thick with no stomata (Proctor 2012). These delicate ferns measure less than 20 cm in height, so desiccation and recovery represent a more adaptive strategy than homoiohydry. How the xylem refills in poikilohydric ferns is unclear but desiccation-tolerant taxa such as *Selaginella lepidophylla* (Hook. & Grev.) Spring and numerous cheilanthoid ferns typically have narrow xylem conduits so capillarity may be sufficient to drive hydraulic recovery. Metabolically assisted refilling along the lines of Secchi and Zwieniecki (2011) cannot as yet be ruled out. Given that *S. lepidophylla* has a weakly developed to non-existent root system and still recovers from complete desiccation, root pressure is unlikely to play a role in embolism reversal.

4.3 Ecological Fern Hydraulics: The Role of Polyploidy in Fern Physiology and Hydraulics

No review of fern physiology would be complete without discussion of the role that polyploidy and introgression play in shaping niche partitioning. Polyploidy is a common phylogenetic occurrence across ferns (Manton 1950; Wood et al. 2009). Ecologically, polyploidy in ferns is important as many of our modern temperate landscapes are dominated by polyploid and/or hybrid ferns. In a recent community survey of *Dryopteris* species in a mesic 20 ha plot at the Huyck Preserve in New York, 80 % of species encountered were either polyploids or hybrids, and 85 % of the individuals encountered fell in this category (Watkins unpublished data). Polyploids, especially those of recent origin, can be constrained during reproduction for lack of compatible cytotypes (Levin 1975). In theory, this could result in reduced matings relative to parents. Yet, polyploids persist and often come to dominate in some cases (but see Buggs and Pannell 2007). One potential mechanism of this maintenance is ecophysiological niche differentiation across ploidy levels. If polyploids are more physiologically capable (e.g. greater hydraulic conductance, higher photosynthetic rates and increased spore production) than diploids, the former could attain an equal or greater ecological footing (DeBodt et al. 2005; Fawcett et al. 2009; Coate et al. 2011; Testo et al. 2014). Whereas some studies have shown this relationship (e.g. Kao and Parker 2010; Warner and Edwards 1993; Li et al. 1996, 2009), others have found no differences or the opposite pattern (e.g. Baack and Stanton 2005). Our understanding of how physiological differences relate to ploidy is limited. One of the most thorough efforts to date has been Sessa's (2012, 2014) work on the genus *Dryopteris*. In a series of studies the authors found that allopolyploids were frequently transgressive in field ranges, yet they failed to find any relationship of carbon relations with ploidy and no evidence of introgression in physiological characters in allopolyploid hybrids.

Some authors have speculated that xylem attributes vary with plant ploidal levels. Stebbins (1971) observed that cell size increases with ploidy, and Pockman and Sperry (1997) suggested that ploidy-driven differences in vessel size may drive population distributions of *Larrea tridentata* in the American southwest. Hence, a priori predictions are that the xylem of polyploid ferns will comprise wider and longer conduits that are more likely to lead to higher transport rates. Ferns may be the ideal system in which to study the hydraulic consequences of polyploidy, and recent studies by Watkins (unpublished data) have examined xylem structure and function in the wood fern *Dryopteris*, a common genus in temperate North American forests. The genus comprises 11 species that include six diploid taxa and five allopolyploid taxa that vary in ploidy. In a survey of tracheid lengths from herbarium sheets of ten taxa (6–9 individuals each), there was a general increase in length with ploidal level (see tracheid length in Fig. 1.9a). Diploid species and triploid hybrids produced similar tracheid lengths while tetraploids, pentaploids and hexaploids were significantly longer (Fig. 1.9a). In an effort to evaluate the association of hydraulics with ploidal level and/or parent:hybrid relationships, Watkins surveyed xylem conductance in five taxa in northern regions of New York state. *Dryopteris cristata* (4*n*) hybridizes with *D. intermedia* (2*n*) to produce the 3*n* *D.*×*bootii*. *Dryopteris intermedia* also hybridizes with 4*n* *Dryopteris carthusiana* to produce another sterile hybrid, 3*n* *D.*×*triploidea*. K_s was evaluated across parents and hybrids and *D.* × *bootii* was significantly higher and thus transgressive in K_s relative to either parent (see K_s in Fig. 1.9b), *Dryopteris*×*triploidea* had similar values of K_s to both parents. Curiously, *Dryopteris*×*bootii* tends to be quite rare and frequently limited to wet swampy sites; whereas *D.* × *triploidea* can be abundant in some forests and in this upstate New York site was more common than 4*n* *D. carthusiana*. The mechanism controlling the distributions of hybrids and diploid/polyploidy ferns is unknown, yet the data presented here suggest that greater exploration of the role of ploidy and hybrid origin is warranted.

4.4 Comparative Hydraulics of Epiphytic and Terrestrial Ferns

Our understanding of fern ecology has grown significantly in recent years (Ranker and Haufler 2008; Mehltreter et al. 2010). Of particular interest in the study of fern biology are the functional differences between epiphytic and terrestrial ferns (Hietz and Briones 1998; Watkins and Cardelus 2012, 2009; Watkins et al. 2007a, b, c). Like many other lineages, ferns experienced a major radiation following the Cretaceous. A significant proportion of the diversity of this radiation is of epiphytic ferns that rapidly entered the novel angiosperm-dominated canopy (Watkins and Cardelus 2012 and references therein). Comparative studies of both gametophytes and sporophytes of epiphytic and terrestrial ferns have revealed unique suites of functional characters associated with each group. Watkins et al. (2007a, b) have shown that the gametophytes of epiphytic ferns are significantly longer lived, are less likely to recruit from the gametophyte stage into the sporophyte stage and are more stress tolerant than terrestrial species (Watkins et al. 2007a; Pittermann et al. 2013). Studies on comparative sporophyte hydraulics have shown that epiphytic ferns have significantly reduced leaf and xylem K_s and increased P_{50} values relative to terrestrial species. Watkins et al. (2010) and Watkins and Cardelus (2012) have suggested that increased P_{50} values may have come at the expense of hydraulic conductivity, and that the reduced stipe lengths commonly observed in epiphytic ferns may be a direct response to reduce path lengths to water movement. A survey of total vascular area in the stipes of several tropical species of terrestrial, epiphytic, root climbing and hemiepiphytic ferns shows that terrestrial species produce more total vascular area than other growth forms and that epiphytic species produce the least area. This pattern holds when total leaf area is taken into account. These data suggest that significant and unappreciated gross differences exist in the vascular system of different fern growth forms.

Diameter measurements of hundreds of individual tracheids from more than two dozen species reveal a fundamental difference in the frequency of tracheid size classes (see Fig. 1.10). Epiphytic and hemiepiphytic species have similar tracheid diameter distributions with a broad normal distribution with many small diameter tracheids (Watkins, unpublished data). On the opposite end of the spectrum, terrestrial species can be characterized by having relatively few tracheids with small diameters and large proportion of conduits with large diameters. On average, terrestrial species produced individual tracheids that were twice as large in diameter compared to epiphytic species (Fig. 1.10). Whereas pit membrane attributes play a critical role in cavitation resistance, reduced pit area, and thus small-diameter tracheids, have been shown to limit the spread of air from one conduit to another (Brodersen et al. 2014). The epiphytic habitat in the canopy is frequently drought stressed, and such conditions could select for reduced tracheid diameters in an effort to avoid cavitation. The four hemiepiphytic species surveyed were all primary, as defined by Canestraro et al. (2014), establishing first as epiphytes and then connecting and never losing connection to the forest floor. These four species had similar diameter distributions as epiphytic taxa, yet all fronds sampled were from mature individuals that were rooted in the terrestrial soil matrix.

Earlier isotopic and nutrient analyses suggest that both water and nutrient relations in hemiepiphytic ferns are more similar to terrestrial than epiphytic species (Watkins et al. 2007c). Based on phylogenetic evidence, hemiepiphytic species can arise from terrestrial, epiphytic or terrestrial root-climbing ancestors. Terrestrial root-climbers are outwardly similar to hemiepiphytes (in fact this group is frequently misidentified as hemiepiphytes) but never lose terrestrial connections (Canestraro et al. 2014). These species were more similar to terrestrial species than hemiepiphytic taxa. Interestingly, Canestraro et al. (2014) have recently shown that photosynthetic rates can vary between terrestrial and climbing stages of some root-climbers, yet not in others. There is a need to understand better the changes in functional anatomy and ecophysiology across taxa that change life forms. Such information will be critical in our quest to understand the evolution of epiphytic ferns and the radiation of lineages into novel habitats. Our sampling is small and the phylogeny of the surveyed species too poorly understood to make detailed categorizations. However, it is possible that hemiepiphytic species, even when derived from terrestrial ancestors, are canalized in tracheid anatomy. The selective pressures induced by drought may overwhelm pressures for increased growth and reproduction. Terrestrial root-climbers may gain the same benefit as hemiepiphytes (i.e. increased height that results in exploration of higher light levels, increased spore dispersal distances) without the added complexities of an epiphytic stage. Clearly more work needs to be done on the comparative xylem anatomy of these taxa.

5 Evolutionary Patterns in Fern Xylem Structure

Vascular arrangements in fern and lycophytes have fascinated botanists since the nineteenth century and even today, variation in stelar patterns is useful for taxonomic identification (Bower 1923; Ogura 1972). An interest in developmental processes rather than functional attributes motivated earlier studies, but the desire to document stelar structure in a broader evolutionary context (including fossil taxa) was the common thread of Bower's and Ogura's discourses. Beck et al. (1982) contributed a seminal paper that not only explored the known vascular arrangements of stems and rhizomes across extinct and living land plants but also formalized stelar nomenclature. When it comes to understanding the evolutionary and functional significance of stelar structures, modern phylogenetic approaches allow us to build on these contributions with greater resolution and thus extend the realm of possibilities; we are now able to map structural and physiological attributes onto the fern phylogeny to document whether stelar patterns originated independently in response to common selective pressures, whether the observed arrangements appear randomly or whether directional selection favoured specific phenotypes over others. These evolutionary patterns are not mutually exclusive and separate clades may reveal different trends.

Here we have documented stelar patterns in frond petioles (stipes) from 182 fern species. Cross sections were obtained from a variety of sources, including new hand-sections from fresh or herbarium material, published sources such as Ogura (1972) and Bower (1923), and online repositories (e.g. plantsystematics.org). From these data, we identified a consensus stele type for each currently recognized family and mapped this across the family-level phylogeny (see Fig. 1.11). Formal reconstructions have yet to be conducted, but several key patterns emerge from this phylogenetic analysis.

First, it is evident that some of the early diverging families, such as Psilotaceae and Equisetaceae, possess single vascular strands (protosteles) in their highly reduced leaves, which are developmentally homologous with the fronds of higher pteridophytes. The familiar *Psilotum nudum* stems also have protosteles but the vascular arrangement in equisetalean stems appears to have undergone some degree of duplication, as multiple strands of primary xylem and associated lacunae encircle the stem cylinder. Plants are modular organisms, so duplication or extension of the protostele may have also occurred in the Ophioglossaceae and Marattiaceae, leading to polycyclic steles in the latter (Schneider et al. 2002). Iteration of vascular strands is consistent with the greater complexity of these two lineages: *Ophioglossum* develops one lateral megaphyll while the polycyclic Marattiales develop multiple fronds that may be up to five-pinnate. The Marattiales have some of the largest fronds among the pteridophytes.

C-shaped vascular strands with variously curling ends are present in the Osmundaceae, Gleicheniaceae, Dipteridaceae and Matoniaceae (Fig. 1.11). Superficially these resemble an incomplete solenostele. As previously mentioned, the Hymenophyllaceae have adopted a poikilohydric strategy and thus exhibit highly reduced xylem. A return to a central, protostelic vascular strand is evident in the schizaeoid clade, which includes the climbing Lygodiaceae, as well as the Anemiaceae and Schizaeaceae. A cylindrical vascular bundle may be adaptive in the Lygodiaceae because torsional and compressive stresses are more equally distributed than in an asymmetrical stele. As in the Hymenophyllaceae, minimized xylem is evident in the aquatic ferns belonging to Marsileaceae and Salviniaceae that require little or no water transport.

The mostly arborescent families comprising the tree fern clade (Thyrsopteridaceae, Loxsomataceae, Culcitaceae, Plagiogyriaceae, Cibotiaceae, Cyatheaceae, Dicksoniaceae and Metaxyaceae) are generally characterized by large fronds, and this is reflected in their complex, often highly divided stelar structures that may resemble folded ribbons. As in the Marattiaceaeles, the fronds are frequently highly divided so the many iterative bundles of these distantly related lineages may represent convergent evolution.

The steles of the early-diverging polypod families (Saccolomataceae, Cystodiaceae, Lindsaeaceae, Lonchitidaceae, Dennstaedtiaceae and Pteridaceae) are described generally as an omega shape, with some variation in the Cystodiaceae and Lonchitidaceae. Several species with large, multi-pinnate leaves belong to the Dennstaedtiaceae and Pteridaceae, so folded vascular ribbons are also present in the petioles (Fig. 1.2e; Vasco et al. 2013). Fronds of some Pteridaceae such as the genus

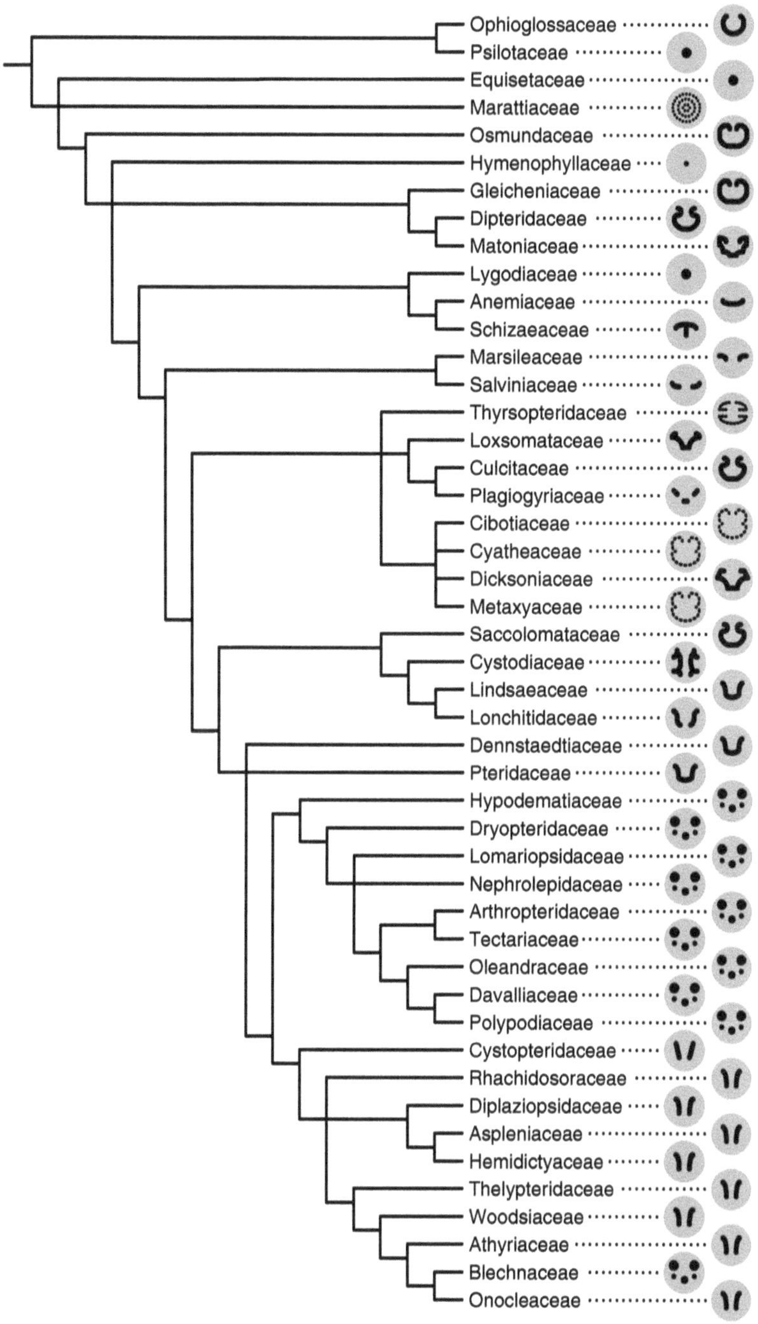

Fig. 1.11 Mapping petiole stelar types onto the fern phylogeny. The phylogenetic relationships and classification presented outside of polypod ferns are based on Smith et al. (2006). Those within eupolypods I follow Liu et al. (2014) and those within eupolypods II follow Rothfels et al. (2012). Other relationships shown reflect those in Schuettpelz and Pryer (2009), with the classification based on Christenhusz et al. (2011). Stelar types represent a consensus for each family from available cross sections. See text for details

Pteris, can exceed 2 m in length but unlike the highly divided dictyosteles of similarly sized tree fern fronds, *Pteris* petiolar steles are usually single stranded. The recently discovered hybrid *Pteris* × *caridadiae* is an interesting exception because its vascular strand is interrupted despite maintaining the overall omega shape of its parent plants (Testo et al. 2015).

Distinct vascular patterns arise in the two most recently derived clades on the fern phylogeny: the eupolypods I (Hypodematiaceae, Dryopteridaceae, Lomariopsidaceae, Nephrolepidaceae, Arthropteridaceae, Tectariaceae, Oleandraceae, Davalliaceae, Polypodiaceae) and eupolypods II (Cystopteridaceae, Rhachidosoraceae, Diplaziopsidaceae, Aspleniaceae, Hemidictyaceae, Thelypteridaceae, Woodsiaceae, Athyriaceae, Blechnaceae, Onocleaceae). The steles in these groups are dictyostelic as they always comprise at least two major vascular bundles. Several smaller strands accompany the two large bundles in the eupolypod I lineage and in the Blechnaceae of the eupolypod II clade (Figs. 1.2c, d and 1.11). So reliable are these patterns that they are frequently used for the purpose of identification.

Understanding how selection acted on stelar patterns in ferns requires consideration of paleoclimate change as well as biome assembly in tropical and temperate regions. Recent phylogenetic approaches have revealed compelling trends in fern diversity that largely point to the post-Cretaceous speciation and radiation of epiphytic taxa. Global climatic shifts may have contributed to this pattern: recent studies of adaptive radiations in monocot, succulent and arid-adapted conifer lineages are consistent with Cenozoic aridification (Edwards et al. 2010; Arakaki et al. 2011; Pittermann et al. 2012). However, equally relevant are several analyses that suggest diversification of fern epiphytes occurred in response to the expansion of complex, angiosperm-dominated forests (Schneider et al. 2004; Schuettpelz and Pryer 2009; Watkins and Cardelus 2012). Release from competition on the crowded forest floor and an increase in light availability may have driven the move to the canopy, but selection for increased cavitation resistance (greater P_{50}), thicker leaf cuticles, and desiccation tolerance in both sporophytes and gametophytes may have allowed epiphytic ferns to gain traction in what is generally a dryer and more climatically variable habitat than in the understory (Watkins et al. 2007a; Dubuisson et al. 2009; Schuettpelz and Pryer 2009; Watkins and Cardelus 2012). Although the speciation of epiphytes stands out in the post-Cretaceous evolutionary trajectory of ferns, the concurrent diversification of eupolypod I terrestrial ferns also warrants a closer look; the formation of temperate forests and drier biomes may have selected for specific physiological attributes associated with greater drought tolerance in non-tropical taxa.

The exclusive presence of dictyosteles in the petioles of more derived fern lineages suggests that xylem structure was under strong selection for drought resistance during the Cenozoic. Sectored dictyosteles in the eupolypods I and II comprise at least two vascular strands, sometimes more (Figs. 1.3 and 1.11). Wider conduits occupy the larger strands, so a broad range of tracheid sizes would ensure that alternative pathways for water movement are available should transport in one of the large bundles be hampered by embolism (Brodersen et al. 2012). In contrast, the mostly integrated C- or omega-shaped stelar arrangements in the early-derived fern groups may increase the probability of air rapidly spreading throughout the xylem due to close contact between conduits. Selection for more numerous narrower tracheids in epiphytic taxa would drive increased cavitation resistance due to greater conduit redundancy but also because reduced pit area will lower the likelihood of air-seeding (Wheeler et al. 2005; Christman et al. 2009; Brodersen et al. 2014). A broader sampling across the fern phylogeny is needed to gain a more nuanced perspective on the evolution of xylem structure and function in ferns, particularly those in the eupolypod I and II lineages.

Other vascular attributes may factor into the success of eupolypod ferns. In a phylogenetically extended survey, Hernandez-Hernandez et al. (2012) indicated an increased presence of the circumendodermal band in more derived ferns. Resembling a thick casparian strip, this band encapsulates the xylem and phloem tissues (Fig. 1.2b). A number of hypotheses have been proposed to explain the presence of this structure. For example, this waxy layer may increase frond structural rigidity (Hernandez-Hernandez et al. 2012), an important trait in many epiphytes in which fronds are displayed at steep angles and vulnerable to damage by falling canopy debris. Epiphytic ferns may also be capable of foliar water absorption, so biomechanical needs may trump xylem-mediated water transport in humid habitats. Foliar absorption is a critical water-use strategy for plants in the redwood forest understory, especially for *P. munitum* (Kaulf.) C. Presl which derives almost all of its water from the frequent fog events that blanket the redwood forests of northern California (Limm et al. 2009; Limm and Dawson 2010). The aggregate of vascular and leaf traits that underpin the physiology of eupolypod ferns is a topic of much interest and sure to reveal novel physiological and structural solutions to life in the canopy.

6 Concluding Remarks

There are many possibilities for research on the physiology, ecology, evolution and paleobotany of seed-free vascular plants. At the organismal scale, it is not unreasonable to propose that the absence of critical features such as bifacial vascular cambia and lateral branching restricted the phenotypic diversity of ferns and lycopods. Indeed, the body plan of seed-free tracheophytes couldn't be more different than that of eutracheophytes. Though incomplete, the fossil record shows this to be true; with an arborescent canopy, ancient seed-free plants such as the progymnosperms explored a broader morphospace than modern pteridophytes (Meyer-Berthaud et al. 1999, 2000). However, the limits and possibilities of phenotypic variation in ferns have not been considered to the degree that they have in woody plants. For example, Niklas' adaptive walks through the plant kingdom mimic evolutionary processes, and provide multiple morphological solutions that balance conflicting needs of reducing surface area whilst enhancing mechanical stability, light capture, and spore dispersal (Niklas 1992, 1999). The model relies on the predictable but spatially variable bifurcation of branches such that model outputs result in highly variable yet structurally conserved phenotypes, many of which closely resemble today's angiosperm and conifer flora (Niklas 1999). The branching assumptions would require adjustment to reflect growth patterns of modern ferns but the model does provide a solid framework for coupling vascular anatomy with frond patterns in the context of basic physiology and reproduction.

The interplay of form and function should not only be limited to extant taxa. Michael Cichan's work with woody Carboniferous plants was prescient; not only was he able to infer developmental patterns of unifacial vascular cambia from fossils, but he applied the standard Hagen–Poiseuille model of fluid flow to model water transport in several groups of lianoid and arborescent plants (Cichan 1985a, b, 1986). Wilson and colleagues took this approach one step further by modelling both transport efficiency and cavitation resistance in a range of taxa including *Asteroxylon mackiei* (see illustration in Fig. 1.6) and *Medullosa*, a Carboniferous seed fern (Wilson et al. 2008; Wilson and Knoll 2010; Wilson and Fischer 2010). Their models suggest that unusually wide and long conduits coupled with permeable membranes may have supported high conductivities in the woody *Medullosa*, but by the same token, this suite of traits likely rendered these plants vulnerable to cavitation. Judging by the anatomical similarity of their xylem, especially the bordered pit structure, the progymnosperms such as *Archaeopteris-Callixylon* (see micrographs in Fig. 1.7) may have had similar vascular physiology. New methods such as X-ray synchrotron tomography have recently revealed the xylem and even the inter-conduit pit structure of an early Devonian woody plant, showing conduits in excess of 70 μm in diameter (Strullu-Derrien et al. 2014). The suggestion that this ancient plant was vulnerable to embolism agrees with the opinions of Wilson (2013), who proposed that high transport efficiency evolved many times over the past 400 million years whilst cavitation resistance lagged behind. Testing this hypothesis may mostly be a matter of locating appropriate fossil material since technologies that improve specimen visualization are developing quite rapidly.

Seed-free vascular plants are highly diverse with respect to their ecology and morphology so there is much to be learned about the ecophysiological attributes that support this variation. There is little doubt that the dictyosteles of recently derived eupolypod ferns were under strong selective pressures during the Cenozoic but a thorough sampling regime will be necessary to uncover the tempo and evolutionary trajectory of traits such as xylem area, tracheid size, vascular connectivity, cavitation resistance and hydraulic efficiency. Additional measures of allometry, capacitance, stomatal density and leaf venation patterns will greatly enhance our understanding of how co-ordinated hydraulic function evolved in ferns and lycophytes. An alternative and equally interesting possibility is that morphological diversity is not the product of natural selection per se, but rather neutral state transitions that confer neither a cost nor benefit to the plant. Fern sporophytes are incredibly diverse with respect to size and shape, and given the co-occurrence of different frond structures in similar habitats (Watkins et al. 2010; Vasco et al. 2013), the physiological consequence of this variation may be subtle or nil. Instead, selection for mechanical stability, spore dispersal, longevity, gametophyte biology, edaphic type or other aspects of life history strategy may govern leaf shape (Creese et al. 2011; Watkins et al. 2007a, b, c; Pittermann et al. 2013). The physiological attributes of extant ferns and lycophytes should be well understood if we are to make any reasonable inferences about the drivers of sporophyte evolution over deep time, but the vasculature is but one element of a complex suite of traits that contribute to the success of seed-free vascular plants.

Acknowledgments We sincerely thank Dr. Uwe Hacke for the opportunity to contribute to this volume and for his comments on the manuscript. Dr. Robbin Moran's assistance with preparation of Fig. 1.11 is much appreciated. The National Science Foundation is gratefully acknowledged for support of this work (JP, IOS-1258186).

References

Agashe SN (1968) Phloem studies in the pteridophytes, part I. Equisetum. Am Fern J 58:74–77
Alder NN, Pockman WT, Sperry JS, Nuismer S (1997) Use of centrifugal force in the study of xylem cavitation. J Exp Bot 48:665–674
Arakaki M, Christin P-A, Nyffeler R, Lendel A, Eggli U, Ogburn RM, Spriggs E, Moore MJ, Edwards EJ (2011) Contemporaneous and recent radiations of the world's major succulent plant lineages. Proc Natl Acad Sci U S A 108:8379–8384
Baack EJ, Stanton ML (2005) Ecological factors influencing tetraploid speciation in snow buttercups (*Ranunculus adoneus*): niche differentiation and tetraploid establishment. Evolution 59:1936–1944
Barghoorn ES (1964) Evolution of cambium in geologica time. In: Zimmermann MH (ed) The formation of wood in forest trees. Academic, New York, pp 3–17
Beck CB (1970) The appearance of gymnospermous structure. Biol Rev 45:379–400
Beck CB (2010) An introduction to plant structure and development. Cambridge University Press, Cambridge, p 459
Beck CB, Schmid R, Rothwell G (1982) Stelar morphology and the primary vascular system of seed plants. Bot Rev 48:691–815
Boyce C, Cody G, Fogel M, Hazen R (2003) Chemical evidence for cell wall lignification and the evolution of tracheids in early Devonian plants. Int J Plant Sci 165:691–702.
Bower FO (1923) The ferns (Filicales), vol 1, Analytical examination of the criteria of comparison. Cambridge University Press, Cambridge, p 359
Brodersen CR, McElrone AJ, Choat B, Matthews MA, Shackel KA (2010) The dynamics of embolism repair in xylem: in vivo visualizations using high-resolution computed tomography. Plant Physiol 154:1088–1095
Brodersen CR, Lee EF, Choat B, Jansen S, Phillips RJ, Shackel KA, McElrone AJ, Matthews MA (2011) Automated analysis of three-dimensional xylem networks using high-resolution-computed-tomography. New Phytol 191:1168–1179
Brodersen CR, Roark LC, Pittermann J (2012) The physiological implications of primary xylem organization in two ferns. Plant Cell Environ 35:1898–1911
Brodersen C, Jansen S, Choat B, Rico C, Pittermann J (2014) Cavitation resistance in seedless vascular plants: the structure and function of interconduit pit membranes. Plant Physiol 165:895–904
Brodribb TJ, Feild TS (2010) Leaf hydraulic evolution led a surge in leaf photosynthetic capacity during early angiosperm diversification. Ecol Lett 13:175–183
Brodribb TJ, Holbrook NM (2004) Stomatal protection against hydraulic failure: a comparison of co-existing ferns and angiosperms. New Phytol 162:663–670
Brodribb TJ, McAdam SAM (2011) Passive origins of stomatal control in vascular plants. Science 331:582–585

Brodribb TJ, Feild TS, Jordan GJ (2007) Leaf maximum photosynthetic rate and venation are linked by hydraulics. Plant Physiol 144:1890–1898
Buggs RJA, Pannell JR (2007) Ecological differentiation and diploid superiority across a moving ploidy contact zone. Evolution 61:125–140
Calkin HW, Gibson AC, Nobel PS (1985) Xylem water potentials and hydraulic conductances in eight species of ferns. Can J Bot 63:632–637
Canestraro BK, Moran RC, Watkins JE (2014) Reproductive and physiological ecology of climbing and terrestrial *Polybotrya* (Dryopteridaceae) at the La Selva Biological Station, Costa Rica. Int J Plant Sci 175:432–441
Carlquist S, Schneider EL (2007) Tracheary elements in ferns: new techniques, observations and concepts. Am Fern J 97:199–211
Choat B, Pittermann J (2009) New insights into bordered pit structure and cavitation resistance in angiosperms and conifers. New Phytol 182:557–560
Choat B, Lahr EC, Melcher PJ, Zwieniecki MA, Holbrook NM (2005) The spatial pattern of air seeding thresholds in mature sugar maple trees. Plant Cell Environ 28:1082–1089
Christman MA, Sperry JS, Adler F (2009) Testing the "rare pit" hypothesis for xylem cavitation resistance in three species of *Acer*. New Phytol 182:664–674
Cichan MA (1985a) Vascular cambium and wood development in Carboniferous plants. I. Lepidodendrales. Am J Bot 72:1163–1176
Cichan MA (1985b) Vascular cambium and wood development in Carboniferous plants. II. *Sphenophyllum plurifoliatum* Williamson and Scott (Sphenophyllales). Bot Gaz 146:395–403
Cichan MA (1986) Conductance in the wood of selected Carboniferous plants. Paleobiology 12:302–310
Coate JE, Schlueter JA, Whaley AM, Doyle JJ (2011) Comparative evolution of photosynthetic genes in response to polyploid and nonpolyploid duplication. Plant Physiol 155:2081–2095
Coomes DA, Allen RB, Bentley WA et al (2005). The hare, the tortoise and the crocodile: the ecology of angiosperm dominance, conifer persistence and fern filtering. J Ecol 93:918–935.
Creese C, Lee A, Sack L (2011) Drivers of morphological diversity and distribution in the Hawaiian fern flora: trait associations with size, growth form, and environment. Am J Bot 98:1–11
DeBodt S, Maere S, Van de Peer Y (2005) Genome duplication and the origin of angiosperms. Trends Ecol Evol 20:591–597
Domec J-C, Lachenbruch B, Meinzer F, Woodruff D, Warren JM, McCulloh K (2008) Maximum height in a conifer is associated with conflicting requirements for xylem design. Proc Natl Acad Sci U S A 105:12069–12074
Dubuisson J-Y, Schneider H, Hennequin S (2009) Epiphytism in ferns: diversity and history. C R Biol 332:120–128
Edwards DS (1986) *Aglaophyton major*, a non-vascular land-plant from the Devonian Rhynie Chert. Bot J Linn Soc 93:173–204
Edwards EJ, Osborne CP, Stromberg CAE, Smith SA, C4 Grasses Consortium (2010) The origins of C4 grasslands: integrating evolutionary and ecosystem science. Science 328:587–591
Fawcett JA, Maere S, Van de Peer Y (2009) Plants with double genomes might have had a better chance to survive the Cretaceous-Tertiary extinction event. Proc Natl Acad Sci U S A 106:5737–5742
Feild TS, Brodribb TJ, Iglesias A, Chatelet DS, Baresch A, Upchurch GR, Gomez G, Mohr BAR, Coiffard C, Kvacek J, Jaramillo C (2011) Fossil evidence for Cretaceous escalation in angiosperm leaf vein evolution. Proc Natl Acad Sci U S A 108:8363–8366
Fotedar RL, Shah JJ (1975) Phloem structure and development in *Blechnum orientale*. Am Fern J 65:52–60
Gaupels F, Buhts A, Knauer T, Deshmukh S, Waller F, Van Bel AJE, Kogel K-H, Kehr J (2008) Adaptation of aphid stylectomy for analyses of proteins and mRNAs in barley phloem sap. J Exp Bot 59:3297–3306
George L, Bazzaz F (1999) The fern understory as an ecological filter: growth and survival of canopy-tree seedlings. Ecology 80:846–856

Gerrienne P, Gensel PG, Strullu-Derrien C, Lardeux H, Steemans P, Prestianni C (2011) A simple type of wood in two early Devonian plants. Science 33:837

Gullo MAL, Raimondo F, Crisafulli A, Salleo S, Nardini A (2010) Leaf hydraulic architecture and water relations of three ferns from contrasting light habitats. Funct Plant Biol 37:566–574

Hacke UG, Sperry JS, Pockman WT, Davis SD, McCulloh KA (2001) Trends in wood density and structure are linked to prevention of xylem implosion by negative pressure. Oecologia 126:457–461

Hacke U, Sperry J, Wheeler J, Castro L (2006) Scaling of angiosperm xylem structure with safety and efficiency. Tree Physiol 26:689–701

Hacke UG, Sperry JS, Feild TS, Sano Y, Sikkema EH, Pittermann J (2007) Water transport in vesselless angiosperms: conducting efficiency and cavitation safety. Int J Plant Sci 168:1113–1126

Hernandez-Hernandez V, Terrazas T, Mehltreter K, Angeles G (2012) Studies of petiolar anatomy in ferns: structural diversity and systematic significance of the circumendodermal band. Bot J Linn Soc 169:596–610

Hietz P, Briones O (1998) Correlation between water relations and within-canopy distribution of epiphytic ferns in a Mexican cloud forest. Oecologia 114:305–316

Jacobsen AL, Ewers FW, Pratt RB, Paddock WA, Davis SA (2005) Do xylem fibers affect vessel cavitation resistance? Plant Physiol 139:546–556

Jansen S, Choat B, Pletsers A (2009) Morphological variation of intervessel pit membranes and implications to xylem function in angiosperms. Am J Bot 96:409–419

Jensen KH, Liesche J, Bohr T, Schulz A (2012) Universality of phloem transport in seed plants. Plant, Cell Environ 35:1065–1076

Kao RH, Parker IM (2010) Coexisting cytotypes of *Arnica cordifolia*: morphological differentiation and local-scale distribution. Int J Plant Sci 171:81–89

Kenrick P, Crane PR (1991) Water-conducting cells in early fossil land plants: implications for the early evolution of tracheophytes. Bot Gaz 152:335–356

Kenrick P, Crane PR (1997) The origin and early evolution of plants on land. Nature 389:33–39

Kitin PB, Fujii T, Abe H, Funada R (2004) Anatomy of the vessel network within and between tree rings of *Fraxinus lanuginosa* (Oleaceae). Am J Bot 91:779–788

Lens F, Sperry JS, Christman MA, Choat B, Rabaey D, Jansen S (2011) Testing hypotheses that link wood anatomy to cavitation resistance and hydraulic conductivity in the genus *Acer*. New Phytol 190:709–723

Levin DA (1975) Minority cytotype exclusion in local plant populations. Taxon 24:35–43

Li WL, Berlyn GP, Ashton PMS (1996) Polyplolds and their structural and physiological characteristics relative to water deficit in *Betula papyrifera* (Betulaceae). Am J Bot 83:15–20

Li WD, Biswas DK, Xu H, Xu CQ, Wang XZ, Liu JK, Jiang GM (2009) Photosynthetic responses to chromosome doubling in relation to leaf anatomy in Lonicera japonica subjected to water stress. Funct Plant Biol 36:783–792

Limm EB, Dawson TE (2010) *Polystichum munitum* (Dryopteridaceae) varies geographically in its capacity to absorb fog water by foliar uptake within the redwood forest ecosystem. Am J Bot 97:1121–1128

Limm EB, Simonin KA, Bothman AG, Dawson TE (2009) Foliar water uptake: a common water acquisition strategy for plants of the redwood forest. Oecologia 161:449–459

Liu H-M, He L-J, Schneider H (2014). Towards the natural classification of tectarioid ferns: confirming the phylogenetic relationships of Pleocnemia and Pteridrys (eupolypods I). J Syst Evol 52:161–174

Manton I (1950) Problems of cytology and evolution in the Pteridophyta. Cambridge University Press, Cambridge

Marrs RH, Watt AS (2006) Biological flora of the British Isles: *Pteridium aquilinum* (L.) Kuhn. J Ecol 94:1272–1321

Mauseth JD, Fujii T (1994) Resin-casting: a method for investigating apoplastic spaces. Am J Bot 81:104–110

McCulloh K, Sperry JS, Lachenbruch B, Meinzer FC, Reich PB, Voelker S (2010) Moving water well: comparing hydraulic efficiency in twigs and trunks of coniferous, ring-porous, and diffuse-porous saplings from temperate and tropical forests. New Phytol 186:439–450

Mehltreter K, Walker LR, Sharpe JM (2010) Fern Ecology. Cambridge University Press, New York

Meyer-Berthaud B, Scheckler SE, Wendt J (1999) *Archaeopteris* is the earliest known modern tree. Nature 398:700–701

Meyer-Berthaud B, Scheckler SE, Bousquet J-L (2000) The development of *Archaeopteris*: new evolutionary characters from the structural analysis of an Early Famennian trunk from southeast Morocco. Am J Bot 87:456–468

Moran RC (2008) Diversity, biogeography, and floristics. In: Ranker TA, Haufler CH (eds) Biology and evolution of ferns and lycophytes. Cambridge University Press, Cambridge, pp 367–394

Morrow AC, Dute RR (1998) Development and structure of pit membranes in the rhizome of the woody fern *Botrichium dissectum*. IAWA J 19:429–441

Mullendore DL, Windt CW, Van As H, Knoblauch M (2010) Sieve tube geometry in relation to phloem flow. Plant Cell 22:579–593

Nikinmaa E, Sievanen R, Holtta T (2014) Dynamics of leaf gas exchange, xylem and phloem transport, water potential and carbohydrate concentration in a realistic 3-D model tree crown. Ann Bot 114(44):653–666. doi:10.1093/aob/mcu068

Niklas KJ (1985) The evolution of tracheid diameter in early vascular plans and its implications on the hydraulic conductance of the primary xylem strand. Evolution 39:1110–1122

Niklas KJ (1992) Plant biomechanics: an engineering approach to plant form and function. University of Chicago Press, Chicago, p 410

Niklas KJ (1999). Evolutionary walks through a land plant morphospace. J Exp Bot 50:39–52.

Niklas KJ (2000) The evolution of plant body plans—a biomechanical perspective. Ann Bot 85:411–438

Niklas K, Smocovitis V (1983) Evidence for a conducting strand in the early Silurian (Llandoverian) plants: implications for the evolution of the land plants. Paleobiology 9:126–137

Niklas KJ, Speck T (2001) Evolutionary trends in safety factors against wind-induced stem failure. Am J Bot 88:1266–1278

Ogura Y (1972) Comparative anatomy of vegetative organs of the pteridophytes. Gebruder Borntraeger, Berlin, p 502

Page CN (2002) Ecological strategies in fern evolution: a neopteridological overview. Rev Palaeobot Palynol 119:1–33

Panshin AJ, de Zeeuw C (1980) Textbook of wood technology. McGraw-Hill, New York

Pickard WF, Abraham-Shrauner B (2009) A "simplest" steady-state Munch-like model of phloem translocation, with source and pathway and sink. Funct Plant Biol 36:629–644

Pittermann J (2010) The evolution of water transport in plants: an integrated approach. Geobiology 8:112–139

Pittermann J, Sperry JS, Hacke UG, Wheeler JK, Sikkema EH (2005) The torus-margo pit valve makes conifers hydraulically competitive with angiosperms. Science 310:1924

Pittermann J, Sperry JS, Hacke UG, Wheeler JK, Sikkema EH (2006) Mechanical reinforcement against tracheid implosion compromises the hydraulic efficiency of conifer xylem. Plant Cell Environ 29:1618–1628

Pittermann J, Choat B, Jansen S, Stuart SA, Lynn L, Dawson T (2010) The relationships between xylem safety and hydraulic efficiency in the Cupressaceae: the evolution of pit membrane form and function. Plant Physiol 153:1919–1931

Pittermann J, Limm E, Rico C, Christman M (2011) Structure function constraints of tracheid-based xylem: a comparison of conifers and ferns. New Phytol 192:449–461

Pittermann J, Stuart SA, Dawson TE, Moreau A (2012) Cenozoic climate change shaped the evolutionary ecophysiology of the Cupressaceae conifers. Proc Natl Acad Sci U S A 109:9647–9652

Pittermann J, Brodersen C, Watkins JE (2013) The physiological resilience of fern sporophytes and gametophytes: advances in water relations offer new insights into an old lineage. Front Plant Sci. doi:10.3389/fpls.2013.00285

Pockman WT, Sperry JS (1997) Freezing-induced xylem cavitation and the northern limit of *Larrea tridentata*. Oecologia 109:19–27

Proctor MCF (2012) Light and desiccation responses of some Hymenophyllaceae (filmy ferns) from Trinidad, Venezuela and New Zealand: poikilohydry in a light-limited but low evaporation ecological niche. Ann Bot 109:1019–1026

Proctor MCF, Tuba Z (2002) Poikilohydry and homoihydry: antithesis or spectrum of possibilities? New Phytol 156:327–349

Ranker TA, Haufler CH (2008) Biology and evolution of ferns and lycophytes. Cambridge University Press, New York

Raven JA (1984) Physiological correlates of the morphology of early vascular plants. Bot J the Linn Soc 88:105–126

Robinson RC, Sheffield E, Sharpe JM (2010) Problem ferns: their impact and management. In: Mehltreter K, Walker LR, Sharpe JM (eds) Fern ecology. Cambridge University Press, Cambridge, pp 255–322

Rothfels CJ, Sundue MA, Kuo L-Y, Larsson A, Kato M, Schuettpelz E, Pryer KM (2012) A revised family-level classification for eupolypod II ferns (Polypodiidae: Polypodiales). Taxon 61:515–533

Rothwell GW, Karrfalt EE (2008) Growth, development and systematics of ferns: does *Botrychium* S.L. (Ophioglossales) really produce secondary xylem? Am J Bot 95:414–423

Rothwell GW, Stockey RA (2008) Phylogeny and evolution of ferns: a paleontological perspective. In: Ranker T, Hausler C (eds) Biology and evolution of ferns and lycophytes. Cambridge University Press, Cambridge

Rowe N, Speck T (2005) Plant evolutionary forms; an ecological and evolutionary perspective. New Phytol 166:61–72

Rowe N, Isnard S, Speck T (2004) Diversity of mechanical architectures in climbing plants: an evolutionary perspective. J Plant Growth Regul 23:108–128

Salleo S, Lo Gullo MA, Trifilo P, Nardini A (2004) New evidence for a role of vessel-associated cells and phloem in the rapid xylem refilling of cavitated stems of *Laurus nobilis* L. Plant Cell Environ 27:1065–1076

Schneider H, Pryer KM, Cranfill R, Smith AR, Wolf PG (2002) Evolution of vascular plant body plans: a phylogenetic perspective. In: Cronk QCB, Bateman RM, Hawkins JA (eds) Developmental genetics and plant evolution. Taylor & Francis, London, pp 330–364

Schneider H, Schuettpelz E, Pryer KM, Cranfill R, Magallon S, Lupia R (2004) Ferns diversified in the shadow of angiosperms. Nature 428:553–557

Schuettpelz E, Pryer KM (2009) Evidence for a Cenozoic radiation of ferns in an angiosperm-dominated canopy. Proc Natl Acad Sci U S A 106:11200–11205

Schulte PJ, Gibson AC, Nobel P (1987) Xylem anatomy and hydraulic conductance of *Psilotum nudum*. Am J Bot 74:1438–1445

Scoffoni C, Rawls M, McKown A, Cochard H, Sack L (2011) Decline in leaf hydraulic conductance with dehydration: relationship to leaf size and venation architecture. Plant Physiol 156:832–843

Secchi F, Zwieniecki MA (2011) Sensing embolism in xylem vessels: the role of sucrose as a trigger for refilling. Plant Cell Environ 34:514–524

Sessa EB, Zimmer EA, Givnish TJ (2012) Phylogeny, divergence times, and historical biogeography of new world Dryopteris (Dryopteridaceae). Am J Bot 88:730–750

Sessa EB, Givnish TJ (2014) Leaf form and photosynthetic physiology of Dryopteris species distributed along light gradients in eastern North America. Funct Eco 28:108–123

Serbet R, Rothwell GW (1999) *Osmunda cinnamomea* (Osmundaceae) in the Upper Cretaceous of western North America: additional evidence for exceptional species longevity among filicalean ferns. Int J Plant Sci 160:425–433

Sharpe JM, Mehltreter K (2010) Ecological insights from fern population dynamics. In: Mehltreter K, Walker LR, Sharpe JM (eds) Fern ecology. Cambridge University Press, New York, pp 61–139

Smith AR, Pryer KM, Schuettpelz E, Korall P, Schneider H, Wolf PG (2006) A classification for extant ferns. Taxon 55:705–731

Sperry JS (2003) Evolution of water transport and xylem structure. Int J Plant Sci 164:S115–S127

Sperry JS, Hacke UG, Pittermann J (2006) Size and function in conifer tracheids and angiosperm vessels. Am J Bot 93:1490–1500

Stebbins GL (1971) Chromosomal evolution in higher plants. Addison-Wesley, Reading

Strullu-Derrien C, Kenrick P Tafforeau P, Cochard H, Bonnemain J-L, Le Herisse A, Lardeux H, Badel E (2014) The earliest wood and its hydrualic properties document in c. 407-million-year-old fossils using synchrotron microtomography. Bot J Linn Soc 175:423–437

Taylor TN, Taylor EL, Krings M (2009) Paleobotany. The biology and evolution of fossil plants. Academic, Oxford

Testo WL, Watkins JE, Barrington DS (2014) Dynamics of asymmetrical hybridization in North American wood ferns: reconciling patterns of inheritance with gametophyte ecology. New Phytol. doi:10.1111/nph.13213

Testo WL, Watkins JE, Pittermann J, Momin R (2015) *Pteris* X *caridadiae* (Pteridaceae), a new hybrid fern from Costa Rica. Brittonia doi:10.1007/s12228-015-9370-8

Tyree MT, Davis SD, Cochard H (1994) Biophysical perspectives of xylem evolution: is there a tradeoff of hydraulic efficiency for vulnerability to cavitation? IAWA J 15:335–360

Vasco A, Moran RC, Ambrose BA (2013) The evolution, morphology, and development of fern leaves. Front Plant Sci 4:345. doi:10.3389/fpls.2013.00345

Veres JS (1990) Xylem anatomy and hydraulic conductance of Costa Rican Blechnum ferns. Am J Bot 77:1610–1625

Warner DA, Edwards GE (1993) Effects of polyploidy on photosynthesis. Photosynth Res 35:135–147

Watkins JE, Cardelús CL (2009) Habitat differentiation of ferns in a lowland tropical rain forest. Am Fern J 3:162–175

Watkins JE, Cardelus CL (2012) Ferns in an Angiosperm world: Cretaceous radiation into the epiphytic niche and diversification on the forest floor. Int J Plant Sci 173:695–710

Watkins JE, Mack MC, Sinclair T, Mulkey SS (2007a) Ecological and evolutionary consequences of desiccation tolerance in tropical fern gametophytes. New Phytol 176:708–717

Watkins JE, Mack MC, Mulkey SS (2007b) Gametophyte ecology and demography of epiphytic and terrestrial tropical ferns. Am J Bot 94:701–708

Watkins JE, Rundel P, Cardelus CL (2007c) The influence of life form on carbon and nitrogen relationships in tropical rainforest ferns. Oecologia 153:225–232

Watkins JE, Holbrook NM, Zwieniecki MA (2010) Hydraulic properties of fern sporophytes: consequences for ecological and evolutionary diversification. Am J Bot 97:2007–2019

Wheeler JW, Sperry JS, Hacke UG, Hoang N (2005) Intervessel pitting and cavitation in woody Rosaceae and other vesseled plants: a basis for a safety vs. efficiency trade-off in xylem transport. Plant Cell Environ 28:800–812

White RA, Weidlich WH (1995) Organization of the vascular system in the stems of *Diplazium* and *Blechnum* (Filicales). Am J Bot 82:982–991

Wilson JP (2013) Modeling 400 million year of plant hydraulics. Paleontol Soc Pap 19:1–20

Wilson JP, Fischer WW (2010) Hydraulics of *Asteroxylon mackei*, an early Devonian vascular plant, and the early evolution of water transport tissue in terrestrial plants. Geobiology 9:121–130

Wilson JP, Knoll AH (2010) A physiologically explicit morphospace for tracheid-based water transport in modern and extinct seed plants. Paleobiology 36:335–355

Wilson JP, Knoll AH, Holbrook NM, Marshall CR (2008) Modeling fluid flow in *Medullosa*, an anatomically unusual Carboniferous seed plant. Paleobiology 34:472–493

Wood TE, Takebayashi N, Barker MS, Mayrose I, Greenspoon PB, Rieseberg LH (2009) The frequency of polyploid speciation in vascular plants. Proc Natl Acad Sci U S A 106: 13875–13879

Zimmermann M, Tomlinson PB (1974) Vascular patterns in palm stems: variations of the *Raphis* principle. J Arnold Arb 55:402–424

Zimmermann M, Tyree MT (2002) Xylem structure and the ascent of sap. Springer, Berlin

Zwieniecki MA, Melcher PJ, Holbrook NM (2001) Hydraulic properties of individual xylem vessels of *Fraxinus americana*. J Exp Bot 52:257–264

Chapter 2
The Hydraulic Architecture of Conifers

Uwe G. Hacke, Barbara Lachenbruch, Jarmila Pittermann, Stefan Mayr, Jean-Christophe Domec, and Paul J. Schulte

1 Introduction

Conifers survive in diverse and sometimes extreme environments (Fig. 2.1a–f). Piñon-juniper communities are found in semi-arid environments, receiving ca. 400 mm of yearly precipitation (Linton et al. 1998), which is less than half the

U.G. Hacke (✉)
Department of Renewable Resources, University of Alberta,
442 Earth Sciences Building, Edmonton, AB, Canada, T6G 2E3
e-mail: uwe.hacke@ualberta.ca

B. Lachenbruch
Department of Wood Science and Engineering, Oregon State University,
Corvallis, OR 97331, USA

Department of Forest Ecosystems & Society, Oregon State University, Corvallis, OR USA
e-mail: barb.lachenbruch@oregonstate.edu

J. Pittermann
Department of Ecology and Evolutionary Biology, University of California,
Santa Cruz, CA 95064, USA
e-mail: jpitterm@ucsc.edu

S. Mayr
Department of Botany, University of Innsbruck, Sternwartestr. 15, Innsbruck 6020, Austria
e-mail: stefan.mayr@uibk.ac.at

J.-C. Domec
Bordeaux Sciences Agro—INRA, UMR ISPA,
1 cours du Général de Gaulle, Gradignan 33175, France

Nicholas School of the Environment, Duke University, Durham, NC 27708, USA
e-mail: jc.domec@agro-bordeaux.fr

P.J. Schulte
School of Life Sciences, University of Nevada, Las Vegas, NV 89154-4004, USA
e-mail: paul.schulte@unlv.edu

U. Hacke (ed.), *Functional and Ecological Xylem Anatomy*,
DOI 10.1007/978-3-319-15783-2_2

Fig. 2.1 Conifers growing in diverse habitats. (**a**, **b**) *Sequoiadendron giganteum* in the Sierra Nevada mountains of California (photos: B. Pratt). (**c**) *Pinus mugo*, partly covered by snow, at the timberline in Tyrol, Austria (photo: S. Mayr). (**d**) *Pinus cembra* at the alpine timberline in Tyrol, Austria (photo: S. Mayr). (**e**) Boreal mixedwood forest near Whitecourt, Alberta (photo: A. Schoonmaker). The teepee was constructed for a shading experiment. (**f**) *Pinus sylvestris* near Bergen, Norway (photo: A. Jacobsen)

average precipitations received by other coniferous tree species worldwide. *Picea mariana* and *Larix laricina* grow in boreal peatlands where they face fluctuating water tables and cold, often frozen soils (Lieffers and Rothwell 1987). Timberline trees such as *Pinus cembra* experience short growing seasons and cope with frost, winter desiccation, and mechanical challenges (Mayr et al. 2012) (Fig. 2.1c, d). The Cupressaceae family is particularly diverse in terms of the morphology and habitat preference exhibited by its members (Pittermann et al. 2012). While *Taxodium distichum* is adapted to water-logged soils, many *Juniperus* and *Pinus* species are exceptionally drought tolerant. Conifers are "archetypical stress tolerators" (Brodribb et al. 2012), and their xylem structure plays a key role in enabling them to survive in stressful habitats. However, north temperate conifers can also occupy more favorable environments; for example, conifers are an integral component of North American mixed eastern deciduous forests, and the Pacific Northwest landscape is characterized by Douglas-fir and old-growth conifer forests.

The water-conducting cells in the xylem of conifers are tracheids, which are overlapping single-celled hollow conduits, closed at both ends. Water moves through a tracheid's lumen, then passes through a pit pair into the lumen of an adjacent tracheid. Relative to vessel-based xylem of the vast majority of angiosperms, the conifers' ancestral vascular system has been regarded as an Achilles heel that limits the conifers' physiological capacity for growth and competition (Bond 1989; Brodribb et al. 2012). This is understandable: vessels are comprised of stacked, single-celled vessel elements that can be wider than tracheids, such that maximum hydraulic efficiency of angiosperm xylem can be one to two orders of magnitude higher than that of conifers (Ewers 1985). However, when observed through the lens of natural history, conifer xylem could hardly be called a handicap. Indeed, the tracheid-based xylem structure of conifers supports the largest (*Sequoiadendron giganteum*, *Pseudotsuga menziesii*) and tallest (*Sequoia sempervirens*) trees (Fig. 2.1a, b). Bristlecone pine (*Pinus longaeva*) trees are among the oldest known living eukaryotic organisms; several specimens still alive are well over 4,000 years of age (Flanary and Kletetschka 2005).

In this chapter, our goal is to provide a synthesis of what we currently know about the hydraulic architecture of conifers. In the last decade or so, much has been learned about the ecophysiological processes in conifer taxa and how they relate to xylem properties. In future years, we will likely expand our knowledge of topics such as needle and fine root hydraulics, aquaporin function, and phloem transport. While *Arabidopsis thaliana* and other herbaceous species will continue to be widely used as model systems in molecular and developmental biology, it is clear that tall, long-lived woody plants face unique challenges in terms of water and assimilate transport, biomechanics, storage, reproduction, and stress tolerance.

We start with information on how the xylem functions in stems, coarse roots, and then briefly, in needles. The focus is on the structural means by which xylem transports water in conditions with plentiful water, drought, and freezing (Sect. 2). The ecophysiology of a woody plant, however, results from the interaction of the whole plant with the environment, as well as its growth and repair abilities. We therefore review information on within-plant variation in xylem hydraulic properties and link some of the relationships of hydraulic properties to a plant's physiological capabilities (Sect. 3).

2 Conifer Xylem and How It Works

2.1 *Tracheids: The Conduits for Water Transport*

While fiber tracheids and libriform fibers provide structural support in angiosperm xylem, conifer xylem lacks fibers so the tracheids must play a dual role in both transporting water to the canopy, and physically supporting it. More than 90 % of a conifer xylem cross section consists of tracheids (Fig. 2.2a, b). Schulte (2012b) estimated that there is a total number of ~500 million tracheids in the trunks of young Douglas-fir trees. In terms of cell types, conifer wood may therefore appear simple in structure. Closer inspection, however, reveals much variation in cell dimensions, wall thickness, pit structure, and other structural features within a cross section, as well as variation in the ultrastructure of the cell wall itself. Depending on developmental constraints and according to hydraulic and biomechanical tasks, the morphology of tracheids varies considerably, and this variation has direct implications for transport safety and efficiency.

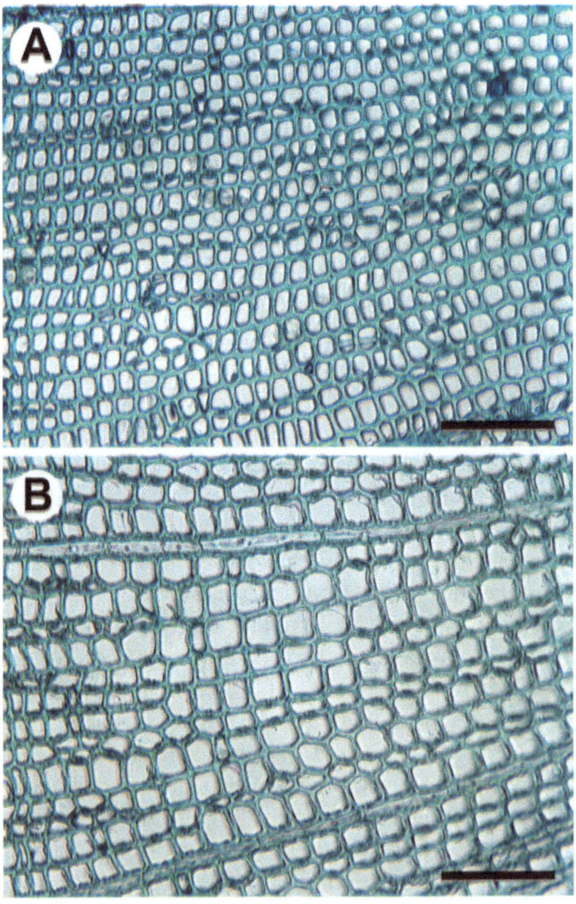

Fig. 2.2 Cross sections showing branch xylem of *Widdringtonia cedarbergensis* (**a**) and *Taxodium mucronatum* (**b**). *W. cedarbergensis* is a xeric species. Its xylem is extremely resistant to cavitation and is characterized by narrow, thick-walled tracheids. *T. mucronatum*, by contrast, is primarily a riparian tree. Tracheids are wider and less reinforced than those of *W. cedarbergensis*. Scale bars = 100 μm. Photos: J. Pittermann

2.1.1 Tracheids and Xylem Structure

Tracheid diameters usually vary between 5 and 80 μm (Tyree et al. 1994; Linton et al. 1998; Pittermann and Sperry 2003; Wilson and Knoll 2010, their Table 1; Schulte 2012b; McCulloh et al. 2014). Diameters tend to be smaller in twigs than in trunks and roots (Bannan 1965; Dunham et al. 2007; Domec et al. 2009). Tracheid diameter is limited by at least two factors. Their tangential expansion during development is likely limited by the width of fusiform initials. Tracheids are formed side by side tangentially. Developing vessel elements, by contrast, can expand at the expense of adjacent fibers and axial parenchyma cells. An indirect limit on tracheid diameter relates to the fact that tracheids, being unicellular, are limited in length. If conduits cannot increase in length as they increase in diameter, their hydraulic conductivity (for a definition of hydraulic conductivity see Tyree and Ewers 1991) will eventually stop increasing (Sperry et al. 2006). For a maximum tracheid length of 5 mm, the model of Hacke et al. (2004) predicted no further increase in tracheid conductivity for an increase in diameter beyond approximately 70–80 μm, which is also close to the observed limit of tracheid diameters.

Tracheid length is usually less than 5 mm (Bannan 1965; Sperry et al. 2006; Dunham et al. 2007; Domec et al. 2008; Schulte 2012b). Longer tracheids are associated with wood that has both higher hydraulic efficiency and higher mechanical properties such as bending strength (Rundel and Stecker 1977; Domec and Gartner 2002a; Rosner 2013), but the relationship with mechanical properties is driven by differences in cell wall properties that are usually associated with cells in positions where cells are longer, and not related to cell length per se (Lachenbruch et al. 2011).

Tracheid length may reflect the length of the fusiform initials in the cambium, but that is unclear. In a typical angiosperm or gymnosperm, xylem cell lengths increase asymptotically along a transect outward from the pith (Panshin and Zeeuw 1980). This relationship, however, is not seen in several circumstances (e.g., roots of *Pseudotsuga menziesii*, Peterson et al. 2007), suggesting that the length of cambial initials does not have total control over tracheid length. The relatively short and narrow tracheids found in extant conifers contrast with the tracheids belonging to *Medullosa*, an extinct, Carboniferous seed fern with a putatively upright habit, similar to modern tree ferns (Taylor et al. 2009). Medullosan tracheids measured up to 28 mm in length and commonly exceeded 150 μm in diameter (Wilson et al. 2008; Wilson and Knoll 2010). These enormous tracheids in the extinct seed fern suggest that short, narrow tracheids in stems of extant gymnosperms are not merely a reflection of developmental constraints, but that tracheid dimensions are under selection for both hydraulic and structural performance. Indeed, the stem tracheids of the late-Devonian pro-gymnosperm *Archaeopteris* bear this out. Fossils show this spore-bearing tree to be a hybrid of conifer and fern attributes whereby the trunk exhibits conifer like, pycnoxylic xylem that supports a canopy of fern-like leaves (see Chap. 1). As in extant conifers, stem tracheids are on average narrower than 30 μm, thereby reflecting the support function, whilst root tracheids frequently reach diameters in excess of 40 μm (Domec et al. 2009, Pittermann, unpublished data). Interestingly, tracheids in fern primary xylem can also be much longer and wider than tracheids of conifers (see Chap. 1), which is presumably possible because fern tracheids play little role in mechanical support of the plant.

As stated above, a volume of coniferous wood is typically about 90–93 % tracheids (Panshin and Zeeuw 1980); the remaining 7–10 % is parenchyma. In contrast to the tracheids, most parenchyma cells maintain their nuclei and cytoplasm until the cells die, at heartwood formation. A characteristic feature of conifer wood is the relatively small amount of parenchyma, particularly axial parenchyma (Evert 2006), compared to the angiosperms, which can have well over 50 % of their xylem volume as parenchyma (although 10–40 % is more typical for commercial temperate-zone hardwoods, Panshin and Zeeuw 1980). The exact function of axial and ray parenchyma is not entirely understood, but they certainly have roles in the transport of chemical signals and other constituents and may contribute to the reversal of embolisms, heartwood formation (Dellus et al. 1997; Taylor et al. 2002; Spicer 2005), and both inducible and constitutive defense. In-depth information can be found in Chap. 8. Xylem parenchyma has three forms in coniferous wood: axial parenchyma, radial (ray) parenchyma, and, in some species, epithelial cells (the cells surrounding resin canals that produce the resin that helps plants defend against biotic agents after injury). We have only rudimentary understanding of the three-dimensional configuration of the parenchyma cells network; in fact, LaPasha and Wheeler (1990) report the unexpected finding that the epithelial cells do not appear to have symplastic connections between the radial and the axially oriented resin canals. Ray parenchyma is in contact with phloem tissue (Fig. 2.3a, b) and thus provides a means by which materials in the phloem can move into the xylem and vice versa. Aquaporin water channels are expressed in ray parenchyma cells (Fig. 2.3a–c), which may facilitate water exchange between ray cells and between rays and axial tracheids. Curiously, only a subset of the conifer species have normal resin canals (those not formed by trauma), and most of these species also have a feature called a ray tracheid, which is a dead cell within the ray that somewhat resembles an axial tracheid in that it has bordered pits, but that has not been shown to transport water (Barnard et al. 2013).

2.1.2 Tracheids and Specific Conductivity

The longer conduits are for a given diameter, the farther apart the end walls are and the less they contribute to the total flow resistance. Examining root and stem wood from 19 conifer species belonging to the Pinaceae, Cupressaceae, Podocarpaceae, and Araucariaceae, Pittermann et al. (2006a) found that total tracheid resistivity was nearly equally divided between lumen and pit, with the passage through the pits contributing approximately 64 % to the total tracheid resistivity. Remarkably, the limitation to hydraulic efficiency caused by passage through the pits was independent of tracheid size in the conifers surveyed. This was associated with proportionality between tracheid diameter and length. The observed tracheid diameter versus length allometry appeared to optimize hydraulic efficiency (Pittermann et al. 2006a).

Mean conduit diameters usually vary significantly between conifer stems and conifer roots, and between conifer stems and angiosperm stems (Fig. 2.4a). Samples were typically 8–12 mm in diameter and mostly represented juvenile xylem. Conifer

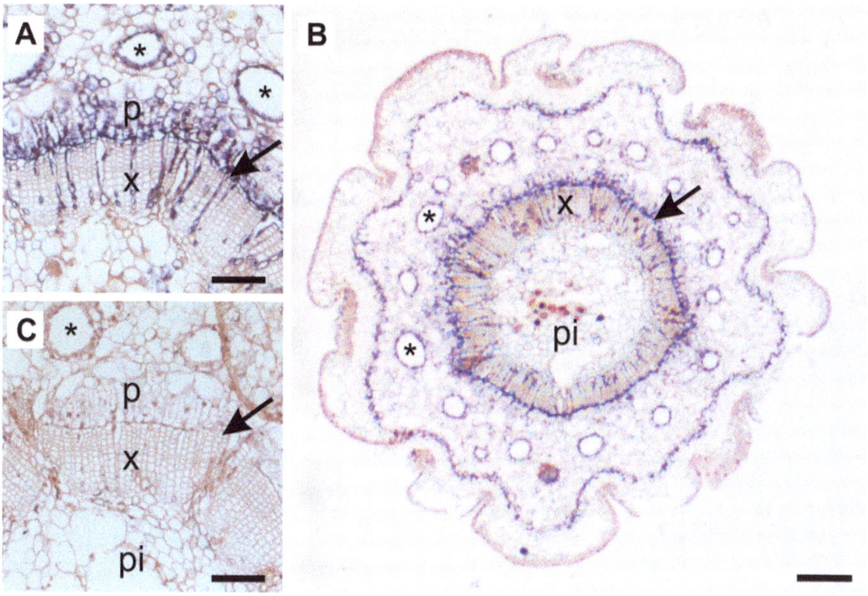

Fig. 2.3 In situ mRNA hybridization of the aquaporin gene *PgPIP1;2* in cross sections of a white spruce (*Picea glauca*) twig. Sections were hybridized with DIG-labeled antisense *PgPIP1;2* RNA probes (**a**, **b**) as described previously (Laur and Hacke 2014). A negative control hybridized with DIG-labeled sense probe is shown in (**c**). Regions of aquaporin expression are indicated by *dark purple* staining. The gene was expressed in xylem rays, the cambial region (*arrows*), the phloem, and in other cortex cells. No staining was found in the negative control. *p* phloem, *pi* pith, *x* xylem. *Asterisks* mark resin ducts. *Bars* in (**a**) and (**c**) = 100 µm, *bar* in (**b**) = 200 µm. Photos: J. Laur

species were from Pinaceae, Cupressaceae, Podocarpaceae, and Araucariaceae. The angiosperm species were eudicot shrubs, trees, and vines from numerous North American habitats and taxa. Conifer stems (mostly branches) had by far the narrowest conduit diameter range. While the median conduit diameter of angiosperm stems and conifer roots was similar, some ring-porous trees and vines had conduits that were much wider than those of conifers (note the log-scale).

Differences in conduit diameter have consequences for transport efficiency, and xylem efficiency in return impacts water supply to the leaf and stomatal conductance (Pittermann et al. 2012). In the data set shown in Fig. 2.4b, vines and some conifer roots exhibited the highest xylem area-specific conductivities. Remarkably, the median xylem area-specific conductivity at full saturation (K_{smax}) of conifer roots was higher than in the other groups. This is explicable in the light of highly conductive pits and the fact that a large fraction of conifer wood is occupied by conduits owing to the double role of tracheids in transport and mechanical support (Sperry et al. 2012). For a given mean conduit diameter, conifer xylem has slightly higher conductivities than angiosperm xylem in which vessels are dispersed in a fiber-parenchyma matrix (Sperry et al. 2006, their Fig. 3B).

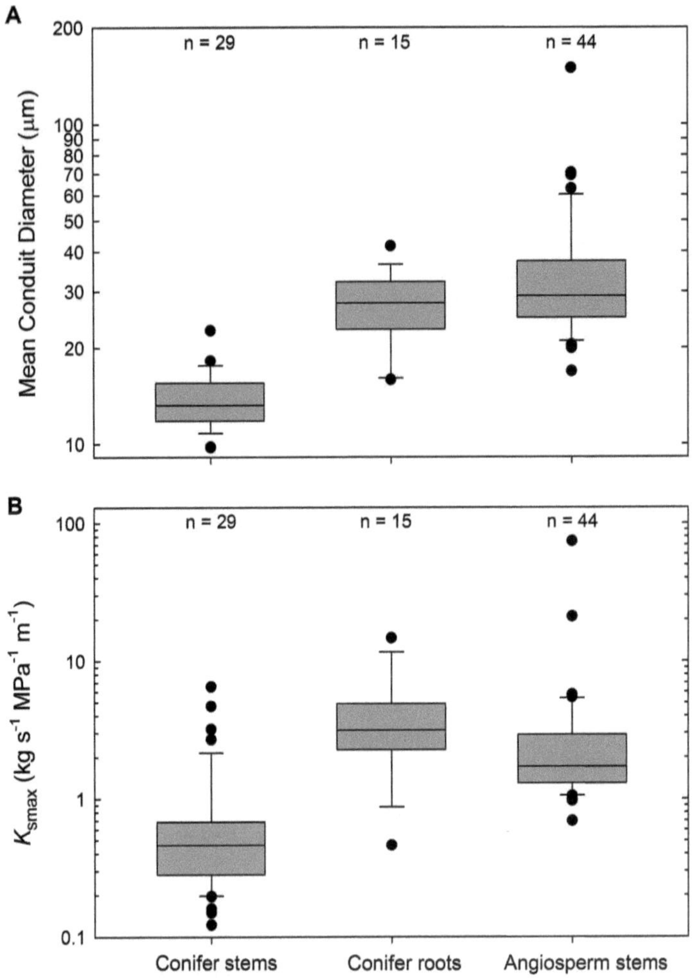

Fig. 2.4 (**a**) Box plots of mean conduit diameters and (**b**) xylem area-specific conductivity (K_{smax}) of conifer stems, conifer roots, and angiosperm stems. The ends of the boxes define the 25th and 75th percentiles. The *middle line* in the boxes represents the median. The error bars (*whiskers*) define the 10th and 90th percentiles. Outliers beyond the whiskers are individually plotted as *circles*. The data is for flushed samples that were at least 14 cm in length, and was taken from various publications (Hacke et al. 2006; Pittermann et al. 2006a; Hacke and Jansen 2009; Schoonmaker et al. 2010; Schreiber et al. 2011)

However, vessels can achieve greater maximum diameters than tracheids, which can ultimately provide greater conductivity per xylem area. Tropical trees reach the greatest maximum transport per stem cross-sectional area since they can combine large vessels with large sapwood areas (Sperry et al. 2012). Nonetheless, in habitats where vessel diameter is constrained by freezing and/or drought stress, both xylem types should exhibit similar K_{smax} values, and this may help conifers compete with angiosperms (Pittermann et al. 2005; Sperry et al. 2008).

2.2 *Bordered Pits: The Tracheid-to-Tracheid Valves for Water Transport*

Because water transport relies on the movement of water under tension (negative hydrostatic pressure), the introduction of gas would result in a breakage of the water column. Therefore, there must be very strong selective pressure for avoiding the breakage of the water column by gas in the tissues (e.g., earlywood) through which transport is needed. Coniferous pits are elegant structures that permit water movement from one cell to another, whilst protecting the cells from air entry in case an adjacent tracheid is gas-filled (see below). Coniferous bordered pits are circular perforations in the tracheid cell wall, paired with a coincident perforation on the adjacent tracheid (and so technically, they are considered as pit pairs). The volume between the two borders is called the pit chamber. The perforation is covered with a torus–margo pit membrane (Fig. 2.5a–d). The number of bordered pits within one earlywood tracheid can vary from approximately 15 to >100 (Evert 2006; Schoonmaker et al. 2010; Schulte 2012b). Pit pairs are largely confined to the radial walls, suggesting a relative ease of water-sharing in the tangential direction around a stem, as opposed to the radial (between growth rings) direction. Pit pairs are most abundant near the tips of the tracheids, which appears to limit the loss of flow resistance across the entire cell (Aumann and Ford 2002). While narrow tracheids have only one row of inter-tracheid pits, wide tracheids (e.g., in roots) often have two rows of pits. Two pit rows may connect a tracheid with two distinct tracheids or just one adjacent conduit (Schulte 2012b). Details of how pits are formed are provided in Chap. 3. The structure and function of these pits represents a unique compromise between requirements for efficient water transport and the need to prevent spread of air from embolized tracheids (Pittermann et al. 2005; Domec et al. 2008). Since tracheid length is limited, pit characteristics are a major determinant of tracheid and whole xylem hydraulic conductivity.

In the center of the pit membrane is a thickened, approximately circular torus, which is impermeable to water. The torus is surrounded by a porous membrane region called the margo, which provides passage for water. When two adjacent tracheids are water-filled, the torus is centrally located in the pit chamber (Fig. 2.6a), allowing water flow between tracheids through the pores of the margo, as influenced by passive movement according to the pressure gradients. Should an air bubble enter one of the tracheids, the membrane will move according to the pressure gradient, such that the torus will block the aperture in the pit border (Fig. 2.6b). This state is called pit aspiration.

Across species and plant organs (roots and stems), pit membrane diameter scaled linearly with both torus and aperture diameters (Fig. 2.7). In Pinaceae species, the torus represented 45–49 % of the pit diameter (Hacke et al. 2004; Domec et al. 2006; Hacke and Jansen 2009). In tall trees, this ratio may vary with height. The tight scaling of pit dimensions is likely to have important functional implications. Conifers have circular bordered pits, and the valve, consisting of the torus and margo, works well to seal such pits. Angiosperms, in contrast, have a much wider range of pit shapes, including the extreme, the scalariform pit, which is relatively

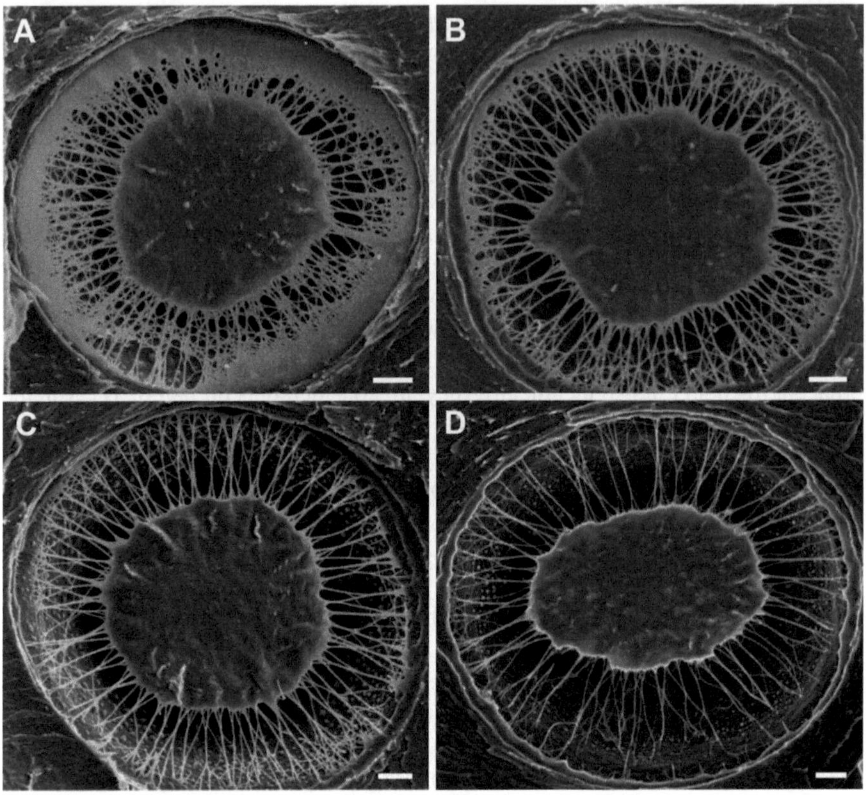

Fig. 2.5 Scanning electron microscopy images showing torus–margo pits of lodgepole pine (*Pinus contorta*) stems. Samples were taken from open-grown (**a–c**) and shaded (**d**) trees as explained in Schoonmaker et al. (2010). The images represent a continuum of margo porosities. Some pits have a dense margo with small pores, particularly near the edge of the membrane (**a**) while other pits have few margo "spokes" and large pores between the margo fibrils. The latter was particularly frequent in shaded trees (**d**). *Bars* = 1 μm. Images: A. Schoonmaker and G. Braybrook

long and narrow. Although there are tori or pseudotori in some angiosperm species (for details, see Chap. 3), these pits lack the highly conductive margo portion and are not at all like those of conifers in conductive ability (Carlquist 2012).

In many cases, the pit membrane can rebound from an aspirated position. This capacity is a prerequisite for the recovery of hydraulic conductivity after embolized tracheids have been refilled. In *Picea abies*, very close agreement has been found between the amount of winter embolism (expressed as percent loss of hydraulic conductivity, PLC) and the ratio of aspirated versus open pits (Mayr et al. 2014). Nearly one-half of all inter-tracheid pits were closed in January and opened as PLC decreased and conductivity was reestablished in spring.

The valve action of torus–margo pits also appears to facilitate embolism repair when the transpiration stream is under tension (Mayr et al. 2014). Under these conditions, refilling in individual tracheids may only be possible if these conduits are

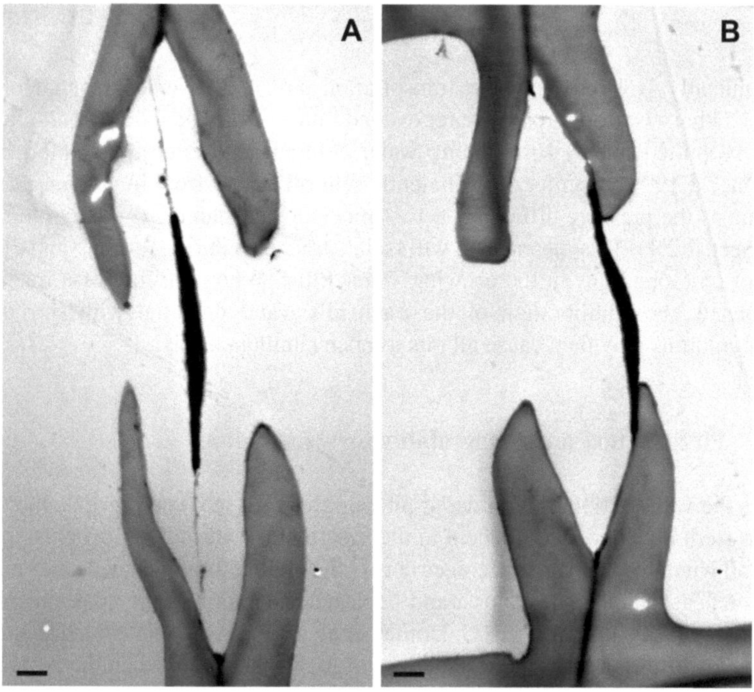

Fig. 2.6 Transmission electron micrographs of pit membranes in roots of *Picea glauca*. The membrane is relaxed (**a**) or aspirated, blocking the aperture (**b**). *Bars* = 1 μm. From Hacke and Jansen (2009)

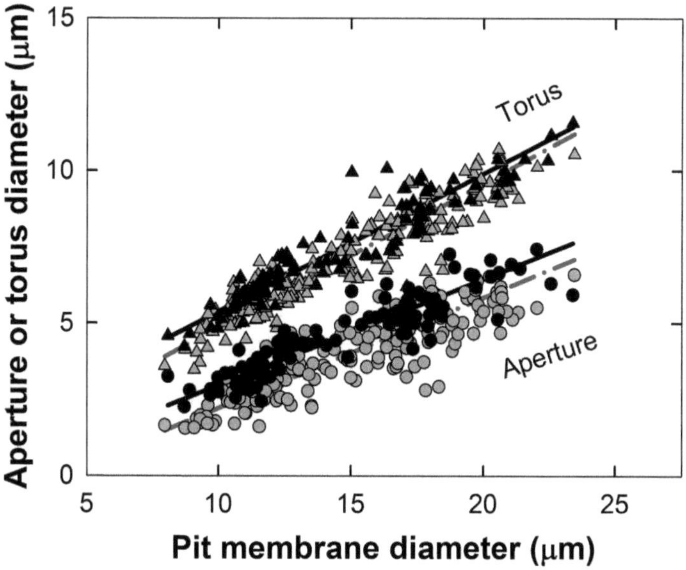

Fig. 2.7 Torus (*triangles*) or pit aperture (*circles*) diameter versus pit membrane diameter for root and stem tracheids. Measurements were either done on surface views of pits (*closed symbols* and *solid lines*) or on cross sections of pits (*gray symbols* and *dashed lines*) as explained previously (Hacke and Jansen 2009). The regression lines are shown for pooled stem and root data. The data were taken from three boreal conifer species (*Abies balsamea*, *Picea glauca*, and *Picea mariana*). From Hacke and Jansen (2009)

hydraulically isolated from the transpiration stream (Zwieniecki and Holbrook 2009). Otherwise, any drop of water exuded into a refilling conduit would immediately be pulled out by surrounding water columns under tension (Holbrook and Zwieniecki 1999). Conifer pits efficiently seal off empty from functional tracheids as long as the pressure difference across pits is lower than the cavitation threshold (see Sect. 2.2.1). This mechanism will isolate tracheids during refilling and dissolution of entrapped gas in the growing water films. When refilling of a tracheid is completed, the equilibration of the tracheid's water potential with surrounding water columns may then cause all pits to open simultaneously.

2.2.1 Pit Structure and Vulnerability to "Air-Seeding"

While the valve action of a bordered pit pair creates a seal that prevents the spread of gas from an embolized tracheid to the functional, water-filled tracheids, the seal will fail when the pressure difference across the pit membrane exceeds the "air-seed pressure," at which air leaks past and nucleates cavitation in the adjacent tracheid (Hacke et al. 2004, their Fig. 1C; Domec et al. 2008). Available evidence suggests that beyond this air-seed pressure, failure of the valve occurs when the torus is displaced from its sealing position, exposing pores in the margo through which gas enters the water-filled tracheid (Sperry and Tyree 1990; Domec et al. 2006). How can differences in cavitation resistance between species and plant organs be explained in the light of pit structure?

Based on the sealing function of the torus, we expect that the size of the torus relative to the pit aperture impacts cavitation resistance. A torus too small for a given aperture diameter may result in greater vulnerability to air-seeding. A torus wider than the optimum reduces margo area and conductivity. One may therefore hypothesize that there is an optimal range of "torus overlap." The term torus overlap may be defined as the fraction of the pit border that is covered by the torus $[(D_t - D_a)/(D_m - D_a)]$, where D_t, D_a, and D_m are torus, aperture, and pit membrane diameter, respectively. Reported values of torus overlap are between 0.21 and 0.38 (Hacke et al. 2004; Hacke and Jansen 2009).

Torus overlap can also be quantified by using the D_t/D_a ratio (Domec et al. 2008). When D_t and D_a were measured in Douglas-fir trees along a height gradient of 85 m, striking trends were found. While D_t remained relatively unchanged (Fig. 2.8a), D_a decreased significantly with increasing height (Fig. 2.8b). As a result, there was an increase in the D_t/D_a ratio with increasing height in both trunk and branch tracheids (Fig. 2.8c). Higher D_t/D_a ratios mean that the pit can sustain a greater pressure difference between an embolized and a water-filled tracheid (ΔP); however, smaller aperture diameters come at the expense of a decline in hydraulic conductivity (Domec et al. 2008).

In 15 species of Cupressaceae, the D_t/D_a ratio also showed a strong relationship with cavitation resistance (expressed as the xylem pressure inducing 50 % loss of hydraulic conductivity, P_{50}). The D_t/D_a ratio was mainly determined by variation in D_a, which decreased nearly twofold with more negative cavitation pressures

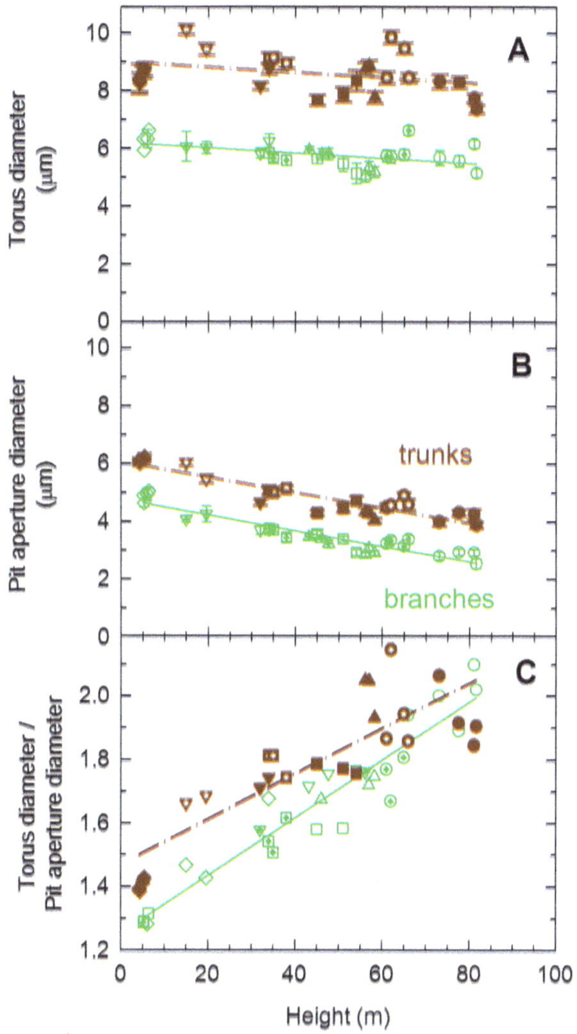

Fig. 2.8 Pit anatomical traits as a function of height in branches (*green*) and trunks (*brown*) of large Douglas-fir trees. While torus diameter (**a**) remained relatively constant with height, pit aperture diameter (**b**) decreased significantly. As a result, the ratio of torus to aperture diameter (**c**) increased with height. Values are means ± 1 standard deviation. From Domec et al. (2008)

(Pittermann et al. 2010). Lower D_a was also observed at higher elevation in *Picea abies*, which is exposed to increasing frost near the timberline (Mayr et al. 2002). Finally, a recent survey reported that torus overlap increased with increasing cavitation resistance (Bouche et al. 2014). Available data from Pinaceae and Cupressaceae therefore indicate that torus overlap is a key determinant of cavitation resistance, and that it is largely driven by variation in D_a.

Aside from torus overlap, cavitation resistance is potentially impacted by a wide variety of other structural pit features. To name a few, torus extensions, torus thickness/flexibility, the presence of pores in the torus, margo properties, the distance the margo must stretch to reach the aperture, and the surface properties of the torus and the internal side of the pit border (smooth or rough surface) have all been implicated

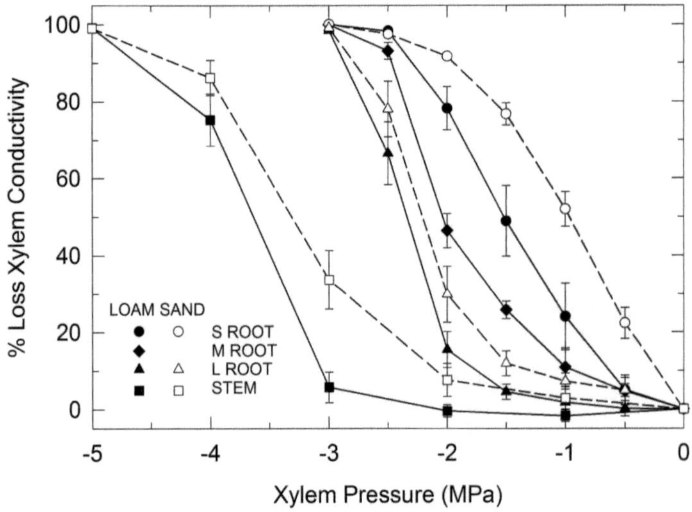

Fig. 2.9 Vulnerability curves showing the percentage loss of hydraulic conductivity as a function of xylem pressure for roots (*S* small 1.6–4.1 mm diameter; *M* medium 4.2–6.4 mm diameter; *L* large 8.2–13 mm diameter) and stems of *Pinus taeda* trees growing in sand (*open symbols*, *dashed lines*) versus loam (*solid symbols* and *line*) soil. Means and standard errors are shown for segments from six trees per site. From Hacke et al. (2000)

in potentially affecting air-seeding (Hacke et al. 2004; Cochard et al. 2009; Hacke and Jansen 2009; Delzon et al. 2010; Schoonmaker et al. 2010; Jansen et al. 2012; Bouche et al. 2014; Plavcová et al. 2013). However, some reports are contradictory (e.g., regarding torus thickness and torus extensions), and it is not always clear whether observed correlations reflect causal relationships. Nonetheless, neither torus extensions nor punctured tori appear to be rare phenomena across Pinaceae species, and it has been hypothesized that such "irregularities" in pit structure can have substantial influence on cavitation resistance (Plavcová et al. 2013). While punctured tori appear to be associated with increased vulnerability to cavitation (Jansen et al. 2012), the impact of torus extensions on ΔP may be species-specific (Pittermann et al. 2010; Schoonmaker et al. 2010).

The xylem of conifer stems typically shows sigmoidal vulnerability curves with a well-defined and highly reproducible cavitation threshold (Fig. 2.9). This may in part be explained by the fact that earlywood tracheids, which contribute to most of the water flow within a ring (Domec and Gartner 2002b) are relatively uniform in terms of size and hydraulic properties compared with a typical population of vessels in an angiosperm stem. Assuming that cavitation resistance is to a large extent determined by pit structure, one may conclude that pit structure in earlywood tracheids is also fairly homogeneous, at least in terms of the structural features of pits that relate to ΔP. This conclusion is consistent with the tight scaling of D_m, D_t, and D_a (Fig. 2.7), but it seems inconsistent with the astonishing variability in margo porosity (Fig. 2.5a–d) and the seemingly random occurrence of features like torus extensions and torus pores that can be observed even within an axis. This could either mean that these features have only minor impact on cavitation resistance or it may be an indication that cavitation resistance is not as much determined by torus and margo structure as previously thought. Cavitation is therefore probably not a direct consequence of the rupture of the margo strands or when air is sucked through the margo pores, but rather related to the capacity of the torus to stay in place when aspirated. These minute mechanical pit features may contribute to the stochasticity of the cavitation response in conifers, and probably reflect some degree of variation associated with wall digestion during tracheid development. The available evidence indicates that selection has favored adjustments in torus thickness, torus diameter, and aperture diameter in a predictable direction that balances cavitation resistance with efficient water flow.

In latewood tracheids, air seeding appears to occur through margo pores without torus aspiration (Domec and Gartner 2002b). Latewood is highly vulnerable to cavitation at normal operating pressures, and may serve more in water storage than in transport (Domec and Gartner 2002b; McCulloh et al. 2014). Similarly, compression wood, which has mainly mechanical function, also showed higher vulnerability to drought-induced cavitation compared to opposite wood (Mayr and Cochard 2003).

2.2.2 Pit Structure and Hydraulic Conductivity

Many conifer trees attain impressive heights (Fig. 2.1b), despite the fact that tracheids are only a few mm in length. Water would have to flow past a minimum of 7,500 pit pairs to get to the top of a 30 m tall tree if tracheids average 4 mm long and water flowed from the tip of one tracheid into the base of the next. Such flow is possible because torus–margo pits are very efficient in terms of water transport (Hacke et al. 2004; Pittermann et al. 2005, 2006a; Sperry et al. 2006; Domec et al. 2008). What is it that makes these pits so efficient? As noted previously (and apparent in Fig. 2.5), the conifer (torus–margo) membrane has much larger pores than does the typical membrane in angiosperms (Pittermann et al. 2005; Choat et al. 2008; Choat and Pittermann 2009). This porous nature of conifer pit membranes was suggested a century ago in drawings by Bailey (1913). Estimates of the resistivity of the margo in conifers fall within a range of 3- to 40-fold lower than found for the pit membrane of angiosperms (Hacke et al. 2004). Choat et al. (2008) reviewed a large number of angiosperm and conifer estimates, also showing a clear distinction in pit membrane resistivity between conifers and angiosperms despite considerable variation.

Modeling studies of flow through conifer pits using a computational fluid dynamics approach (Lancashire and Ennos 2002; Schulte 2012a) can help to distinguish resistance components attributed to various pit structures such as the pit aperture, and the two components of the membrane: the torus providing an obstruction to flow across the pit membrane and the margo. In a study of pits in *Picea mariana* (Schulte 2012a), the margo contributed the majority of the resistance in the modeled pit from a sun-grown tree (63 % of pit resistance). But in the shade-grown case, the margo contributed only 38 % of the total pit resistance. The differences appeared to be due to the presence of wider pores in the shade pit, suggesting that considerable variation in the pit membrane component of resistance is likely to occur among

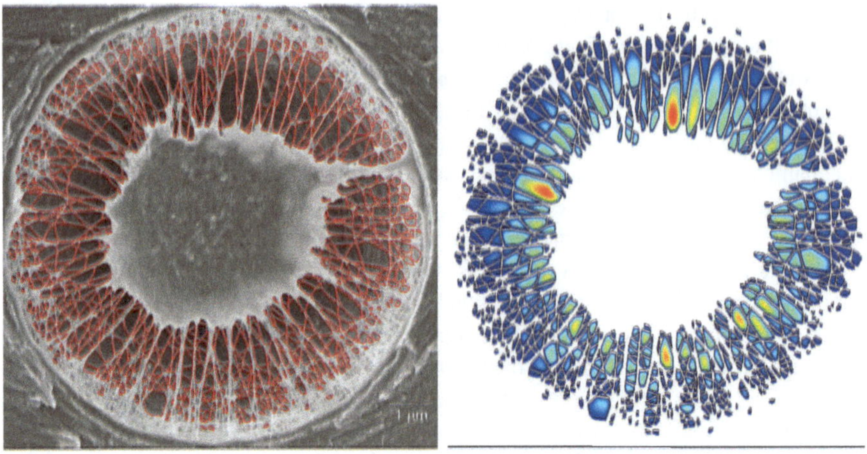

Fig. 2.10 Modeling of pits in *Picea mariana* stems. The *left* image shows an SEM view of a pit membrane (torus and margo) where the pore edges have been drawn as polygons in AutoCAD for import into the fluid dynamics model. The *right* image shows results from the model solution, whereby the color scale (0–7 mm/s as *red*) shows flow velocity within the margo pores. See Schulte (2012a) for details

conifers. Model solutions can be used to estimate the flow rate through individual pores (Fig. 2.10) and the total contribution to flow of each margo pore. Although the wide pores are much fewer in number than narrow pores, they are more significant for flow: for the pits modeled, only a few percent of the margo pores accounted for nearly half of the total flow. As shown in Fig. 2.5a, b, there is a tendency for denser margo structures at the edge of the membrane. Pores tend to be larger closer to the torus, and this has implications for hydraulic conductivity. A characteristic of flow through the margo pores (apparent in Fig. 2.10) is that flow tends to be greater through pores at the inner edge of the margo adjacent to the torus. This suggests that margo structure (the distribution of margo strands and pore size) might be somewhat optimized for providing high flow rate while also providing strength for mechanical support of the torus.

Most of the work described above has been done on Pinaceae and Cupressaceae species. However, not all conifers may demonstrate the functionality and selective directionality of pit attributes that have been found in the north temperate conifers (Pittermann et al. 2006a).

2.3 *Freezing-Induced Embolism*

Wide tracheids, such as those found in the riparian bald-cypress, *Taxodium distichum* commit species to environments without freezing or water stress. Conduit diameter determines vulnerability to freezing-induced embolism; conifers with narrow tracheids are better protected against freezing-induced embolism in cold climates than trees with wider conduits (Pittermann and Sperry 2003). Indeed, boreal latitudes, north-temperate treelines and high deserts such as those in the south-western USA are dominated by members of the Pinaceae (Brodribb et al. 2012). Why are tracheids so advantageous in cold climates?

Freeze–thaw-induced cavitation is a process by which air-bubbles frozen in the xylem sap expand during the thaw, and thus create an embolism in functional xylem conduits. The process begins when during freezing, air is pushed out of the ice crystal lattice, coalesced into bubbles and thus trapped in the frozen xylem sap (Sevanto et al. 2012). During the thaw, as transpiration reintroduces tension into the water column and the xylem regains functionality, the bubbles may either expand and fill the conduit with a mixture of air and water vapor, or they may collapse, with air redissolving back into the xylem sap. Which of the two options occurs is in large part determined by the xylem tension during the thaw (P_x) and the dimensions of the xylem conduits. Over two decades worth of research has shown that xylem conduit dimensions are mechanistically related to species' vulnerability to freeze–thaw cavitation (Ewers 1985; Davis et al. 1999; Pittermann and Sperry 2003).

The classic bubble-expansion mechanism rests on the relationship between P_x and the bubble pressure P^*, which is the combined pressure of the surface tension (t) and the air/water mixture (P_b, assumed to be a relative 0 MPa) of a bubble in liquid sap (Domec 2011). P^* is directly related to the diameter of an air-bubble (D_b) whereby

$$P^* = -4t / D_b + P_b \tag{2.1}$$

For a bubble to expand and nucleate cavitation, P_x must be more negative than P^*, the bubble's equilibrium pressure. Equation 2.1 reminds us that the larger the bubble, the lower the P^* and the greater the likelihood of bubble expansion under mild P_x. Accordingly, conifers with small tracheids were reported to be hardly susceptible to freeze–thaw-induced embolism (Hammel 1967; Sucoff 1969; Sperry and Sullivan 1992; Sperry et al. 1994; Davis et al. 1999; Feild and Brodribb 2001; Feild et al. 2002).

Several studies have shown that species with large-diameter conduits are more vulnerable to freeze–thaw-induced cavitation than those with narrower vessels or tracheids. Presumably, this is because the size of the bubble is directly proportional to the distance that the ice front must travel, so wider conduits lead to larger bubbles. This process is largely independent of conduit length. Angiosperms and conifers with equal-diameter vessels and tracheids show similar losses in conductivity at a given P_x during the thaw. Indeed, conduits wider than 43–44 μm will embolize during a mild P_x of −0.5 MPa (Davis et al. 1999; Pittermann and Sperry 2003). However, this critical diameter is predictably reduced to values below 18 μm under more negative thaw xylem pressures, which have been observed to vary between −1 and −4 MPa in nature (Pittermann and Sperry 2006 and literature cited therein). Aside from *Taxodium distichum* and *Pinus caribaea*, few conifers have branch tracheids that exceed 20 μm in diameter so they are well adapted to high-elevation habitats and the north-temperate and boreal regions where freeze–thaw cycles are a seasonal, or even daily occurrence during the winter season (Feild and Brodribb 2001; Schoonmaker et al. 2010). Tracheids in roots may be above 40 µm in diameter, but they are often insulated by soil and snow, and experience fewer freeze–thaw events than stems.

Repeated freeze–thaw events in conjunction with drought may cause higher loss of conductivity than single freeze–thaw events (Mayr et al. 2003a). Timberline conifers are exposed to more than 100 freeze–thaw cycles per winter; these frost cycles in combination with low water potentials cause up to 100 % loss of conductivity in *Picea abies* and other species (Mayr et al. 2006). Embolism formation upon freeze–thaw events amplified by drought stress was also reported in *Pinus* and *Larix* species (Sparks and Black 2000; Sparks et al. 2001). It is still unclear if increasing conductivity losses on repeated freeze–thaw cycles reflect stochasticity of the underlying bubble-expansion mechanism or if additional processes are involved. Ultrasonic emission analyses indicate that cavitation events are already initiated during freezing (Mayr et al. 2007; Mayr and Sperry 2010) and that minimal temperatures of a frost cycle are of relevance for bubble formation (Charrier et al. 2014).

Freeze–thaw events can cause very complex hydraulic patterns in conifers as observed at the alpine timberline: When the soil and/or trunk base are frozen, water supply is blocked, which can last for months in extreme environments (Fig. 2.1c). Evergreen conifers then have to keep stomata closed during winter and require a cuticular protection which sufficiently minimizes water loss as long as water supply is blocked (Michaelis 1934; Tranquillini 1976). However, overheating of needles, limited cuticular maturation, ice blast, or insufficient stomatal closure cause a slow but permanent and thus substantial decrease in water potential (Mayr et al. 2012). In parallel, freeze–thaw events lead to a segmentation of the trees' hydraulic system, as frozen xylem sections separate needles from branches and branches from the trunk. This segmentation changes in diurnal courses, so that timberline conifers show complex and dynamic patterns in water potential which, in turn, influence the formation of freeze–thaw-induced embolism (Fig. 2.11).

2.4 Phenotypic Plasticity

Much of the data discussed above refers to variation of traits across species. There is less information on intraspecific variability of hydraulic traits. Chapter 4 describes phenotypic plasticity with regards to hydraulic traits in *Populus* genotypes. How much plasticity exists in conifer species? In a study on 12 Scots pine (*Pinus sylvestris*) populations sampled along a climatic gradient in Western Europe, variability across sites was high for K_{smax} and leaf specific xylem conductivity (LSC), but not for P_{50}, midday water potential, mean tracheid diameter, and wood density (Martínez-Vilalta et al. 2009). The stability of midday water potentials (isohydric behavior) was a consequence of reducing the leaf-to-sapwood area ratio (A_L:A_S) with decreasing water availability, and possibly other factors. The study on Scots pine could not resolve how much of the observed variability in hydraulic traits was

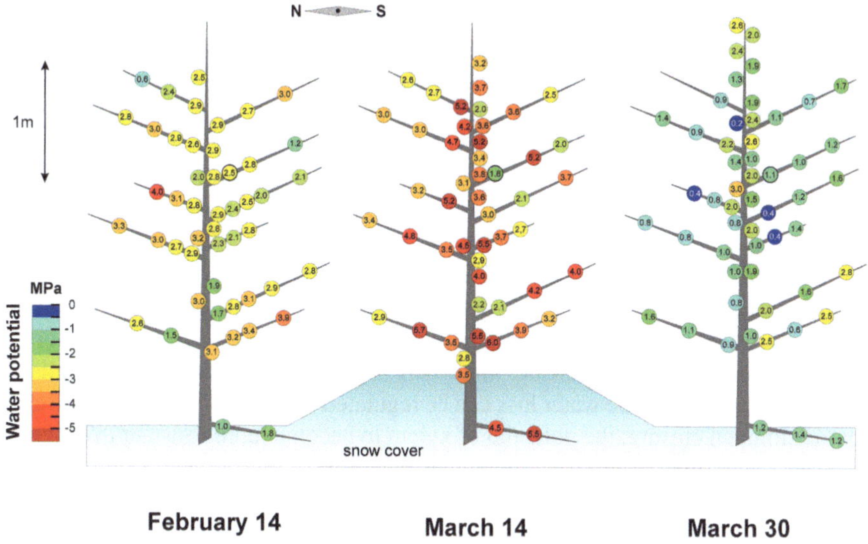

Fig. 2.11 Pattern in water potential during winter 2006 at the alpine timberline (2,035 m). Samples (*end twigs*) were taken from north and south exposed branches of one *Picea abies* tree in February as well as mid- and end of May and measured with the Scholander technique. Modified after Mayr and Charra-Vaskou (2007)

genetic in origin or the result of phenotypic plasticity (Martínez-Vilalta et al. 2009). Common garden experiments are useful in shedding light on this question.

Recent common garden experiments with interior spruce (*Picea glauca* and *Picea engelmannii*) and lodgepole pine (*Pinus contorta*) seed sources studied variation in frost hardiness and seedling growth. Most of the trait variation existed within populations. There was a remarkable degree of genetic diversity in the measured traits, implying a high evolutionary capacity to adapt to new environmental conditions (Liepe 2014). This is an interesting finding, and it raises the question how much genetic diversity exists for key hydraulic and wood anatomical traits, such as P_{50} and mean tracheid diameter.

A study on eight populations of *Pinus canariensis* growing at two common garden sites found considerable variability in P_{50} and other traits, and this was largely the result of phenotypic plasticity (López et al. 2013). Variability in hydraulic traits was also examined in six *Pinus pinaster* populations growing at a mesic and a xeric site (Corcuera et al. 2011). P_{50} and K_{smax} were significantly lower in the xeric site, and these shifts were mainly attributed to phenotypic plasticity. By contrast, wood density was genetically controlled and was not affected by the growing conditions.

Substantial phenotypic plasticity in vulnerability to cavitation and other important hydraulic traits was also found in other studies. For instance, loblolly pine (*Pinus taeda*) trees growing on loam versus sand soil exhibited phenotypic plasticity in P_{50}, root-to-leaf area ratios (A_R:A_L), and other traits (Hacke et al. 2000). Another study on loblolly pine found that nutrient addition caused an over 50 % reduction in A_R:A_L relative to control trees (Ewers et al. 2000). Reduction in A_R:A_L

was primarily due to a large increase in leaf area with improved nutrition. Changes in A_R:A_L are important, because they impact the potential for water uptake during soil drought. Schoonmaker et al. (2010) studied hydraulic acclimation to shading in four boreal conifer species. Open-grown and understory trees differed in many wood anatomical and hydraulic traits. One of the most profound changes occurred at the pit level. Shaded trees had, on average, larger pores in the margo than open-grown trees (Fig. 2.5). More porous pit membranes compensated for narrower tracheids in shaded trees, but also contributed to increased vulnerability to cavitation.

These studies suggest that conifers exhibit considerable potential for acclimating to different environments. Changes may occur at various levels, from shifts in A_R:A_L and A_L:A_S ratios to microscopic changes at the pit level. However, the piñon mortality described in Sect. 3 indicates that there are genetically determined limits to drought tolerance. While water loss can be regulated by the amount of leaf area and/or via stomatal closure, the potential of xylem to become more resistant to drought-induced cavitation during development appears to be rather limited. Limits to phenotypic plasticity may also constrain the height growth of giant conifers (Koch et al. 2004). To the extent that the water potential at the top of tall trees differs from the potential at the base of the tree, plasticity in wood structure and function will be required to adequately supply leaves with water.

2.5 *Needle Structure and Water Movement in Needles*

Needle architecture in conifers has been described as canalized and being limited in venation patterns compared with angiosperm leaves (Brodribb et al. 2012). While the diversity of venation patterns is limited, conifer needles are marvels in terms of tissue complexity and longevity. Huber (1947) pointed out that we can find many different cell types in a small cross section of a pine needle and viewed pine needles as some of the most complex structures in all of plant anatomy. Many structure–function relationships related to water and assimilate transport remain to be explored. In *Pinus longaeva*, needles remain alive on branches for 30 or more years. With advancing needle age, little or no change in the number of xylem cell layers was observed, but there was a significant increase in the layer of phloem cells (Ewers 1982). Hence, needles produced secondary phloem but no secondary xylem. In *Pinus longaeva*, newly produced phloem cells seem to continually replace dying phloem cells. The work of Ewers (1982) also implies that needle xylem can remain functional for many years.

Needles have been reported to be more vulnerable to dehydration-induced declines in hydraulic conductance than stems (Johnson et al. 2009; Johnson et al. 2011; McCulloh et al. 2014). Changes of needle hydraulic conductance may be caused by cycles of embolism and refilling (Woodruff et al. 2007; Johnson et al. 2012), reversible collapse of tracheids (Cochard et al. 2004; Brodribb and Holbrook 2005; Zhang et al. 2014), changes in the membrane permeability of bundle sheath cells (Laur and Hacke 2014), and other changes in the properties of living cells (see Chap. 10).

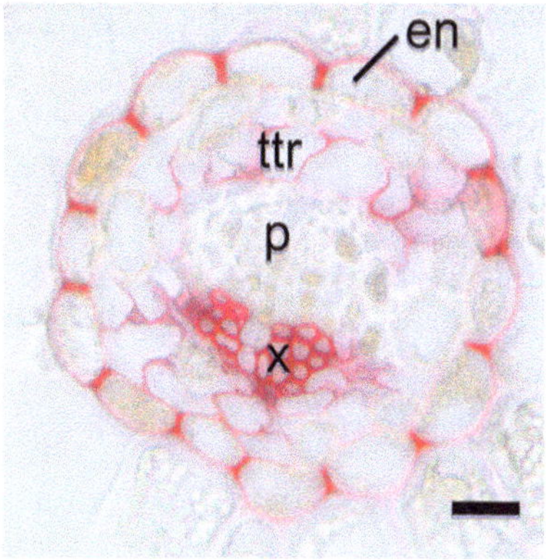

Fig. 2.12 Light microscopy image of a *Picea glauca* needle cross section. The section was stained with phloroglucinol-HCl; lignified cells walls are shown in *red*. Lignin was detected in radial cell walls of the endodermis (*en*), transfusion tracheids (*ttr*), and in xylem tracheids (*x*). *Bar*=20 µm. From Laur and Hacke (2014)

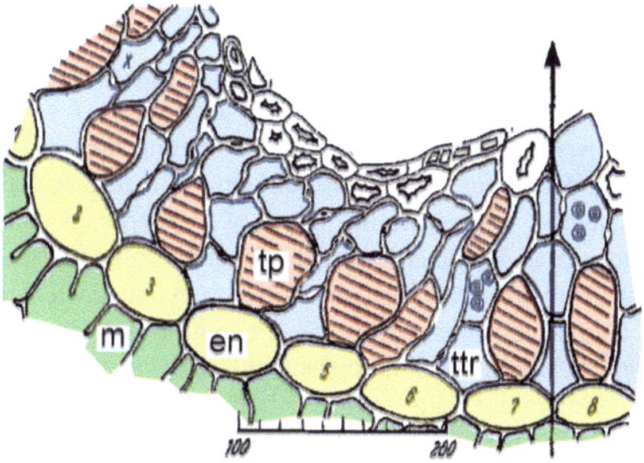

Fig. 2.13 Cross section of a pine needle (*Pinus nigra*). Modified after Huber (1947). Endodermis cells (*yellow*) are numbered (#*1*–*8*). Transfusion tracheids (*blue*, *ttr*) typically face the radial walls of the endodermis cells. Transfusion parenchyma cells (*tp*, *orange*, marked with *hatched lines*) are in contact with the median endodermis cell walls. *m* mesophyll (*green*). The *arrow* points to the center of the needle

As shown in Fig. 2.12, axial needle xylem of *Picea glauca* is surrounded by phloem and transfusion tissue. The work of Zhang et al. (2014) suggests that transfusion tissue may play an important role in needle hydraulics. This tissue includes transfusion tracheids, which serve in water storage and radial transport between the axial xylem and the bundle sheath (Fig. 2.13). Transfusion tracheids of *Taxus baccata* needles undergo reversible collapse in response to dehydration–rehydration cycles, and these volume changes were linked to reversible changes in leaf hydraulic conductance (Zhang et al. 2014). The reversible shrinkage of these cells may impact

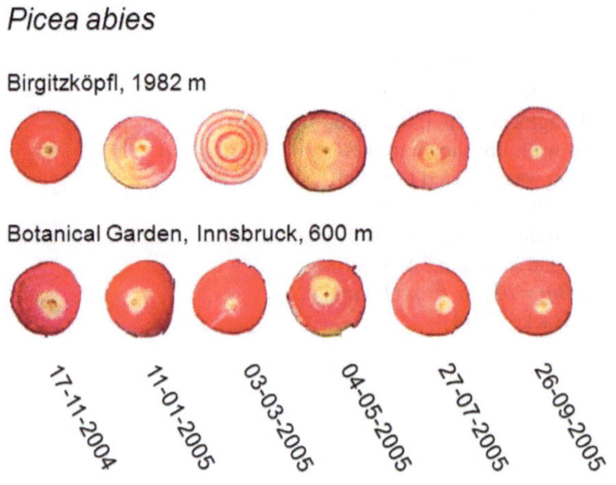

Fig. 2.14 Dye staining of branch cross sections of *Picea abies*. Branches were harvested from trees at low and high elevation in winter 2004/2005 and stained with Phloxine-B. *Red* areas indicate functional xylem, *non-stained* areas are embolized (Schmid and Mayr, unpublished)

the xylem pressure of axial tracheids. More work will be necessary to validate the existing hydraulic and imaging techniques that are suitable for studying needles, and to develop novel empirical methods and modeling approaches for studying needle hydraulics. Water potential measurements on a fine spatial and temporal scale would be helpful in interpreting some of the reported changes in needle hydraulic conductance and in determining the optimal needle transport capacity that minimizes the pressure drop per given investment in xylem (Zwieniecki et al. 2006).

Aquaporins (AQPs) are present in various needle tissues (see Laur and Hacke 2014 for details), including the bundle sheath. These water channels likely play a key role in facilitating radial water exchange between vascular and photosynthesizing needle tissue. Individual AQP genes of *Picea glauca* had different tissue distribution patterns of expression; yet all four of the studied genes showed expression in phloem cells (Laur and Hacke 2014). The prominent occurrence of AQPs in needle phloem suggests a role for water channels in phloem loading/unloading.

Needles are capable of taking up water from fog, intercepted rainfall, or melting snow (e.g., Burgess and Dawson 2004; Breshears et al. 2008a). Needle water uptake appears to play an important role in timberline conifers, which recover hydraulic conductivity in late winter and early spring, even when soils remain frozen. The needles of timberline trees are able to take up water from melting snow, and this water contributes to embolism repair in branches (Mayr et al. 2014). At the time of refilling, AQPs appear in the needle phloem and endodermis indicating a role in water uptake and the refilling process. Figure 2.14 shows a series of dye-perfused stem cross sections of *Picea abies* with functional xylem throughout the year at low elevation but remarkable winter embolism at the timberline. The entire cross section is refilled so that full functionality is restored for the vegetation period. The dye perfusion experiments shown in Fig. 2.14 indicate that conifer stems can recover hydraulic conductivity in late winter and early spring, even when soils remain frozen.

3 Conifer Hydraulic Architecture: How the Xylem Works in the Context of the Whole Tree

The performance of xylem in a live plant depends not only on the properties of the material, but on the distribution of water potentials within the plant at any time and how that affects physiological processes like photosynthesis and growth. The distribution of water potentials, in turn, depends on both the soil water availability and the atmospheric vapor pressure deficit, but also on the plant directly: the amount of xylem (sapwood width, length of the path), the variation in xylem properties within the plant, the size of the root system, the leaf area, capacitance (how much water will come out of storage per MPa water potential), and physiological behavior like the triggers of stomatal opening and closure. Moreover, the environmental factors and some of the plant-related factors will change in a cyclical manner (with seasons and phenology, for example), and others will change on a finer time scale. We have a relatively good understanding of how the plant-related factors should affect hydraulic performance in general. There is still much to learn, however, about the regulation and coordination of processes that contribute to the observed water potentials throughout a plant.

3.1 Leaf Specific Xylem Conductivity

Leaf specific xylem conductivity is a property that incorporates some of the architectural information about the plant into hydraulic information. The LSC is the hydraulic conductivity of a stem segment normalized by the leaf area distal to the segment, rather than by the stem cross-sectional area (Tyree and Ewers 1991). Whereas K_{smax} tells how "efficient" the material is at allowing water to pass, LSC tells how much water transport can occur per leaf area. The higher the LSC, the lower the water potential gradient required to maintain a particular transpiration rate.

In conifers, LSC generally increases greatly with stem diameter (Tyree and Ewers 1991), meaning that branches tend to have lower, LSC than the trunk (Domec et al. 2012). In branches of *Juniperus communis* trees, LSC and K_{smax} were low at the base, increased toward the distal parts and decreased again near the branch tips (Beikircher and Mayr 2008). These patterns are controlled in part by differences in tracheid dimension. One of the consequences of such patterns of LSC is the development of steep water potential gradients in smaller diameter branches (Tyree and Ewers 1991 and literature cited therein). Small distal branches are also the parts of the tree that are the least buffered from environmental conditions (sun, wind, drought), and thus are worthy of more study to better understand how entire plants deal with drought stress, and to better understand the process of branch die-back and death during drought.

A physical map of the water potentials at the apices of all the branches in a tree may have a strong relationship with a physical map showing the amount of extension growth of the stems, with more growth occurring where water is more available.

During the growing season, these ultimate water potentials depend on the flow path's resistance and on the leaf area supplied. We do not have a strong understanding of the feedback between growth and wood development—does a low water potential direct the formation of a bud that will make a small leaf area? Most research on conifers has been conducted in trees with a pyramidal form that results from a dominant central leader and spreading angled branches off the leader. It would be interesting to better understand the relationship of tree form, branch growth, and hydraulic architecture by having more research in coniferous species that do not have this pyramidal form.

3.2 The Transport Network

Within a woody plant, water moves from the root to the stomatal cavity through a network of xylem conduits. Much research attention has focused on the optimal form of that network for transporting water with the least biomass investment. It has long been known that the bole has higher K_{smax} than branches, and that the higher the order of branch, typically the lower the K_{smax} (e.g., Dunham et al. 2007; Domec et al. 2012). This pattern results largely from the positive correlation of tracheid diameter and stem diameter (Petit et al. 2009). McCulloh et al. (2010) concluded that conifers had a pattern of change in stem conductivity with stem size that was intermediate between the steep change in ring-porous hardwoods, and the shallower change in diffuse-porous hardwoods.

Another issue related to the transport network is the pathways by which water moves radially. Assuming that water is delivered primarily to the outer growth ring in roots and that it exits from xylem attached primarily to the outer three or four growth rings in the shoot (Maton and Gartner 2005), but that it ascends the tree through many growth rings, one can ask how that water moves in the radial direction, near both at the source and transpiration ends. In a typical conifer, the bordered pits are mostly on the tracheids' radial walls, which means water can move most easily in the tangential direction (e.g., laterally, within a ring). In some species there is a small amount of tangential wall pitting in the last-formed latewood and the first-formed earlywood (Panshin and Zeeuw 1980; Kitin et al. 2009), presumably facilitating movement between rings. Barnard et al. (2013) used dye to visualize the radial water movement, and found no evidence that water movement occurred through ray tracheids.

3.3 Transport Safety and Drought-Induced Cavitation

Cavitation resistance potentially confers drought tolerance because maintenance of water-filled conduits during drought ensures sufficient supply of water to leaves. The greater the cavitation resistance, the higher the potential for gas exchange and for extracting water from a drying soil (Sperry et al. 1998). Across Cupressaceae species, the P_{50} of stems was correlated with the minimum growing-season precipitation of the species' native range (Pittermann et al. 2012). Similar relationships have also been found for southern hemisphere conifers (Brodribb and Hill 1999).

A number of case studies on the piñon-juniper communities of the south-western USA revealed how cavitation resistance of conifers relates to habitat and gas exchange. Utah juniper (*Juniperus osteosperma*) dominates in the lower, drier sites while Colorado piñon (*Pinus edulis*) is more common as precipitation increases with elevation. Juniper is generally considered as the more drought-tolerant species, which agrees with juniper having more resistant xylem than piñon (Linton et al. 1998). Tracheid diameters in juniper were narrower than in piñon, and the hydraulically weighted tracheid diameter was correlated with cavitation resistance.

The two species showed contrasting dynamics in terms of xylem pressure and gas exchange. *P. edulis* exhibited pronounced stomatal closure during drought, which allowed plants to maintain high xylem pressures. Due to its relatively vulnerable xylem, piñon appeared to be operating near its hydraulic limits (Linton et al. 1998). These differences in water relations between the two species may explain why piñon disappears from piñon-juniper woodlands at lower, drier elevations.

The Linton et al. (1998) study revealed that co-occurring plants may not only differ in their resistance to cavitation, but also in their strategies for avoiding cavitation. The latter may be just as important as cavitation resistance per se for predicting how species may cope with climate change and associated changes in moisture availability and how co-occurring species influence each other under drought-stress conditions. An isohydric strategy might fail when neighboring plants risk high water losses and a reduction of soil water potentials.

A 4-year drought (from 2000 to 2003), which was associated with anomalously high temperatures, triggered extensive mortality of *P. edulis* (Breshears et al. 2005). The die-off likely relates to the greater xylem vulnerability of piñon relative to juniper, and the water use strategies that are associated with these differences in cavitation resistance. While *P. edulis* exhibited higher sap flux than *J. osteosperma* under favorable conditions, *P. edulis* used isohydric stomatal control to avoid cavitation during periods of water shortage. *J. osteosperma*, by contrast, sacrificed high sap flux rates for greater resistance to cavitation and exhibited anisohydric stomatal regulation (West et al. 2008). The high resistance to drought-induced cavitation in *Juniperus* species helps explain the survival of junipers during recent extreme droughts and why they have been able to expand into drier environments across the western and central USA over the last century (Willson et al. 2008).

Prolonged stomatal closure in *P. edulis* in response to persistent water stress may lead to a depletion of carbon reserves (Plaut et al. 2012; Sevanto et al. 2014). This may make trees susceptible to beetle attack, and may eventually lead to mortality (Breshears et al. 2008b; West et al. 2008). Drought-stressed trees may also lack the carbon reserves that are required to reestablish hydraulic function after a drought (Brodribb and Cochard 2009), i.e., trees may not be able to refill embolized xylem and grow new fine roots (Plaut et al. 2012). In addition, it remains to be tested if vulnerability to cavitation increases in conifer xylem that previously cavitated and refilled (so-called "cavitation fatigue," Hacke et al. 2001a).

The studies on the piñon-juniper woodlands indicate that it is useful to interpret cavitation resistance in the context of stomatal behavior. As shown for piñon, trees may not necessarily die from outright hydraulic failure (McDowell et al. 2013; Sevanto et al. 2014), but rather from the consequences that arise from the need to avoid hydraulic failure. Accumulation of hydraulic deterioration as observed in trembling aspen (Anderegg et al. 2013) may also be important in conifers. Knowing the vulnerability curves (or the P_{50}) of different species and plant organs is a prerequisite for understanding and predicting mortality patterns. Recent research comparing traits of *Pseudotsuga menziesii* trees after a severe drought in France showed slightly higher wood densities in the survivors (Ruiz Diaz Britez et al. 2014). These results are consistent with research showing that clones of the same species that had higher density also had more negative values of P_{50} (Dalla-Salda et al. 2009).

What structural features are associated with greater cavitation resistance and why does the P_{50} of conifer stems scale with the minimum xylem pressure measured in plants under natural conditions (Choat et al. 2012)? While more cavitation-resistant xylem will be required in dry habitats, it may not be advantageous in mesic habitats. There are at least two costs associated with resistant xylem. First, species that are resistant to drought-induced cavitation tend to have narrower tracheids and lower K_{smax} than species with vulnerable xylem (Fig. 2.15a). We caution, however, that the diameter-vulnerability trends are not always observed in other data sets. Second, greater cavitation resistance was associated with greater wood density, an excellent correlate of wood strength (Fig. 2.15b) (again, not always observed in other data sets), which is based on respective carbon investments.

A priori, one may not expect to find either of these relationships. Wood density reflects earlywood and latewood densities weighted by their respective proportion within a ring, whereas the hydraulic properties of a sample are mainly determined by the earlywood (Domec et al. 2009). Further, it is generally assumed that cavitation resistance is primarily determined by properties of inter-tracheid pits, and it is not clear how pit properties would scale with wood density and tracheid diameter. Why then have correlations between P_{50}, wood density, and tracheid diameter emerged in many datasets (Hacke et al. 2004; Pittermann et al. 2006b, 2012; Domec et al. 2009; Rosner 2013)?

A link between P_{50}, conduit reinforcement, and wood density can be explained in the light of cohesion-driven water transport. The more resistant the xylem is to cavitation, the more negative the xylem pressure the plant can sustain, and the stronger the tracheid walls must be. Bending stresses arise in the double wall between water- and air-filled tracheids (Hacke et al. 2001b). The thicker the double wall (t) relative to its span (b), the greater the reinforcement against collapse. One may therefore expect conduit reinforcement (t/b) to scale with the pressure range tracheids must cope with. This pressure range is limited by cavitation. A riparian *Taxodium* tree has vulnerable xylem, wide tracheids, and will require less conduit reinforcement than *Widdringtonia cedarbergensis*, which has extremely resistant xylem ($P_{50}=-11.3$ MPa, Pittermann et al. 2010). Comparing xylem cross sections of these taxa demonstrates that this expectation is justified (Fig. 2.2).

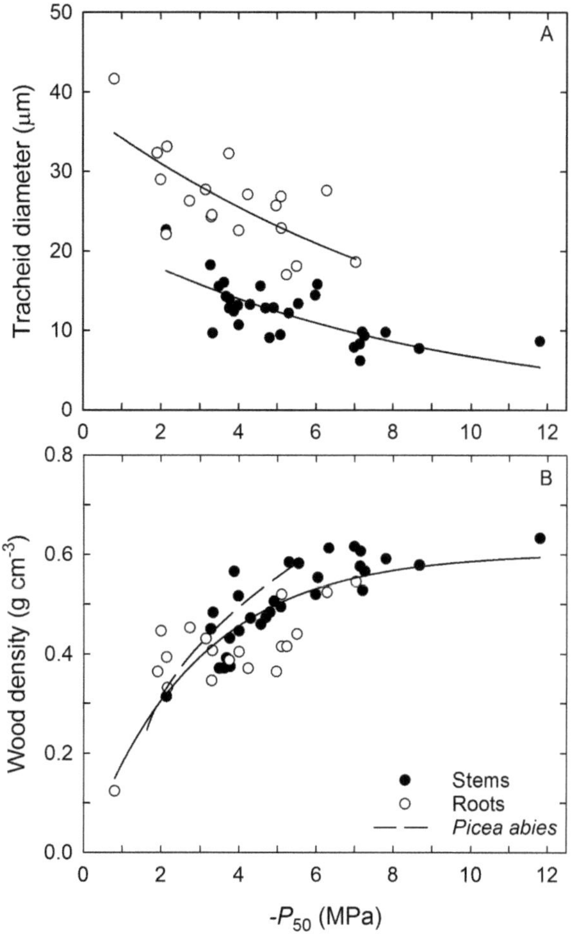

Fig. 2.15 Mean tracheid diameters (**a**) and wood densities (**b**) of roots (*open symbols*) and stems (*closed symbols*) of northern hemisphere conifer species as a function of their cavitation resistance (expressed as the xylem pressure inducing 50 % loss of hydraulic conductivity, P_{50}). The data was taken from various publications (Hacke et al. 2001b); Pittermann et al. 2006b; Hacke and Jansen 2009; Schoonmaker et al. 2010). The *dashed curve* in (**b**) is a regression of data presented by Rosner et al. (2014, their Fig. 2a), and is based on samples taken at various heights from 5- to 50-year-old Norway spruce (*Picea abies*) trees growing in Sweden, Norway, and Germany. See Rosner et al. (2014) for details

In Pinaceae and Cupressaceae species, the significant increase in conduit reinforcement and wood density with more negative P_{50} translated into decreasing tracheid diameters with more negative P_{50}. This is because the variation in *t*/*b* in the data set of Pittermann et al. (2006b) was mainly driven by variation in tracheid diameter rather than wall thickness (but see Bouche et al. 2014). Trends in tracheid diameter impact transport efficiency, which explains why attaining greater mechanical reinforcement comes at the expense of reduced K_{smax} (Pittermann et al. 2006b, 2012). These structure–function trade-offs likely explain the correlations shown in Fig. 2.15.

However, Pittermann et al. (2006b, 2012) found that southern hemisphere species were quite different from the northern hemisphere conifer species featured in Fig. 2.15. The wood of Podocarpaceae, Araucariaceae, and southern hemisphere Cupressaceae appears to be overbuilt for bearing transport stress, making their xylem expensive and hydraulically inefficient. Functionally, this means that for an equivalent P_{50}, conduit diameters and *t*/*b* ratios are much higher in these taxa than in north temperate conifer endemics, reminding us that xylem attributes are somewhat decoupled from transport efficiency in these trees (Pittermann et al. 2006a, b). This suggests that selection for efficient water transport is not as strong in these taxa. Given that the sampled trees occupied humid, mesic habitats in which water limitation is unlikely, it is possible that selection acted on xylem attributes that are more related to rot resistance and longevity. In contrast, tracheid walls in the needles of *Pinus* and *Podocarpus* species were found to be insufficiently reinforced and thus to show cell wall deformation before cavitation (Cochard et al. 2004; Brodribb and Holbrook 2005).

The wood density versus P_{50} relationship has far-reaching consequences due to the significance of wood density for various aspects of tree physiology (for a discussion of some of these, see Pittermann et al. 2006b; Lachenbruch and McCulloh 2014). To the extent that wood density is related to growth and drought sensitivity, the data shown in Fig. 2.15b has potential implications for tree breeding. Rosner et al. (2014) discussed if wood density can be used as a screening trait for drought sensitivity in Norway spruce. They concluded that density is a highly predictive trait for cavitation resistance of trunkwood in Norway spruce (see the dashed curve in Fig. 2.15b). However, they cautioned that the relationship between density and P_{50} is only valid when mechanical demands do not interact with hydraulic demands. Rosner et al. (2014) pointed out that rapid growth is often a principal criterion for tree breeding. Selecting for growth may lead to lower transport safety, at least in species like Norway spruce, in which growth rate is inversely related to wood density. In the light of global climate change it may be safer to select for higher densities, even at the cost of slower short-term growth under favorable conditions.

3.4 *Variations of Xylem Safety Across Plant Organs and with Height*

How does cavitation resistance vary across plant organs? Several hypotheses predict how xylem structure should change along the root-to-branch water transport pathway to optimize tree hydraulic efficiency and protection from embolism (Comstock and Sperry 2000; McCulloh and Sperry 2005), but there are only a few studies that have sampled more than two points along the entire pathway (Domec and Gartner 2002b; McElrone et al. 2004; Burgess et al. 2006; Dunham et al. 2007). Small diameter roots are typically more vulnerable than small diameter stems and branches (Fig. 2.9, Sperry and Ikeda 1997; Hacke et al. 2000; McElrone et al. 2004; Domec et al. 2009), consistent with roots experiencing less negative xylem pressure than above-ground plant organs. In some juniper species, growing in very dry habitats, however, vulnerability does not differ between roots and stems (Linton et al. 1998; Willson et al. 2008).

There are considerable differences in hydraulic properties between small and large roots. Smaller roots of loblolly pine were more vulnerable than larger roots (Fig. 2.9). This trend is consistent with findings of Dunham et al. (2007), who

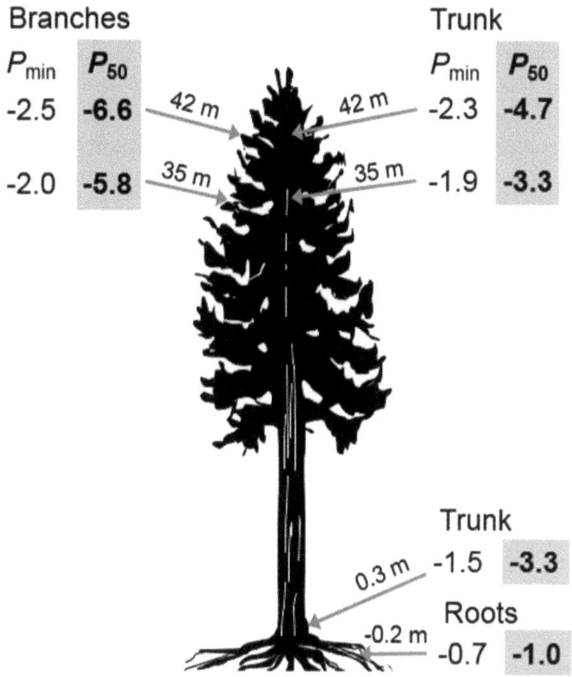

Fig. 2.16 Illustration of the variation in minimum in situ xylem pressure (P_{min}) and xylem pressure causing 50 % loss of hydraulic conductivity (P_{50}) in a large Douglas-fir tree. Both P_{min} and P_{50} become more negative with increasing height and increasing lateral distance from the trunk. Based on Domec et al. (2009) and Dunham et al. (2007)

reported that small/young Douglas-fir roots embolized much more readily than the large structural roots close to the trunk. Roots may be very variable in terms of xylem anatomy and hydraulic properties. This variability is also observed along the trunk (Schulte 2012b) and between branches of different heights. For instance, the mean tracheid diameter in branches of a 90 m *Sequoiadendron giganteum* was correlated with xylem pressure measured at dawn; branches from higher parts of the crown exhibited more negative xylem pressure as well as narrower tracheids (Rundel and Stecker 1977). The authors concluded that the advantages of narrow tracheids under water stress conditions relate to properties of thin-walled tubes in resisting strong mechanical tensions. Studies on giant *Sequoia sempervirens* and *Pseudotsuga menziesii* trees found that air-seeding pressures became substantially more negative with increasing height, a trend observed in branches (Koch et al. 2004; Burgess et al. 2006; Domec et al. 2008) and trunk xylem (Domec et al. 2008). This trend is expected based on the progressive drop in xylem pressure with height (Koch et al. 2004). Much remains to be learned about this variation and how it relates to the water use strategy of a plant.

In large Douglas-fir trees, trunk xylem was more vulnerable and had wider and longer tracheids than branch xylem sampled at the same height (Domec and Gartner 2002a; Dunham et al. 2007) (Fig. 2.16). Large roots (cambial age ~42 years) were very similar to samples taken from the trunk (cambial age ~52 years) in terms of xylem anatomy and hydraulic properties, including P_{50} (Dunham et al. 2007). Conifer branches generally appear to be hydraulically safer than the trunk and small roots.

In *Picea abies*, the leader shoot was found to be more resistant to drought-induced embolism than nearby side branches (Mayr et al. 2003b). This trend is consistent with the need for more embolism-resistant xylem as the pressure drops along the flow path from the roots to the leaves. In each of these studies the decrease in P_{50} in the stems came at the cost of higher flow resistance than in the roots. The safety factor for xylem failure by air-seeding may be defined as the ratio of P_{50} to the minimum summer xylem pressure for a particular plant part (P_{50}/P_{min}). In Douglas-fir and Ponderosa pine, the safety factor was minimal in roots and intermediate in the trunk. Branches had the highest P_{50}/P_{min} ratio (Domec and Gartner 2002a; Domec et al. 2009). This agrees with a study on four conifer species growing in the Pacific Northwest region of North America (McCulloh et al. 2014). The study also reported that the vast majority of hydraulic conductance was lost by midday in needle-bearing shoots while essentially no embolism was predicted to occur in small diameter branches and trunks.

Closer inspection of Fig. 2.16 also reveals that the bases of the trees had lower safety factors for hydraulic failure than top parts. However, roots and trunk bases may exhibit relatively small fluctuations in xylem pressure because of buffering mechanisms related to redistribution of water in the soil and plant tissues (Domec et al. 2009; McCulloh et al. 2014). Small, vulnerable roots may regrow or may be refilled depending on soil moisture availability (Domec et al. 2004).

3.5 Conclusions

As we increase our understanding of conifer xylem, we come to appreciate both its simplicity and its elegance. The tissue contains only two basic cell types: tracheids (mainly axial, some radial) and parenchyma (mainly radial, some axial). It has only three main features that affect water transport: its aspect ratio, its cell wall thickness relative to lumen diameter, and its bordered pit frequency, geometry, and biomechanics. With small alterations in these three factors the material can orchestrate extremely different performance in terms of water transport, embolisms resistance, and capacitance. These small alterations are part of the adaptive traits that allow different species to occupy such disparate habitats as the edge of the permafrost, the high desert of the interior western USA, equatorial South America, and maritime zones in southern Europe. Likewise, these small alterations are also found, in some cases, in different ecotypes of the same species. Furthermore, within a species the same traits can be altered to confer adaptive changes as the functional role alternates between properties conferred by earlywood, and then latewood; in comparison of organs like branch tips, roots, and boles; and as a the plant grows from small to large.

The pits themselves are highly specialized valves that, through their varied within-plant frequency and geometry appear to control not only the spread of embolism (which would disable the water transport system), but the prioritization of delivery of water throughout a tree. Small changes in pit characteristics affect earlywood and latewood function, tangential transport within a single ring, radial transport of water from one growth ring to the next, the speed with which water is used by a tree, and which parts of a tree are hydraulically favored in times of low water availability. Pit characteristics vary by habitat and plant part. Thus, even though conifer wood, at first glance, is a rather uniform material, at the microscopic and functional levels, it is quite varied and gives rise to important differences in capabilities of the woody parts of a plant, which can have a major effect on plant performance.

Lastly at the level of the whole plant, there are several traits beyond wood structure that affect conifer hydraulic performance: the plant-controlled factors that contribute to transpiration (leaf area, stomatal control, cuticular water loss, capacitance), conductance (sapwood area, K_{smax}, pathlength for water transport), and water acquisition (root surface area, absorptive capacity). These traits can differ greatly among species, or by habitat for the same species. As shown above, the understanding of a plant's hydraulic architecture cannot be complete without these considerations, as well as some knowledge of non-plant controlled factors (e.g., pattern of water availability and vapor pressure deficit, and soil water-release characteristics).

There remain many intriguing gaps in our knowledge about how conifer hydraulic function works. There are questions of co-ordination during development (e.g., how are leaf-to-sapwood area ratios controlled) and co-ordination of physiological processes (how does stomatal control of transpiration at various water potentials relate to embolism formation in the roots, stems, and in foliage). There are signaling questions and questions of the extent to which traits can compensate for one another to deliver similar performances. These questions must affect our initial attempts to glibly state what a trait is "for" (e.g., what is the adaptive significance of a certain morphology) or our broad-brush attempts to answer questions of biogeography (e.g., why are there very few pines in the Coast Range of Oregon). Many fundamental advances will be forthcoming still. Currently, advances are resulting not only from experimental ecophysiological projects, but also from functional trait analyses of large datasets that make use of correlations that may not be directly causal. Advances may also result from the use of materials to be developed by molecular biology programs that make novel trait combinations or that knock out one set of capacities, thus calling into question what compensations will occur and how these compensations affect performance. The current and anticipated advances in understanding of conifer hydraulic architecture are likely to shed light on plant adaptation, habitat use, range of conditions an organisms or a species may occupy and the magnitude of changes that can be accommodated by individuals with various levels of traits. Such work will likely be of much value to land and natural resource managers, and will be of use for people working in a variety of areas such as tree breeding, development of urban landscape plant materials, bioremediation, greenhouse gases mitigation, and production of biomass of a variety of qualities for pulp, timber, and energy production.

Acknowledgments Thanks to Sabine Rosner for sharing her data on Norway spruce. Thanks also to Peter Schmid for providing dye staining pictures. U. Hacke gratefully acknowledges support from the Canada Research Chair Program, NSERC, and the Canada Foundation for Innovation. J.-C. Domec was supported by NSF-IOS (award 2011-46746) and by the USDA-AFRI (awards 2011-68002 and 2012-00857). S. Mayr was supported by the Austrian Science Fund (FWF), project I826-B25.

References

Anderegg WRL, Plavcová L, Anderegg LDL, Hacke UG, Berry JA, Field CB (2013) Drought's legacy: multi-year hydraulic deterioration underlies widespread aspen forest die-off and portends increased future risk. Glob Chang Biol 19:1188–1196

Aumann CA, Ford ED (2002) Parameterizing a model of Douglas fir water flow using a tracheid-level model. J Theor Biol 219:431–462

Bailey IW (1913) The preservative treatment of wood–II. The structure of the pit membranes in the tracheids of conifers and their relation to the penetration of gases, liquids, and finely divided solids into green and seasoned wood. J For 11:12–20

Bannan MW (1965) The length, tangential diameter and length/width ratio of conifer tracheids. Can J Bot 43:967–984

Barnard DM, Lachenbruch B, McCulloh KA, Kitin P, Meinzer FC (2013) Do ray cells provide a pathway for radial water movement in the stems of conifer trees? Am J Bot 100:322–331

Beikircher B, Mayr S (2008) The hydraulic architecture of *Juniperus communis* L. ssp. communis: shrubs and trees compared. Plant Cell Environ 31:1545–1556

Bond W (1989) The tortoise and the hare: ecology of angiosperm dominance and gymnosperm persistence. Biol J Linn Soc 36:227–249

Bouche PS, Larter M, Domec J-C, Burlett R, Gasson P, Jansen S, Delzon S (2014) A broad survey of hydraulic and mechanical safety in the xylem of conifers. J Exp Bot 65(15):4419–4431

Breshears DD, Cobb NS, Rich PM, Price KP, Allen CD, Balice RG, Romme WH, Kastens JH, Floyd ML, Belnap J, Anderson JJ, Myers OB, Meyer CW (2005) Regional vegetation die-off in response to global-change-type drought. Proc Natl Acad Sci U S A 102:15144–15148

Breshears DD, McDowell NG, Goddard KL, Dayem KE, Martens SN, Meyer CW, Brown KM (2008a) Foliar absorption of intercepted rainfall improves woody plant water status most during drought. Ecology 89:41–47

Breshears DD, Myers OB, Meyer CW, Barnes FJ, Zou CB, Allen CD, McDowell NG, Pockman WT (2008b) Tree die-off in response to global change-type drought: mortality insights from a decade of plant water potential measurements. Front Ecol Environ 7:185–189

Brodribb TJ, Cochard H (2009) Hydraulic failure defines the recovery and point of death in water-stressed conifers. Plant Physiol 149:575–584

Brodribb T, Hill RS (1999) The importance of xylem constraints in the distribution of conifer species. New Phytol 130:365–372

Brodribb TJ, Holbrook NM (2005) Water stress deforms tracheids peripheral to the leaf vein of a tropical conifer. Plant Physiol 137:1139–1146

Brodribb TJ, Pittermann J, Coomes DA (2012) Elegance versus speed: examining the competition between conifer and angiosperm trees. Int J Plant Sci 173:673–694

Burgess SSO, Dawson TE (2004) The contribution of fog to the water relations of *Sequoia sempervirens* (D. Don): foliar uptake and prevention of dehydration. Plant Cell Environ 27:1023–1034

Burgess SSO, Pittermann J, Dawson TE (2006) Hydraulic efficiency and safety of branch xylem increases with height in *Sequoia sempervirens* (D. Don) crowns. Plant Cell Environ 29:229–239

Carlquist S (2012) Monocot xylem revisited: new information, new paradigms. Bot Rev 78:87–153

Charrier G, Charra-Vaskou K, Kasuga J, Cochard H, Mayr S, Améglio T (2014) Freeze-thaw stress: effects of temperature on hydraulic conductivity and ultrasonic activity in ten woody angiosperms. Plant Physiol 164:992–998

Choat B, Pittermann J (2009) New insights into bordered pit structure and cavitation resistance in angiosperms and conifers. New Phytol 182:557–560

Choat B, Cobb AR, Jansen S (2008) Structure and function of bordered pits: new discoveries and impacts on whole-plant hydraulic function. New Phytol 177:608–625

Choat B, Jansen S, Brodribb TJ, Cochard H, Delzon S, Bhaskar R, Bucci SJ, Feild TS, Gleason SM, Hacke UG, Jacobsen AL, Lens F, Maherali H, Martinez-Vilalta J, Mayr S, Mencuccini M, Mitchell PJ, Nardini A, Pittermann J, Pratt RB, Sperry JS, Westoby M, Wright IJ, Zanne AE (2012) Global convergence in the vulnerability of forests to drought. Nature 491:752–755

Cochard H, Froux F, Mayr S, Coutand C (2004) Xylem wall collapse in water-stressed pine needles. Plant Physiol 134:401–408
Cochard H, Hölttä T, Herbette S, Delzon S, Mencuccini M (2009) New insights into the mechanisms of water-stress-induced cavitation in conifers. Plant Physiol 151:949–954
Comstock JP, Sperry JS (2000) Tansley review no. 119. Some theoretical considerations of optimal conduit length for water transport in plants. New Phytol 148:195–218
Corcuera L, Cochard H, Gil-Pelegrin E, Notivol E (2011) Phenotypic plasticity in mesic populations of *Pinus pinaster* improves resistance to xylem embolism (P50) under severe drought. Trees 25:1033–1042
Dalla-Salda G, Martinez-Meier A, Cochard H, Rozenberg P (2009) Variation of wood density and hydraulic properties of Douglas-fir (*Pseudotsuga menziesii* (Mirb.) Franco) clones related to a heat and drought wave in France. Forest Ecol Manag 257:182–189
Davis SD, Sperry JS, Hacke UG (1999) The relationship between xylem conduit diameter and cavitation caused by freeze-thaw events. Am J Bot 86:1367–1372
Dellus V, Mila I, Scalbert A, Menard C, Michon V, Herve du Penhoat CLM (1997) Douglas-fir polyphenols and heartwood formation. Phytochemistry 45:1573–1578
Delzon S, Douthe C, Sala A, Cochard H (2010) Mechanism of water-stress induced cavitation in conifers: bordered pit structure and function support the hypothesis of seal capillary-seeding. Plant Cell Environ 33:2101–2111
Domec J-C (2011) Let's not forget the critical role of surface tension in xylem water relations. Tree Physiol 31:359–360
Domec J, Gartner B (2002a) Age-and position-related changes in hydraulic versus mechanical dysfunction of xylem: inferring the design criteria for Douglas-fir wood structure. Tree Physiol 22:91–104
Domec JC, Gartner BL (2002b) How do water transport and water storage differ in coniferous earlywood and latewood? J Exp Bot 53:2369–2379
Domec JC, Warren JM, Meinzer FC, Brooks JR, Coulombe R (2004) Native root xylem embolism and stomatal closure in stands of Douglas-fir and ponderosa pine: mitigation by hydraulic redistribution. Oecologia 141:7–16
Domec JC, Lachenbruch B, Meinzer FC (2006) Bordered pit structure and function determine spatial patterns of air-seeding thresholds in xylem of Douglas-fir (*Pseudotsuga menziesii*; Pinaceae) trees. Am J Bot 93:1588–1600
Domec JC, Lachenbruch B, Meinzer FC, Woodruff DR, Warren JM, McCulloh KA (2008) Maximum height in a conifer is associated with conflicting requirements for xylem design. Proc Natl Acad Sci U S A 105:12069–12074
Domec JC, Warren JM, Meinzer FC, Lachenbruch B (2009) Safety factors for xylem failure by implosion and air-seeding within roots, trunks and branches of young and old conifer trees. IAWA J 30:101–120
Domec J-C, Lachenbruch B, Pruyn ML, Spicer R (2012) Effects of age-related increases in sapwood area, leaf area, and xylem conductivity on height-related hydraulic costs in two contrasting coniferous species. Ann For Sci 69:17–27
Dunham S, Lachenbruch B, Ganio L (2007) Bayesian analysis of Douglas-fir hydraulic architecture at multiple scales. Trees 21:65–78
Evert RF (2006) Esau's plant anatomy: meristems, cells, and tissues of the plant body: their structure, function, and development, 3rd edn. Wiley-Interscience, Hoboken
Ewers FW (1982) Secondary growth in needle leaves of *Pinus longaeva* (bristlecone pine) and other conifers: quantitative data. Am J Bot 69:1552–1559
Ewers FW (1985) Xylem structure and water conduction in conifer trees, dicot trees, and lianas. IAWA Bull 6:309–317
Ewers BE, Oren R, Sperry JS (2000) Influence of nutrient *versus* water supply on hydraulic architecture and water balance in *Pinus taeda*. Plant Cell Environ 23:1055–1066
Feild TS, Brodribb T (2001) Stem water transport and freeze-thaw xylem embolism in conifers and angiosperms in a Tasmanian treeline heath. Oecologia 127:314–320
Feild TS, Brodribb T, Holbrook NM (2002) Hardly a relict: freezing and the evolution of vesselless wood in Winteraceae. Evolution 56:464–478

Flanary B, Kletetschka G (2005) Analysis of telomere length and telomerase activity in tree species of various life-spans, and with age in the bristlecone pine *Pinus longaeva*. Biogerontology 6:101–111

Hacke UG, Jansen S (2009) Embolism resistance of three boreal conifer species varies with pit structure. New Phytol 182:675–686

Hacke UG, Sperry JS, Ewers BE, Ellsworth DS, Schäfer KVR, Oren R (2000) Influence of soil porosity on water use in *Pinus taeda*. Oecologia 124:495–505

Hacke UG, Stiller V, Sperry JS, Pittermann J, McCulloh KA (2001a) Cavitation fatigue. Embolism and refilling cycles can weaken the cavitation resistance of xylem. Plant Physiol 125:779–786

Hacke UG, Sperry JS, Pockman WT, Davis SD, McCulloh KA (2001b) Trends in wood density and structure are linked to prevention of xylem implosion by negative pressure. Oecologia 126:457–461

Hacke UG, Sperry JS, Pittermann J (2004) Analysis of circular bordered pit function—II. Gymnosperm tracheids with torus-margo pit membranes. Am J Bot 91:386–400

Hacke UG et al (2006) Scaling of angiosperm xylem structure with safety and efficiency. Tree Physiol 26:689–701

Hammel H (1967) Freezing of xylem sap without cavitation. Plant Physiol 42:55–66

Holbrook NM, Zwieniecki MA (1999) Embolism repair and xylem tension: do we need a miracle? Plant Physiol 120:7–10

Huber B (1947) Zur Mikrotopographie der Saftströme im Transfusionsgewebe der Koniferennadel. Planta 35:331–351

Jansen S, Lamy JB, Burlett R, Cochard H, Gasson P, Delzon S (2012) Plasmodesmatal pores in the torus of bordered pit membranes affect cavitation resistance of conifer xylem. Plant Cell Environ 35:1109–1120

Johnson DM, Meinzer FC, Woodruff DR, McCulloh KA (2009) Leaf xylem embolism, detected acoustically and by cryo-SEM, corresponds to decreases in leaf hydraulic conductance in four evergreen species. Plant Cell Environ 32:828–836

Johnson D, McCulloh K, Meinzer F, Woodruff D, Eissenstat D (2011) Hydraulic patterns and safety margins, from stem to stomata, in three eastern US tree species. Tree Physiol 31:659–668

Johnson DM, McCulloh KA, Woodruff DR, Meinzer FC (2012) Evidence for xylem embolism as a primary factor in dehydration-induced declines in leaf hydraulic conductance. Plant Cell Environ 35:760–769

Kitin P, Fujii T, Abe H, Takata K (2009) Anatomical features that facilitate radial flow across growth rings and from xylem to cambium in *Cryptomeria japonica*. Ann Bot 103:1145–1157

Koch GW, Sillett SC, Jennings GM, Davis SD (2004) The limits to tree height. Nature 428:851–854

Lachenbruch B, McCulloh KA (2014) Traits, properties, and performance: how woody plants combine hydraulic and mechanical functions in a cell, tissue, or whole plant. New Phytol 204:747–764

Lachenbruch B, Moore JR, Evans R (2011) Radial variation in wood structure and function in woody plants, and hypotheses for its occurrence. In: Meinzer FC, Lachenbruch B, Dawson TE (eds) Size- and age-related changes in tree structure and function, vol 4. Springer, Dordrecht, pp 121–164. doi:10.1007/978-94-007-1242-3_5

Lancashire JR, Ennos AR (2002) Modelling the hydrodynamic resistance of bordered pits. J Exp Bot 53:1485–1493

LaPasha C, Wheeler E (1990) Resin canals in *Pinus taeda*. Longitudinal canal lengths and interconnections between longitudinal and radial canals. IAWA Bull 11:227–238

Laur J, Hacke UG (2014) Exploring *Picea glauca* aquaporins in the context of needle water uptake and xylem refilling. New Phytol 203(2):388–400

Lieffers VJ, Rothwell RL (1987) Rooting of peatland black spruce and tamarack in relation to depth of water-table. Can J Bot 65:817–821

Liepe KJ (2014) Genetic variation in lodgepole pine and interior spruce: adaptation to climate and implications for seed transfer. University of Alberta, Edmonton

Linton MJ, Sperry JS, Williams DG (1998) Limits to water transport in *Juniperus osteosperma* and *Pinus edulis*: implications for drought tolerance and regulation of transpiration. Funct Ecol 12:906–911

López R, de Heredia UL, Collada C, Cano FJ, Emerson BC, Cochard H, Gil L (2013) Vulnerability to cavitation, hydraulic efficiency, growth and survival in an insular pine (*Pinus canariensis*). Ann Bot 111:1167–1179

Martínez-Vilalta J, Cochard H, Mencuccini M, Sterck F, Herrero A, Korhonen J, Llorens P, Nikinmaa E, Nolè A, Poyatos R (2009) Hydraulic adjustment of Scots pine across Europe. New Phytol 184:353–364

Maton C, Gartner BL (2005) Do gymnosperm needles pull water through the xylem produced in the same year as the needle? Am J Bot 92:123–131

Mayr S, Charra-Vaskou K (2007) Winter at the alpine timberline causes complex within-tree patterns of water potential and embolism in *Picea abies*. Physiol Plant 131:131–139

Mayr S, Cochard H (2003) A new method for vulnerability analysis of small xylem areas reveals that compression wood of Norway spruce has lower hydraulic safety than opposite wood. Plant Cell Environ 26:1365–1371

Mayr S, Sperry JS (2010) Freeze–thaw-induced embolism in *Pinus contorta*: centrifuge experiments validate the "thaw-expansion hypothesis" but conflict with ultrasonic emission data. New Phytol 185:1016–1024

Mayr S, Wolfschwenger M, Bauer H (2002) Winter-drought induced embolism in Norway spruce (*Picea abies*) at the Alpine timberline. Physiol Plant 115:74–80

Mayr S, Gruber A, Bauer H (2003a) Repeated freeze-thaw cycles induce embolism in drought stressed conifers (Norway spruce, stone pine). Planta 217:436–441

Mayr S, Rothart B, Dämon B (2003b) Hydraulic efficiency and safety of leader shoots and twigs in Norway spruce growing at the alpine timberline. J Exp Bot 54:2563–2568

Mayr S, Hacke U, Schmid P, Schwienbacher F, Gruber A (2006) Frost drought in conifers at the alpine timberline: xylem dysfunction and adaptations. Ecology 87:3175–3185

Mayr S, Cochard H, Améglio T, Kikuta SB (2007) Embolism formation during freezing in the wood of *Picea abies*. Plant Physiol 143:60–67

Mayr S, Schmid P, Beikircher B (2012) Plant water relations in alpine winter. In: Lütz C (ed) Plants in Alpine regions. Springer, Vienna, pp 153–162

Mayr S, Schmid P, Laur J, Rosner S, Charra-Vaskou K, Daemon B, Hacke UG (2014) Uptake of water via branches helps timberline conifers refill embolized xylem in late winter. Plant Physiol 164:1731–1740

McCulloh KA, Sperry JS (2005) Patterns in hydraulic architecture and their implications for transport efficiency. Tree Physiol 25:257–267

McCulloh K et al (2010) Moving water well: comparing hydraulic efficiency in twigs and trunks of coniferous, ring-porous, and diffuse-porous saplings from temperate and tropical forests. New Phytol 186:439–450

McCulloh KA, Johnson DM, Meinzer FC, Woodruff DR (2014) The dynamic pipeline: hydraulic capacitance and xylem hydraulic safety in four tall conifer species. Plant Cell Environ 37:1171–1183

McDowell NG, Fisher RA, Xu C, Domec J, Hölttä T, Mackay DS, Sperry JS, Boutz A, Dickman L, Gehres N (2013) Evaluating theories of drought-induced vegetation mortality using a multimodel–experiment framework. New Phytol 200:304–321

McElrone AJ, Pockman WT, Martínez-Vilalta J, Jackson RB (2004) Variation in xylem structure and function in stems and roots of trees to 20 m depth. New Phytol 163:507–517

Michaelis P (1934) Ökologische Studien an der alpinen Baumgrenze. IV. Zur Kenntnis des winterlichen Wasserhaushaltes. Jahrb Wiss Bot 80:169–247

Panshin AJ, Zeeuw CD (1980) Textbook of wood technology. McGraw-Hill, New York

Peterson MG, Dietterich HR, Lachenbruch B (2007) Do Douglas-fir branches and roots have juvenile wood? Wood Fiber Sci 39:651–660

Petit G, Anfodillo T, De Zan C (2009) Degree of tapering of xylem conduits in stems and roots of small *Pinus cembra* and *Larix decidua* trees. Botany 87:501–508

Pittermann J, Sperry J (2003) Tracheid diameter is the key trait determining the extent of freezing-induced embolism in conifers. Tree Physiol 23:907–914

Pittermann J, Sperry JS (2006) Analysis of freeze-thaw embolism in conifers. The interaction between cavitation pressure and tracheid size. Plant Physiol 140:374–382

Pittermann J, Sperry JS, Hacke UG, Wheeler JK, Sikkema EH (2005) Torus-margo pits help conifers compete with angiosperms. Science 310:1924

Pittermann J, Sperry JS, Hacke UG, Wheeler JK, Sikkema EH (2006a) Inter-tracheid pitting and the hydraulic efficiency of conifer wood: the role of tracheid allometry and cavitation protection. Am J Bot 93:1265–1273

Pittermann J, Sperry JS, Wheeler JK, Hacke UG, Sikkema EH (2006b) Mechanical reinforcement of tracheids compromises the hydraulic efficiency of conifer xylem. Plant Cell Environ 29:1618–1628

Pittermann J, Choat B, Jansen S, Stuart SA, Lynn L, Dawson TE (2010) The relationships between xylem safety and hydraulic efficiency in the Cupressaceae: the evolution of pit membrane form and function. Plant Physiol 153:1919–1931

Pittermann J, Stuart SA, Dawson TE, Moreau A (2012) Cenozoic climate change shaped the evolutionary ecophysiology of the Cupressaceae conifers. Proc Natl Acad Sci 109:9647–9652

Plaut JA, Yepez EA, Hill J, Pangle R, Sperry JS, Pockman WT, McDowell NG (2012) Hydraulic limits preceding mortality in a piñon-juniper woodland under experimental drought. Plant Cell Environ 35:1601–1617

Plavcová L, Jansen S, Klepsch M, Hacke UG (2013) Nobody's perfect: can irregularities in pit structure influence vulnerability to cavitation? Front Plant Sci 4:453

Rosner S (2013) Hydraulic and biomechanical optimization in Norway spruce trunkwood—a review. IAWA J 34:365–390

Rosner S, Světlík J, Andreassen K, Børja I, Dalsgaard L, Evans R, Karlsson B, Tollefsrud MM, Solberg S (2014) Wood density as a screening trait for drought sensitivity in Norway spruce. Can J For Res 44:154–161

Ruiz Diaz Britez M, Sergent A-S, Martinez Meier A, Bréda N, Rozenberg P (2014) Wood density proxies of adaptive traits linked with resistance to drought in Douglas fir (*Pseudotsuga menziesii* (Mirb.) Franco). Trees 28:1289–1304

Rundel PW, Stecker RE (1977) Morphological adaptations of tracheid structure to water stress gradients in the crown of *Sequoiadendron giganteum*. Oecologia 27:135–139

Schoonmaker AL, Hacke UG, Landhausser SM, Lieffers VJ, Tyree MT (2010) Hydraulic acclimation to shading in boreal conifers of varying shade tolerance. Plant Cell Environ 33:382–393

Schreiber SG, Hacke UG, Hamann A, Thomas BR (2011) Genetic variation of hydraulic and wood anatomical traits in hybrid poplar and trembling aspen. New Phytol 190:150–160

Schulte PJ (2012a) Computational fluid dynamics models of conifer bordered pits show how pit structure affects flow. New Phytol 193:721–729

Schulte PJ (2012b) Vertical and radial profiles in tracheid characteristics along the trunk of Douglas-fir trees with implications for water transport. Trees 26:421–433

Sevanto S, Holbrook NM, Ball MC (2012) Freeze/thaw-induced embolism: probability of critical bubble formation depends on speed of ice formation. Front Plant Sci 3:107

Sevanto S, McDowell NG, Dickman LT, Pangle R, Pockman WT (2014) How do trees die? A test of the hydraulic failure and carbon starvation hypotheses. Plant Cell Environ 37:153–161

Sparks JP, Black RA (2000) Winter hydraulic conductivity end xylem cavitation in coniferous trees from upper and lower treeline. Arct Antarct Alp Res 32:397–403

Sparks JP, Campbell GS, Black RA (2001) Water content, hydraulic conductivity, and ice formation in winter stems of *Pinus contorta*: a TDR case study. Oecologia 127:468–475

Sperry JS, Ikeda T (1997) Xylem cavitation in roots and stems of Douglas-fir and white fir. Tree Physiol 17:275–280

Sperry JS, Sullivan JEM (1992) Xylem embolism in response to freeze-thaw cycles and water stress in ring-porous, diffuse-porous, and conifer species. Plant Physiol 100:605–613

Sperry JS, Tyree MT (1990) Water-stress-induced xylem embolism in three species of conifers. Plant Cell Environ 13:427–436

Sperry JS, Nichols KL, Sullivan JEM, Eastlack SE (1994) Xylem embolism in ring-porous, diffuse-porous, and coniferous trees of northern Utah and interior Alaska. Ecology 75:1736–1752
Sperry JS, Adler FR, Campbell GS, Comstock JP (1998) Limitation of plant water use by rhizosphere and xylem conductance: results from a model. Plant Cell Environ 21:347–359
Sperry JS, Hacke UG, Pittermann J (2006) Size and function in conifer tracheids and angiosperm vessels. Am J Bot 93:1490–1500
Sperry JS, Meinzer FC, McCulloh KA (2008) Safety and efficiency conflicts in hydraulic architecture: scaling from tissues to trees. Plant Cell Environ 31:632–645
Sperry JS, Smith DD, Savage V, Enquist BJ, McCulloh KA, Reich PB, Bentley LP, von Allmen EI (2012) A species-level model for metabolic scaling in trees I. Exploring boundaries to scaling space within and across species. Funct Ecol 26:1054–1065
Spicer R (2005) Senescence in secondary xylem: heartwood formation as an active developmental program. In: Holbrook NM, Zwieniecki MA (eds) Vascular transport in plants. Elsevier Academic Press, Oxford, pp 457–475
Sucoff E (1969) Freezing of conifer xylem and the cohesion-tension theory. Physiol Plant 22:424–431
Taylor AM, Gartner BL, Morrell JJ (2002) Heartwood formation and natural durability—a review. Wood Fiber Sci 34:587–611
Taylor EL, Taylor TN, Krings M (2009) Paleobotany: the biology and evolution of fossil plants. Academic, San Diego
Tranquillini W (1976) Water relations and alpine timberline. In: Lange OL, Kappen L, Schulze E-D (eds) Water and plant life. Ecological studies, vol 19. Springer, Berlin, pp 473–491
Tyree MT, Ewers FW (1991) Tansley review no. 34: the hydraulic architecture of trees and other woody plants. New Phytol 119:345–360
Tyree MT, Davis SD, Cochard H (1994) Biophysical perspectives of xylem evolution—is there a tradeoff of hydraulic efficiency for vulnerability to dysfunction. IAWA J 15:335–360
West AG, Hultine KR, Sperry JS, Bush SE, Ehleringer JR (2008) Transpiration and hydraulic strategies in a pinon-juniper woodland. Ecol Appl 18:911–927
Willson CJ, Manos PS, Jackson RB (2008) Hydraulic traits are influenced by phylogenetic history in the drought-resistant, invasive genus *Juniperus* (Cupressaceae). Am J Bot 95:299–314
Wilson JP, Knoll AH (2010) A physiologically explicit morphospace for tracheid-based water transport in modern and extinct seed plants. Paleobiology 36:335–355
Wilson JP, Knoll AH, Holbrook NM, Marshall CR (2008) Modeling fluid flow in *Medullosa*, an anatomically unusual Carboniferous seed plant. Paleobiology 34:472–493
Woodruff DR, McCulloh KA, Warren JM, Meinzer FC, Lachenbruch B (2007) Impacts of tree height on leaf hydraulic architecture and stomatal control in Douglas-fir. Plant Cell Environ 30:559–569
Zhang Y-J, Rockwell FE, Wheeler JK, Holbrook NM (2014) Reversible deformation of transfusion tracheids in *Taxus baccata* is associated with a reversible decrease in leaf hydraulic conductance. Plant Physiol 165(4):1557–1565
Zwieniecki MA, Holbrook NM (2009) Confronting Maxwell's demon: biophysics of xylem embolism repair. Trends Plant Sci 14:530–534
Zwieniecki MA, Stone HA, Leigh A, Boyce CK, Holbrook NM (2006) Hydraulic design of pine needles: one-dimensional optimization for single-vein leaves. Plant Cell Environ 29:803–809

Chapter 3
Development, Structure, and Function of Torus–Margo Pits in Conifers, Ginkgo and Dicots

Roland R. Dute

Abbreviations

A	Pit aperture
B	Pit border
C	Pit cavity
CO	Corona of torus
CM	Cell membrane
D	Desmotubule
DI	Dictyosome
ER	Endoplasmic reticulum
L	Cell lumen
M	Margo
MC	Median cavity
MT	Microtubule
P	Pustular region of torus
PA	Parenchyma cell
PD	Plasmodesma
PL	Protective layer of parenchyma cell wall
PS	Plastid
R	Ring of parallel microfibrils in torus
T	Torus
TP	Torus pad
V	Vesicle

R.R. Dute, B.S., M.S., Ph.D. (✉)
Department of Biological Sciences, Auburn University, Auburn, AL 36849, USA
e-mail: duterol@auburn.edu

U. Hacke (ed.), *Functional and Ecological Xylem Anatomy*,
DOI 10.1007/978-3-319-15783-2_3

1 Introduction

Tracheary elements are cells specialized for long-distance transport of water (Esau 1977). They are connected by bordered pit pairs. "Each pit pair consists of a modified wall, known as a pit membrane, flanked on either side by a pit border containing an opening (or aperture) that provides access to an adjoining cell lumen (Dute et al. 2004)." Thus, water passes from the lumen of one cell through an aperture into a pit cavity and through the pit membrane into the adjoining pit cavity, then through the other pit aperture into the lumen of the neighboring cell (Fig. 3.1a, left) (Dute et al. 2004). Pit membranes allow transfer of water from cell to cell but at the same time impede movement of air embolisms (Pittermann et al. 2005).

Pit membranes are of two basic types (Pittermann et al. 2005). One type, found in most gymnosperms, consists of a permeable margo and impermeable torus (Fig. 3.1b). Water flows through the margo, but in the presence of an embolism, the pit membrane is displaced or aspirated such that the torus blocks an aperture (Fig. 3.1a, right). In contrast, pit membranes of most dicotyledons are homogeneous; that is, they lack a torus but the membrane pores are of a smaller size than the margo pores of gymnosperms (Fig. 3.1c) (Pittermann et al. 2005). The smaller pores reduce water flux, but also make passage for air bubbles more difficult (Dute et al. 2001).

In 1978 Ohtani and Ishida discovered a third type of pit membrane in dicotyledons in which features of the first two types are combined. Thus, the fine mesh of the typical dicot pit membrane is associated with a torus (Fig. 3.1d).

The goal of this chapter is to describe development, form, and function of torus-bearing intervascular pit membranes in seed plants (angiosperms and gymnosperms).

2 Anatomy and Ontogeny: Dicotyledons

Table 3.1 provides a list of those dicot species containing torus-bearing pit membranes in their woods. The table was published originally by Dute et al. in (2010a), but has been updated with information from recent articles (Dute et al. 2010b, 2011; Sano et al. 2013). A glance at the table indicates that the occurrence of torus-bearing pit membranes is homoplastic. Thus, the presence of multiple mechanisms for torus development in dicots should come as no surprise.

2.1 Oleaceae

Torus ontogeny in *Osmanthus* (Oleaceae), in particular that in *O. americanus*,[1] has been studied at some length. Tori are common in the genus, being found in species from five of six sections investigated (Dute et al. 2010b). Studies of *O. americanus* and *O. armatus* indicate the presence of tori in stem, root, and leaf

[1] Recently retransferred to *Cartrema* by Nesom (2012). This species has been retained within *Osmanthus* in this chapter.

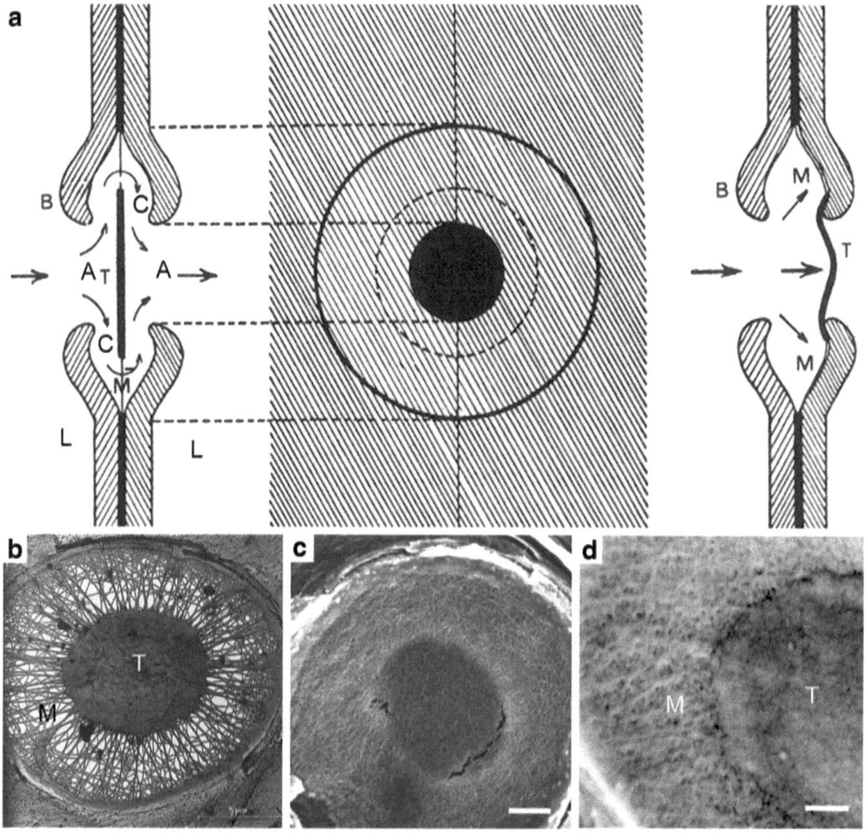

Fig. 3.1 (**a**) Diagrammatic representation of form and function of bordered pit pairs in gymnosperms. At *left*, *arrows* denote passage of water from one cell lumen to the next through the permeable portion of the pit membrane (*M*) separating two bordered pits. At *right* the pit membrane of the pit pair is displaced and the central impermeable torus of the pit membrane inhibits passage of air bubbles between cells. The *central* diagram is the bordered pit pair in face view. (**b**) A face view of the pit membrane of *Tsuga canadensis*, typical of conifers and *Ginkgo*, in which an impermeable torus is encircled by a porous margo. (**c**) A pit membrane of *Daphne jezoensis*, typical of that of dicotyledons. The *central dark circle* is not a torus but rather an image of the subtending pit aperture. Note the tightly woven nature of the cellulose microfibrils over the entire membrane surface (the *cracks* are induced by heat of the electron beam). (**d**) A pit membrane of *Osmanthus americanus*, which represents a combination of the first two types. *Bars*: **b** = 5 μm; **c** = 1 μm; **d** = 1 μm. Diagram (**a**) and photo (**b**) from Zimmermann (1983); photo (**c**) from Dute et al. (1992); photo (**d**) from Dute and Rushing (1987)

(Dute and Rushing 1987; Dute and Elder 2011; Dute et al. 2012a, b). Tori are absent from the vasculature of the flower, but later develop in the fruit stalk and in the xylem of the drupaceous fruit (Dute et al. 2013). When viewed with TEM, the torus thickening of the pit membrane consists of two distinct pads separated by a layer of wall material (Fig. 3.2a) (Dute and Rushing 1987). The latter represents the compound middle lamella, whereas the pads are deposited later in cell ontogeny when the respective pit borders are complete or nearly so (Dute and Rushing 1988).

Table 3.1 Dicot species containing torus-bearing pit membranes in their woods

Family	Species	Citation
Oleaceae	*Chionanthus retusa*	Rabaey et al. (2008a)
	Osmanthus americanus	Dute and Rushing (1987)
	O. aurantiacus	Ohtani (1983)
	O. fortunei	Ohtani and Ishida (1978), Ohtani (1983)
	O. fragrans	
	O. heterophyllus	
	O. insularis	Ohtani (1983)
	O. rigidus	Ohtani (1983)
	O. serratulus	Rabaey et al. (2006)
	O. suavis	
	Picconia azorica	Dute et al. (2008), Rabaey et al. (2008a)
	P. excelsa	
Thymelaeaceae	*Daphne acutiloba*	Dute et al. (1992)
	D. alpina	
	D. altaica	
	D. arbuscula	
	D. arisanensis	Dute et al. (1996)
	D. aurantiaca	
	D. bholua	
	D. blagayana	Dute et al. (1992)
	D. X burkwoodii	
	D. caucasica	
	D. cneorum	Dute et al. (1990)
	D. collina	Dute et al. (1992)
	D. ericoides	Dute et al. (1996)
	D. genkwa	Dute et al. (1992)
	D. glomerata	
	D. gnidioides	Dute et al. (1996)
	D. gnidium	Dute et al. (1992)
	D. jasminea	Dute et al. (1996)
	D. kiusiana	Ohtani and Ishida (1978), Ohtani (1983)
	D. laureola	Dute et al. (1992)
	D. miyabeana	Ohtani and Ishida (1978), Ohtani (1983)
	D. odora	
	D. oleoides	Dute et al. (1992)
	D. papyracea	
	D. petraea	Dute et al. (1996)
	D. retusa	Dute et al. (1992)
	D. stapfii	Dute et al. (1996)
	D. striata	Dute et al. (1992)
	D. tangutica	
	Diarthron arenaria	Dute et al. (2001)
	D. lessertii	
	D. magakjanii	
	D. stachyoides	
	D. turcomanica	
	Stellera chamaejasme (?)	
	Thymelaea villosa	
	Wikstroemia albiflora	Dute et al. (1996)
	W. kudoi	
	W. pauciflora	
	W. yakushimensis	

(continued)

Table 3.1 (continued)

Family	Species	Citation
Rosaceae	*Cercocarpus intricatus*	Jansen et al. (2007)
	C. ledifolius	
	C. montanus	
	C. traskiae	
Ulmaceae	*Planera aquatica*	Dute et al. (2004)
	Ulmus alata	Wheeler (1983)
	U. americana	Jansen et al. (2004)
	U. campestris	Czaninski (1979)
	U. carpinifolia	Jansen et al. (2004)
	U. coritana	
	U. cornubiensis	
	U. davidiana	
	U. diversifolia	
	U. effusa	
	U. fulva	
	U. glabra	
	U. japonica	
	U. laciniata	
	U. macrocarpa	
	U. montana	
	U. parvifolia	
	U. pedunculata	
	U. plotii	
	U. scabra	
	U. thomasii	Wheeler (1983)
	Zelkova acuminata	Jansen et al. (2004)
	Z. crenata	
	Z. cretica	Jansen et al. (2007)
	Z. serrata	
Cannabaceae	*Celtis australis*	Jansen et al. (2007)
	C. laevigata	Wheeler (1983)
	C. occidentalis	
	C. reticulata	
Schisandraceae	*Schisandra chinensis*	Sano et al. (2013)

Pad deposition is associated with a plexus of microtubules and membranous vesicles (Dute and Rushing 1988) (Fig. 3.2b, c). The microtubules approach the surface of a growing pad at different angles. It is hypothesized that the vesicles, containing wall material, move along the microtubules to the torus surface (Dute and Rushing 1988). The microtubule plexus is not directly associated with those microtubules responsible for pit border deposition. Completion of a torus pad is followed by cellular lysis of the adjoining cytoplasm.

Careful observation of torus structure in *Osmanthus* (using SEM or TEM of carbon replicas) indicates the presence of a fibrillar component as well as

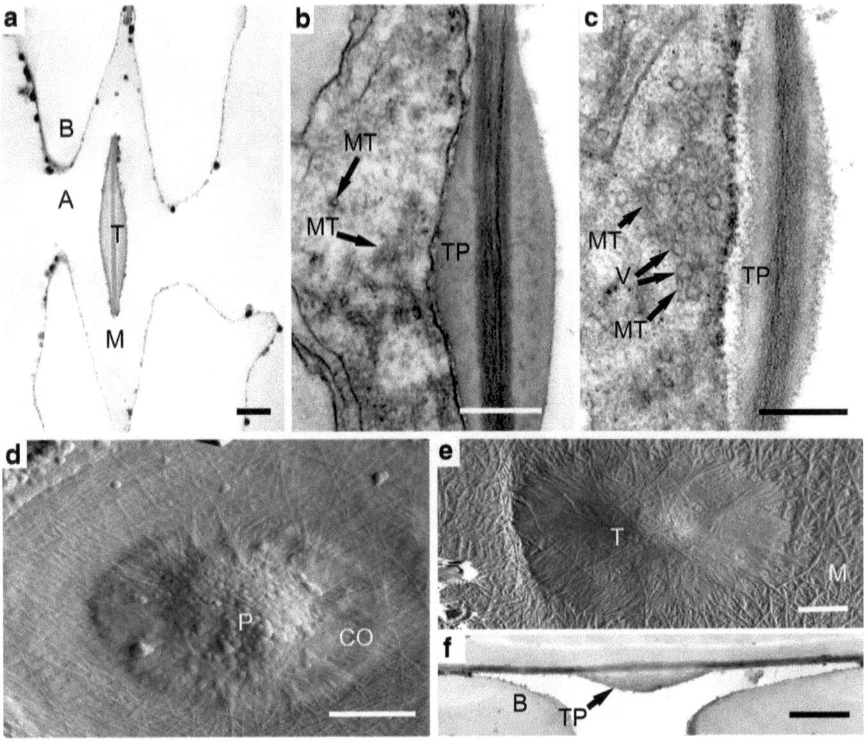

Fig. 3.2 Oleaceae. (**a**) Sectional view of a mature, intervascular pit pair of *Osmanthus americanus*. The torus consists of two pads separated by a compound middle lamella. (**b, c**) Detail of torus pad formation showing both microtubules and dictyosome vesicles. (**d**) Pit membrane whose torus has both corona and a pustular region. (**e**) Torus whose pustular region has been chemically removed exposing the microfibrils beneath. (**f**) A blind pit with a torus pad positioned facing the lumen of the water-conducting cell. *Bars*: **a** = 0.5 μm; **b** = 0.3 μm; **c** = 0.3 μm; **d** = 1 μm; **e** = 0.5 μm; **f** = 1 μm. Photo (**a**) from Dute and Rushing 1987; photos (**b, c, f**) from Dute and Rushing (1988); photos (**d, e**) from Dute and Elder (2011)

amorphous substances (Ohtani and Ishida 1978). Both components can be easily visualized (Fig. 3.2d) using the superior resolution provided by the atomic force microscope (AFM) (Dute and Elder 2011). Observations of air-dried or solvent-dried pit membranes show tori to consist of a central pustular region surrounded by a corona or fringe (Fig. 3.2d). Treatment with acidified sodium chlorite (Ohtani and Ishida 1978; Dute and Elder 2011) removes the pustules and exposes parallel sets of microfibrils crossing one another in the center of the torus (Fig. 3.2e). These microfibrils are responsible for a radially oriented component of microfibrils within the margo (Ohtani and Ishida 1978; Ohtani 1983). It is clear that the pustular layer is the last phase of torus synthesis. The pustular surface stains deeply with toluidine blue O in *O. armatus* (Dute, unpublished results) and reacts with $KMnO_4$ in *O. americanus* (Coleman et al. 2004). Thus, the torus pads contain some lignin.

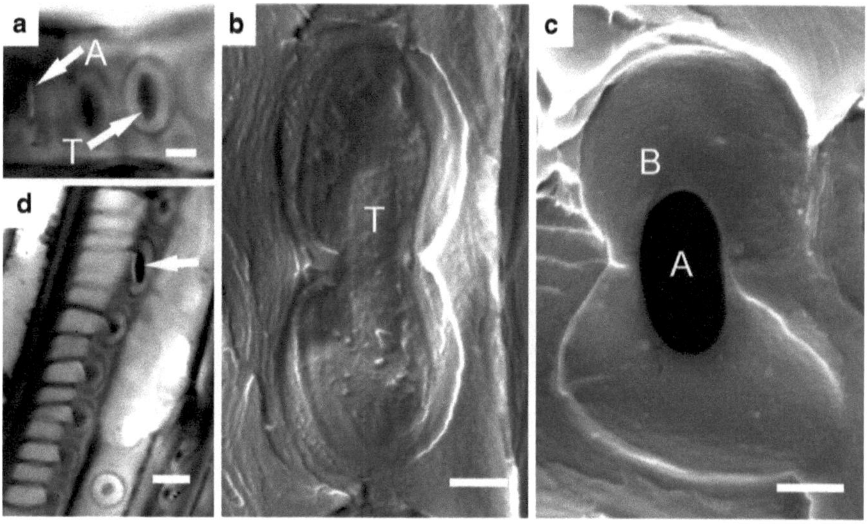

Fig. 3.3 Oleaceae. (**a**) Aberrant pits in a petiole of *O. armatus*. Notice the elongate torus and aperture. (**b**) Elongate torus in what appears to be two fused pits in a root of *O. americanus*. (**c**) Pit border and aperture resulting from two fused pits in the root of *O americanus*. (**d**) *Arrow* denotes elongate torus in a pit from the root of *O. armatus*. *Bars*: **a**=3.4 μm; **b**=2 μm; **c**=2 μm; **d**=5 μm. Photo (**a**) from Dute et al. (2012b); photos (**b–d**) from Dute et al. (2012a)

Within a given bordered pit pair, completion of torus pad deposition and subsequent hydrolysis of matrix material from the margo begins first in the older of two adjacent cells. Cellulose microfibrils of a pit membrane remain undamaged and are easily visualized when viewing the membrane surface with either AFM or SEM (Fig. 3.1d). When viewed in thin section using TEM, however, the mature margo is often practically invisible (Fig. 3.2a).

A given tracheary element acts independently of its neighbors and is programmed to deposit a torus pad on any surface that is enclosed by a pit border, even a blind pit (Fig. 3.2f) (Dute and Rushing 1988).

Beck et al. (1982), during a study of *Callixylon* wood, hypothesized that torus-bearing pit membranes require circular apertures in order to act effectively as valves. Torus-bearing intervascular pit membranes of *Osmanthus* are generally associated with circular to somewhat elliptical pit apertures, but exceptions do exist. Petioles of *Osmanthus armatus* contain tracheary elements with tori. At the boundary of the primary and secondary xylem, both pit apertures and tori are elongate (Fig. 3.3a) (Dute et al. 2012b). Similar instances of elongate tori and apertures exist in roots of *O. americanus* (Fig. 3.3b, c) and *O. armatus* (Fig. 3.3d) (Dute et al. 2012a). In some cases (e.g., Fig. 3.3b, c) pit borders fused during development, resulting in a pit border with a single, elongate aperture. The pit membrane of such a fused pit is also elongate (rather than circular), and the torus pad that is deposited corresponds to the aperture shape. Since torus pad deposition does not occur in

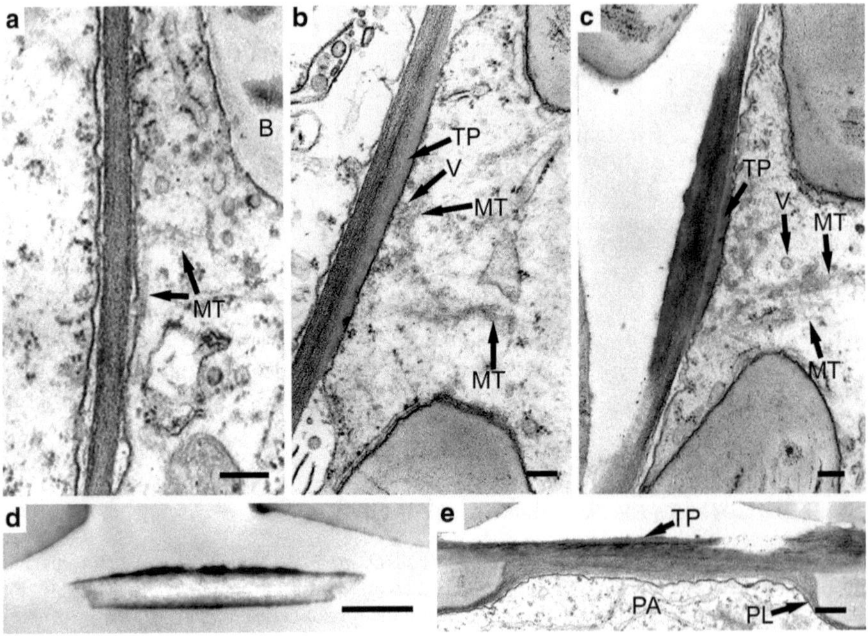

Fig. 3.4 Thymelaeaceae. (**a–c**) Steps in the development of a torus pad in the immature pit membrane of *Daphne odora*. Torus construction is associated with both dictyosome vesicles and masses of microtubules. (**d**) Torus of *D. odora* stained with $KMnO_4$ for detection of lignin. (**e**) Half-bordered pit pair with a torus pad deposited only on the side of the water-conducting cell (*D. odora*). *Bars*: **a**=0.2 μm; **b**=0.2 μm; **c**=0.2 μm; **d**=0.5 μm; **e**=0.5 μm. Photos (**a–c**, **e**) from Dute et al. (1990); Photo (**d**) from Coleman et al. (2004)

Osmanthus until the pit border is complete (or nearly so), one could argue that aperture shape somehow controls torus shape. It would be interesting to observe the quality of seal during aspiration in such aberrant pits.

2.2 Thymelaeaceae

Torus-bearing membranes of *Daphne* (Thymelaeaceae) develop in a similar, but not identical, manner to that of *Osmanthus* (Dute et al. 1990). Membrane thickening at the site of the torus is not initiated until the pit border is near or at completion. Aggregation of microtubules near the site of the future torus begins at this time (Fig. 3.4a). Development of a microtubule plexus follows (Fig. 3.4b). Vesicular deposition of torus material into the developing torus pad at this time is quite distinct (Fig. 3.4c). It appears as if the secreted material impregnates the already existing pit membrane and also accumulates as a surface layer (Fig. 3.4c, d). The secreted material stains positive for lignin (Fig. 3.4d). A torus pad is routinely deposited on the tracheary element side of half-bordered pit pairs (Fig. 3.4e) and blind pits (Dute et al. 1990; Coleman et al. 2004).

2.3 *Rosaceae*

In species of *Cercocarpus* (Rosaceae) torus pads are constructed after completion of the pit borders, but neither microtubule plexus nor mass of vesicles is present (Fig. 3.5a) (Dute et al. 2010a). Torus pads are deposited on the tracheary element pit membrane surface of half-bordered pit pairs (Fig. 3.5b) and blind pits.

2.4 *Ulmaceae/Cannabaceae*

Certain species of the Ulmaceae/Cannabaceae also form tori that are best developed in intervascular pit pairs of latewood tracheary elements (Wheeler 1983; Dute and Rushing 1990; Dute et al. 2004; Jansen et al. 2004). Torus ontogeny has been carefully followed only in *Ulmus alata*, *Celtis laevigata*, and *C. occidentalis* (Dute and Rushing 1990). Torus thickenings in *Ulmus* and *Celtis* occur early in pit membrane ontogeny (before initiation of the secondary wall) and their development is not associated with a microtubule plexus (Fig. 3.6a) (Dute and Rushing 1990). Thickenings are viewed as elaborations of the primary walls of the pit membrane. Tests for lignin in mature tori have proven equivocal (Wheeler 1983; Coleman et al. 2004). AFM images of mature tori in *U. alata* show microfibrils crossing the surface (Fig. 3.6b) in contrast to the situation in *Osmanthus* and *Cercocarpus* (Dute and Elder 2011).

2.5 *Schisandraceae*

Sano et al. (2013) recently observed torus-bearing pit membranes in wood of *Schisandra chinensis*. This report is the first example of tori found in basal angiosperms rather than in core eudicots. Further study of structure and development of these pit membranes is called for.

2.6 *Miscellanea*

Microfibril orientation in the margo of dicot pit membranes has already been discussed relative to *Osmanthus*. This genus possesses the most organized arrangement of radial microfibrils among the torus-bearing dicot genera (Ohtani and Ishida 1978; Dute and Elder 2011). Wheeler (1983) has shown a good correlation between radial microfibrils and the presence of a torus in *Ulmus* and *Celtis* (Fig. 3.7a). Pit membranes of many torus-bearing *Daphne* species, by contrast, had few (Dute et al. 1990) or no (Ohtani and Ishida 1978) radial microfibrils. But torus-bearing pit membranes in aspirated examples of *Daphne aurantica* and in *Wikstroemia* of the same family do have a distinct radial component to the margo (Dute et al. 1996). It

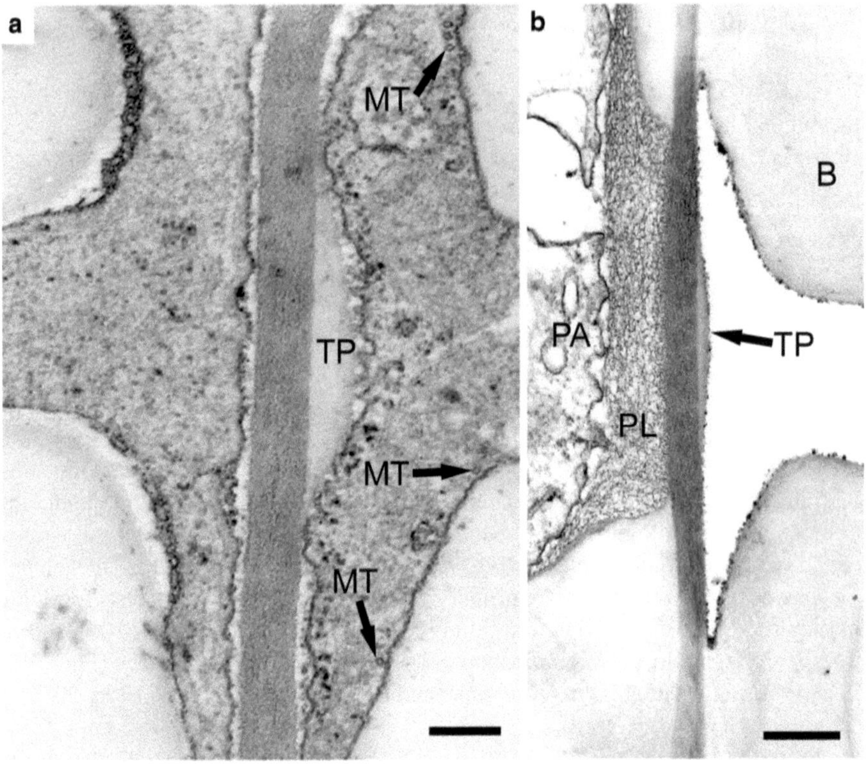

Fig. 3.5 Rosaceae. Torus pad development in *Cercocarpus betuloides*. (**a**) Development in a bordered pit pair. Only one pad is developed; the other cell of the pit pair is younger and a pad has yet to be initiated. (**b**) Half-bordered pit pair in which a torus pad is associated with only the tracheary element (water-conducting cell) side of the pit membrane. *Bars*: **a**=0.2 μm; **b**=0.5 μm. Photos from Dute et al. (2010a)

seems as if this feature varies within the family. When present, radial microfibrils are, in all dicot genera studied, on or near the surface of the pit membrane (Fig. 3.7b) and cover the remainder of the margo, which consists of tightly woven, randomly arranged microfibrils similar to that of intervascular pit membranes of most dicots (Dute and Elder 2011).

Plasmodesmata are the exception rather than the rule in intervascular pit membranes of woody dicots (Barnett 1982, but see Sect. 2.7 below); therefore, discussion of these cytoplasmic channels will be postponed until discussion of the intertracheid pit membranes of gymnosperms, where such structures are common.

2.7 *Pseudotori*

Pseudotori are thickenings associated with pit membranes of tracheary elements (including fiber tracheids) in a number of woody taxa including Rosaceae, Ericaceae, and Oleaceae (Jansen et al. 2007; Rabaey et al. 2008a). The following ontogenetic

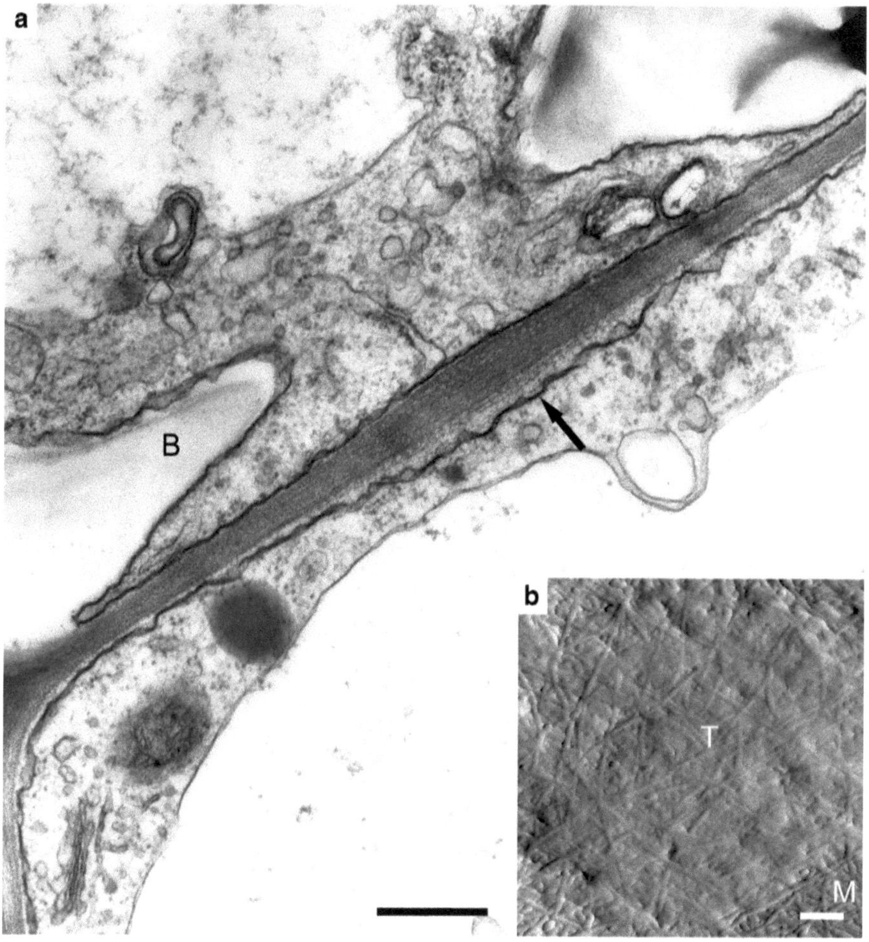

Fig. 3.6 Ulmaceae/Cannabaceae. Pit membrane development between two immature bordered pits in *Celtis laevigata*. The torus thickening indicated with an *arrow* is under construction. That particular cell has yet to build its pit border, whereas the other (older) cell has not only a completed its torus thickening, but has nearly completed its pit border. (**b**) Torus surface in *Ulmus alata* traversed by microfibrils. *Bars*: **a** = 0.5 μm; **b** = 0.25 μm. Photo (**a**) from Dute and Rushing (1990); photo (**b**) from Dute and Elder (2011)

sequence is a summarization of a number of published studies (Parameswaran and Liese 1973, 1981; Parameswaran and Gomes 1981; Barnett 1982, 1987; Jansen et al. 2007; Rabaey et al. 2008a, b; Plavcová and Hacke 2011). In intervascular pit membranes, torus-like structures are formed by thickening of the primary wall layers. These thickenings are often eccentric rather than centrally located on the pit membrane. More than one thickening can occur on a given membrane. These thickenings often contain large numbers of plasmodesmatal channels that are focused toward the middle lamella (Fig. 3.8a). This stage is followed by deposition of pads or caps of wall material over both surfaces of the thickening (Fig. 3.8b). The plasmodesmatal channels might or might not extend into the pads. With dissolution of

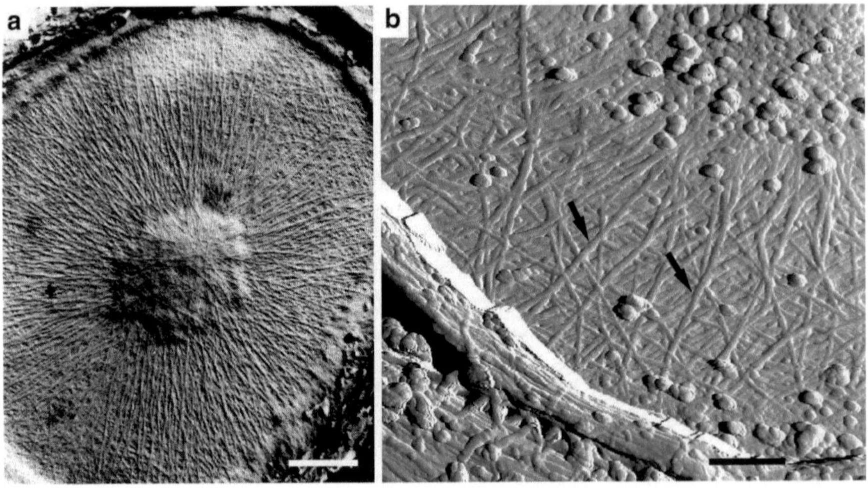

Fig. 3.7 (**a**) Surface of *Celtis reticulata* pit membrane showing torus associated with radial microfibrils traversing the margo. (**b**) Detail of margo of *Osmanthus armatus* showing radial microfibrils (*arrows*) at or near the surface. *Bars*: **a** = 1 μm; **b** = 0.5 μm. Photo (**a**) from Wheeler (1983); photo (**b**) from Dute and Elder (2011)

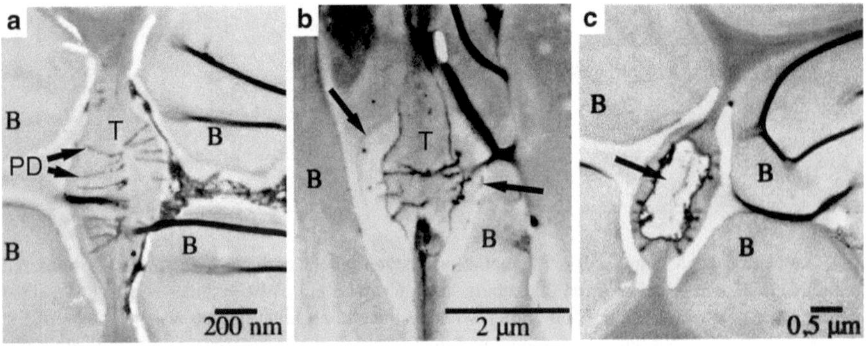

Fig. 3.8 Developmental stages of pseudotori in *Malus yunnanensis* (**a, c**) and *Pittosporum tenuifolium* (**b**). (**a**) Early developmental stage with thickened primary walls and numerous plasmodesmata. (**b**) Caps or pads (*arrows*) subsequently deposited on surfaces of primary walls. (**c**) Loss of matrix material from primary wall thickenings (*arrow*). *Bars*: **a** = 200 nm; **b** = 2 μm; **c** = 0.5 μm. All photographs from Rabaey et al. (2008b)

the cytoplasms, matrix material is lost from the original pit membrane thickening, whereas the pads generally remain intact (Fig. 3.8c). In cases where the pit membrane connects a tracheary element and a parenchyma cell, the membrane surface of the latter is covered by a protective wall layer rather than a pad.

There is no evidence that pseudotori are involved in embolism blockage as are regular tori. The eccentric location of most pseudotori on pit membranes would preclude this possibility. Their actual function is unknown, although the large number of plasmodesmata would imply intercellular communication when the cells are still alive.

3 Anatomy and Ontogeny: Conifers and *Ginkgo*

Traditionally, intertracheid pit membranes of gymnosperms (with the exception of cycads) are thought to consist of a thickened, central torus surrounded by a permeable, fibrillar margo (Fig. 3.1b). Early work showed considerable structural diversity to exist within mature pit membranes of gymnosperms (Bauch et al. 1972). In this section of the chapter, I shall provide an overview of two different pathways of intertracheary pit membrane development in gymnosperms using *Abies firma* (representing Pinaceae) versus *Metasequoia glyptostroboides* (Cupressaceae) and *Ginkgo biloba* as examples[2]. Additional information from other species and genera is included when helpful.

Intervascular pit membranes of gymnosperms not uncommonly contain complex branched plasmodesmata (Fig. 3.9a) (Murmanis and Sachs 1969; Thomas 1969, 1972; Fengel 1972; Fujikawa and Ishida 1972; Timell 1979; Dute 1994; Sano et al. 1999; Dute et al. 2008; Roth-Nebelsick et al. 2010; Jansen et al. 2012). In *Ginkgo*, *Abies*, and *Metasequoia*, branched plasmodesmata are present in the earliest developmental stages (Dute 1994; Dute et al. 2008). Plasmodesmata exist at sites of both future torus and margo. Structurally, plasmodesmata consist of branches (half plasmodesmata) extending from median cavities to the lumen surface (Fig. 3.9b). Plasmodesmatal branches are lined by the cell membrane and contain distinct desmotubules (Fig. 3.9b, c). Those channels within the torus region of the pit membrane elongate as the torus walls thicken. Not all tori within the wood of a specimen sample contain plasmodesmata, but in those that do, the branches (channels) show uneven (clustered) distribution (Fig. 3.9d) (Dute et al. 2008; Jansen et al. 2012). Ultrathin sections frequently show plasmodesmata as entering but one surface and passing only to the median cavity or central nodule region (Fig. 3.9b). Such half plasmodesmata were referred to as “ectoplasmic strands” in pit membranes of *Pinus radiata* by Barnett and Harris (1975). However, serial sections in pit membranes of *Ginkgo* (Dute 1994) and fortuitous sections from conifers (Jansen et al. 2012) indicate that at least some median cavities are connected by plasmodesmatal channels to both pit membrane surfaces.

[2] *Ephedra*, a xeromorphic Gnetalean gymnosperm also has well-developed tori. Pit membrane structure in this genus is dealt with in a recent publication (Dute et al. 2014).

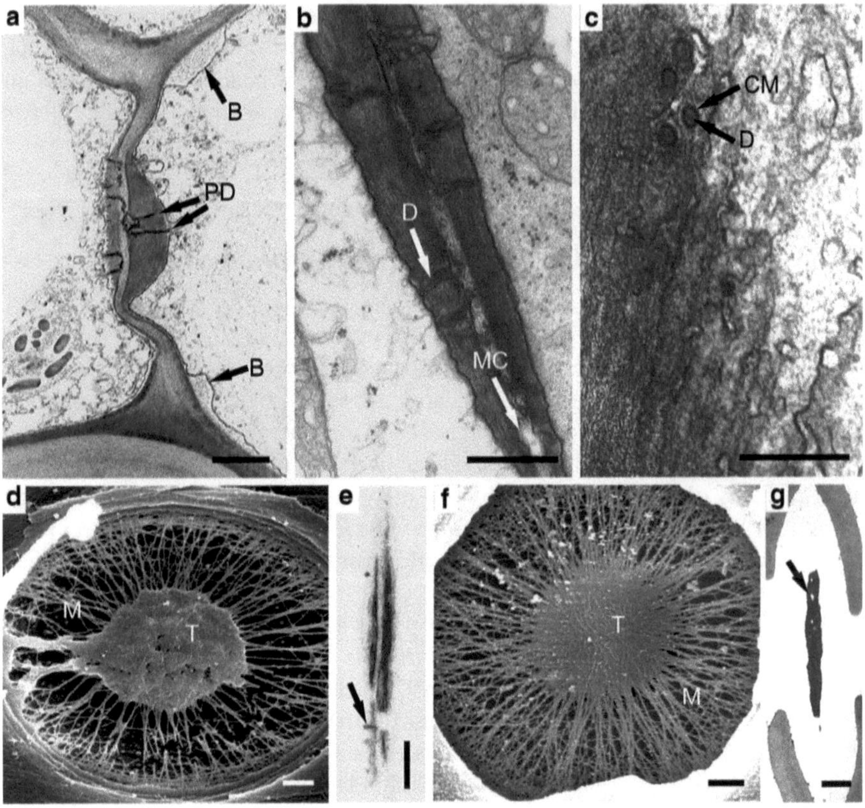

Fig. 3.9 Conifers and *Ginkgo*. (**a**) Immature pit membrane with developing torus thickenings in *Abies firma*. Note the pit borders have just begun to development on the cell at *right*. Pit membranes in these gymnosperms commonly contain plasmodesmata. (**b**) Sectional view through an immature torus of *Ginkgo*. Plasmodesmatal channels contain desmotubules and are associated with median cavities in the middle lamella of the pit membrane. (**c**) Paradermal section of developing pit membrane in *Ginkgo*. Plasmodesmatal channels are seen as *circles*, each lined by the cell membrane and containing a desmotubule. (**d**) A mature pit membrane of *A. firma* with visible plasmodesmatal channels in the torus. (**e**) Mature torus in *Ginkgo* in the process of losing its matrix material. *Arrow* indicates a remnant plasmodesma. (**f**) Mature pit membrane of *Metasequoia glyptostroboides* in which no plasmodesmata are visible. (**g**) Section through a mature torus of *A. firma* in which cavities associated with plasmodesmata and median cavities are still evident. *Bars*: **a** = 0.5 μm; **b** = 0.5 μm; **c** = 0.25 μm; **d** = 1 μm; **e** = 0.5 μm; **f** = 1 μm; **g** = 1 μm. Photos (**a**, **d**, **f**, **g**) from Dute et al. (2008); photos (**c**, **e**) from Dute (1994); photo (**b**) from Dute

The fate of plasmodesmata varies with the genus in question. In both *Ginkgo* and *Metasequoia* maturation of the tracheid is associated with loss of matrix material from both torus and margo. Figure 3.9e shows remnants of at least one plasmodesma in a torus of *Ginkgo* that is undergoing matrix removal. Thin sections of later stages show no evidence of plasmodesmata. It could be argued that the channels are still present in some form, but are simply not stained. However, SEM of torus

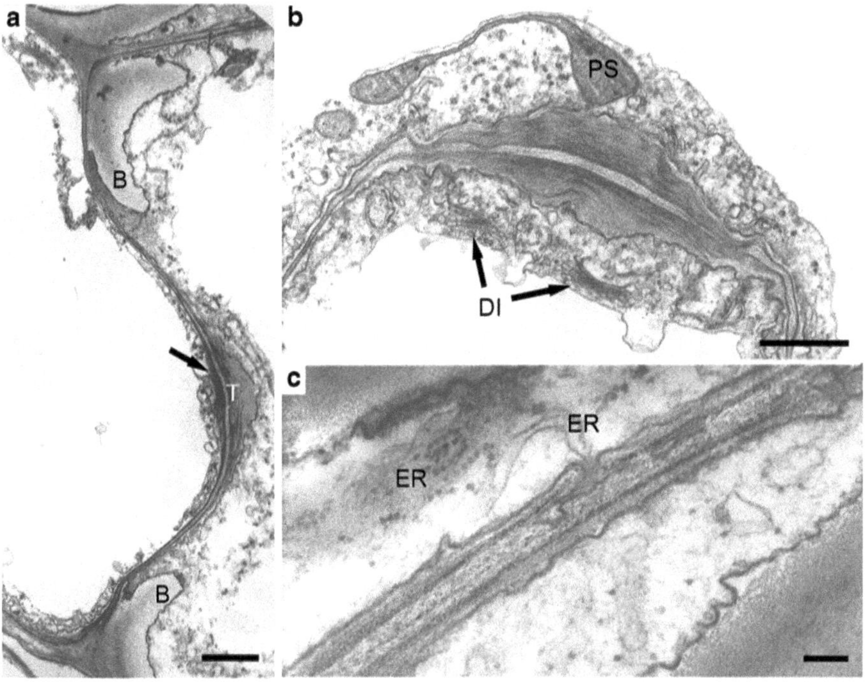

Fig. 3.10 *Ginkgo*. (**a**) Wall thickening of the torus region (*arrow*) occurs before initiation of a pit border in water-conducting cells (tracheids). (**b**) Organelles associated with torus thickening include dictysomes and plastids. (**c**) Plasmodesma in the future margo portion of an immature pit membrane. A strand of endoplasmic reticulum (*ER*) enters the channel and gives rise to a desmotubule. *Bars*: **a** = 1 μm; **b** = 1 μm; **c** = 0.25 μm. All photos from Dute (1994)

surfaces of *Ginkgo* and *Metasequoia* provide no evidence for such structures (Fig. 3.9f). In contrast, in other genera such as *Abies*, the torus retains its matrix during maturation and plasmodesmatal remain evident (Fig. 3.9d, g). For *Metasequoia* and *Ginkgo*, it is hypothesized that matrix removal leads to a realignment of the cellulose microfibrils and disappearance of plasmodesmata (Dute et al. 2008).

Torus thickening begins before initiation of the pit border (Figs. 3.9a and 3.10a) (Thomas 1972; Imamura and Harada 1973; Dute 1994; Dute et al. 2008). In fact, it appears as if torus thickness has reached its maximum by the time of border initiation. Various cytoplasmic membranes and organelles (including dictyosomes) are associated with the developing torus (Fig. 3.10b). Strands of ER, often normal to the torus surface, are noted during various ontogenetic stages of the pit membrane. This observation is not surprising given the number and prominence of plasmodesmata in both torus and margo of the pit membrane. Occasional thin sections show strands of tubular ER entering plasmodesmatal channels, no doubt becoming desmotubules within the plasmodesmata (Fig. 3.10c). No microtubule plexus such as is found in *Osmanthus* or *Daphne* is present in either conifers or *Ginkgo*.

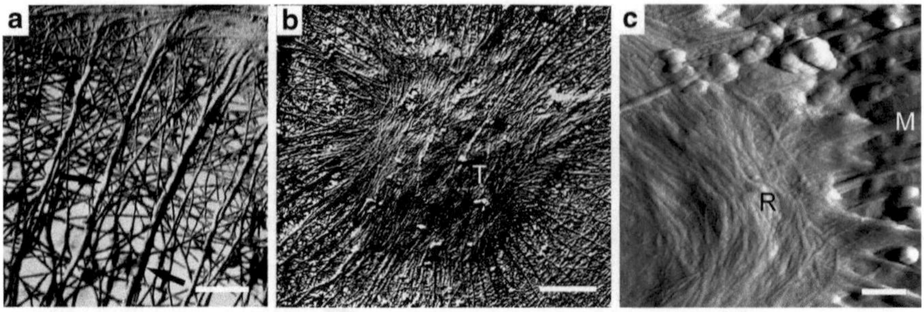

Fig. 3.11 *Pinus*. (**a**) Detail of margo from a mature pit membrane. The surface consists of large radially oriented microfibrils subtended by smaller diameter microfibrils in a reticulate arrangement. (**b**) Face view of a torus thickening caused by crossing of radially oriented microfibrils at the center of the pit membrane. (**c**) A detail of the periphery of a torus in *P. taeda* showing concentric microfibrils. *Bars*: **a** = 0.25 μm; **b** = 1 μm; **c** = 0.5 μm. Photos (**a**, **b**) from Thomas (1970); photo (**c**) from Dute and Elder (2011)

Torus and margo consist of both fibrillar and matrix material. Careful work on *Pinus* and *Taxodium* by Thomas (1968, 1970, 1972) paints the following picture. Randomly oriented microfibrils are covered on both pit membrane surfaces by layers of thicker (larger diameter), radially oriented microfibrils (Fig. 3.11a). The latter cross one another in the center of the membrane and initiate thickening at the site of the torus (Fig. 3.11b). In Pinaceae, but in neither Cupressaceae nor *Ginkgo*, a ring of parallel microfibrils is deposited at the torus periphery (Fig. 3.11c). Finally, incrusting material is added to the torus. Enzyme hydrolysis studies show the torus of conifers to have a high concentration of pectin, whereas the margo (before removal of the matrix material at maturation) contains large amounts of hemicellulose (Imamura et al. 1974).

As pit membrane ontogeny continues, development of the pit border occurs. Late in pit border development access to the pit aperture on the lumen side is blocked by a plastid. This blockage was noted in *Ginkgo* (Fig. 3.12a) (Dute 1994) and later in both *Abies firma* and *Metasequoia* (Fig. 3.12b) (Dute et al. 2008). A similar observation for *Abies balsamea* was made by Timell (1979). The metabolic importance of this developmental stage is unknown.

Completion of the pit border is followed by autolysis during which cell contents are lost and variable amounts of matrix/incrusting materials are removed from the pit membranes. In *Abies* and other Pinaceae (Dute et al. 2008; Hacke and Jansen 2009; Jansen et al. 2012) matrix material is removed from the margo but not from the torus. Although left intact, the torus attracts more heavy metal stain, becoming more electron opaque with the TEM (Fig. 3.9g). In contrast, *Ginkgo* (Dute 1994), *Metsequoia* and *Taxodium* lose most or all of the matrix material from the torus as well (Fig. 3.12c, d). This process causes the torus to lose some of its thickness (Thomas 1972; Dute 1994). Matrix loss from the torus in these cases takes longer than loss from the margo due to the greater thickness of the former. Plasmodesmata are not loci for matrix removal. Mature pit membranes without incrusting material/matrix show radial

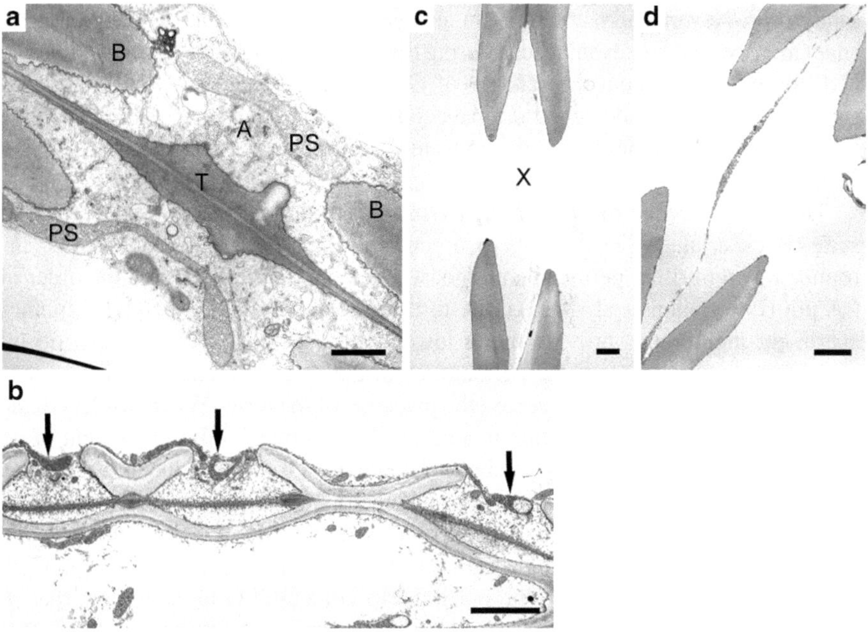

Fig. 3.12 Conifers and *Ginkgo*. Plastid occlusions of developing pit apertures in *G. biloba* (**a**) and *Metasequoia glyptostroboides* (*arrows*) (**b**). Mature bordered pit pairs in *G. biloba* (**c**) and *M. glyptostroboides* (**d**). A pit membrane is present in (**c**) at the region marked with an *X*, but can only be observed by "overexposing" the image. *Bars*: **a** = 1 μm; **b** = 5 μm; **c** = 1 μm; **d** = 1 μm. Photos (**a**, **c**) from Dute (1994); photos (**b**, **d**) from Dute et al. (2008)

microfibrils from the margo traversing the surface of the torus (Fig. 3.9f). Dute et al. (2008) attempted to explain the differential loss of matrix material from different species by suggesting either a difference in torus composition or in autolytic enzymes.

4 Function and Response to Environment: Conifers and *Ginkgo*

The functions of homogeneous and torus-bearing intervascular pit membranes are the same—to facilitate passage of water, but to inhibit passage of air bubbles. These functions are enabled differently in pit membranes of conifers versus angiosperms.

Conifers can be found in a number of environmental situations, but by and large, they are able to endure moisture stress (Rushforth 1987). In his tome entitled *Pinus*, Mirov (1967) details the xeromorphic adaptations within that particular genus—I would include torus-bearing pit membranes among them. Pines and other conifers can withstand water stress caused by drought, high evaporation, low temperatures, and sandy soils. Nevertheless, conifers can also grow in other habitats where their xeric adaptations would be advantageous during the occasional drought (Mirov 1967).

Little is known about the ecology of *Ginkgo*. The tree grows well in exposed sites in soil that provides good drainage yet has ample moisture-holding capacity (He et al. 1997). Although sensitive to water-logging, the tree is resistant to drought (He et al. 1997).

The pit membrane of conifers and *Ginkgo*, considered to represent a derived state (Beck et al. 1982), consists of a central, relatively impermeable torus surrounded by a fibrillar, permeable margo, whose pore diameters are on the order of 0.1 μm (Pittermann et al. 2005). Due to these relatively large openings, gymnosperm pit membranes provide much lower resistance to water flux than do pit membranes of angiosperms (Pittermann et al. 2005) even though membrane surface area in the former is reduced by the presence of the torus. When displaced, the torus should occlude the aperture in a pit border and thus inhibit movement of air embolisms into a neighboring tracheid. The quality of seal produced by an aspirated torus is dependent upon a number of factors, among them: torus thickness and depth of pit chamber. As explained by Hacke and Jansen (2009), thinner (and thus more flexible) tori provide a tighter seal against the pit aperture. Also, depth of the pit chamber is inversely proportional to vulnerability to cavitation; that is, the greater the deflection required by the pit membrane, the more likely it is that the seal will be imperfect. Also, there is some evidence that plasmodesmata traversing the torus can make it "leaky" and allow passage of air bubbles (Jansen et al. 2012).

Calculations show that a constancy of torus diameter versus membrane and aperture diameter leads to maximum pit conductivity per membrane area. Torus overlap is, according to Hacke et al. (2004), "the fraction of the pit border width that is covered by the torus" and is defined by $(D_t - D_a)/(D_m - D_a)$, where D_t equals the torus diameter, D_a the aperture diameter, D_m the membrane diameter. Torus overlap for 17 species of gymnosperms proved to fall between 0.21 and 0.38. Values that are above or below this range lead to reduced flow rate through the pit membranes (Hacke et al. 2004). Permeability of the pits, in turn, limits the saturating length of tracheids that is needed to maximize conductivity through the cells as, according to the authors, "a longer length provides no significant gain in conductivity" (the resistance of the pits becomes rate limiting—see also Lancashire and Ennos 2002) "and a shorter length sacrifices conductivity."

5 Function and Response to Environment: Dicotyledons

A brief mention must be made at this time regarding pit membrane structure of intervascular pits of most angiosperms. According to many authors, these pit membranes lack a torus and consist of randomly crossing or of a reticulate texture of microfibrils with pores measuring in the nms (Pittermann et al. 2005), although recent studies show that some pit membranes can have larger pores (Jansen et al. 2009). This arrangement of microfibrils is the surface layer and typically is seen in undamaged pit membranes with the SEM. In fact, such pit membranes often are multilayered with the inner layer(s) consisting of microfibrils in a parallel arrangement (Schmid and Machado 1968; Dute et al. 1992; Sano 2005; Jansen et al. 2009). The small pores result in high resistance to water flux through the membrane. This is compensated by the great length of angiosperm vessels when compared to gymnosperm tracheids.

The function of torus-bearing membranes of conifers and *Ginkgo* is well established. The same can be said for that of the homogeneous pit membranes in angiosperms. What, however, can we make of torus-bearing pit membranes in dicots where a centrally located torus is surrounded by a margo whose microfibrillar arrangement is identical to that of homogeneous pit membranes? It would seem that a combination of reticulately arranged microfibrils and loss of conducting pit membrane surface by the presence of a torus would cut water flow from cell to cell to an unacceptable level. Added to this is the fact that tori in angiosperms have evolved a number of times (Sano et al. 2013; Table 3.1 this chapter). In other words, tori in the angiosperms could have evolved different times for different reasons under different selective pressures. Little experimental data exist for these species, and one is forced to rely on morphological correlations.

Recent trials at North Carolina State University noted superior performance during the drought of 2007 for specimens of *Chionanthus retusus* (Oleaceae), *Osmanthus fragrans* (Oleaceae), and *Ulmus alata* (Ulmaceae), all three with torus-bearing pit membranes (Bilderback 2007).

It was stated earlier in the chapter that a correlation exists between pit membranes with a torus and a circular pit aperture. According to Beck et al. (1982), such a membrane would be adaptive in providing efficient valve action under conditions inducing pit membrane aspiration. This correlation was made with regard to progymnosperms and conifers but, despite the exceptions noted earlier, this is true also for torus-bearing dicot pit membranes. In contrast, according to Beck et al., a less specialized homogeneous pit membrane would be adaptive in pits with slit-like apertures. However, Wright (1928) notes that "the presence of a round pore (aperture) does not necessarily entail development of a torus." A study of selected species of the Thymelaeceae (Dute et al. 2001) showed the circularity ratio of pit apertures (ratio of long axis to short axis) of *Daphne gnidioides* (torus-bearing) to be 0.71 and of *Pimelea prostata*, *P. arenaria*, and *Gnidia caffra* (all non-torus bearing) to be 0.3, 0.33, and 0.27, respectively. However, *Dirca palustris* (also non-torus bearing) varied in three specimens from 0.63 to 0.78. In the Ulmaceae, the mean circularity ratio for *Planera aquatica* (torus-bearing) was 0.76 (Dute et al. 2004). Jansen et al. (2004) observed two herbarium specimens of *Ulmus lanciniata* with tori present in small diameter tracheary elements, but not in those cells whose diameter was greater than 20 μm. The pit aperture of the former had a circularity ratio of 0.85, whereas the circularity ratio of the latter was 0.61. It seems as if circular apertures and circular tori are correlated in dicotyledons.

5.1 *Oleaceae*

More is known about torus structure, development, location within a plant and species distribution in *Osmanthus* than in any other genus of angiosperm. All major sections of this genus have torus-bearing members except *Notosmanthus* (Dute et al. 2010b). This section differs from the rest of the genus as regards inflorescence characters and chemistry, and probably should be removed (Dute et al. 2010b).

Twelve of thirteen torus-bearing species of *Osmanthus* studied by Dute et al. (2010b) were diffuse porous and one was semi-ring porous. This situation is in contrast to that mentioned in the Ulmaceae (Jansen et al. 2004), where torus presence was associated with ring porosity.

The center of diversity for *Osmanthus* is Southeast Asia, where over 80 % of the species are located (Xiang et al. 2008). *Osmanthus armatus*, for example, is associated with a monsoon climate (Shen and Zhang 2000). The plants grow in thickets on slopes (Wu and Raven 1996) and the species is considered to be drought tolerant and adapted to warm conditions (Shen and Zhang 2000). Perusal of collecting site information for various Asian species (Green 1958) often shows plants growing on dry, steep slopes. Thus, *Osmanthus* species are associated with xeric situations. An exception to this observation is *O. americanus* which, while torus-bearing, is found in low, wet situations in the southeastern United States (Green 1958).

Osmanthus shrubs and trees are evergreen (Xiang et al. 2008). Dute et al. (2012b), in a study of tori in leaves of *O. amaratus* noted, "Tori are considered to be xeromorphic features which, along with thick cuticle and sclereids, are advantageous for the perennial growth habit of leaves." Evergreen leaves are expensive to make, and to repay construction costs must remain attached to the stem for a prolonged period (q.v. discussion in Dute et al. 2012b). Such leaves must be able to survive times of stress, which includes water stress. Thus, there is a clear adaptive advantage to isolating air bubbles in leaf veins.

Wood anatomy (Baas et al. 1988) and sequencing data (Wallander and Albert 2000) show *Osmanthus* and *Picconia* to be closely related, probably sister genera. It should come as no surprise then to learn that the two species of *Picconia* (*P. azorica* and *P. excelsa*) have torus-bearing intervascular pit membranes (Dute et al. 2008; Rabaey et al. 2008a). Both species grow in Macaronesia, *P. excelsa* is on the Canary Islands and Madeira, and *P. azorica* in the Azores (Humphries 1979; Gomes 1998; Arteaga et al. 2006; Ferreira et al. 2011). *Picconia* is considered "a xerophytic, evergreen tree genus" (Ferreira et al. 2011). A recent molecular study has clearly confirmed *Picconia*'s two species as distinct (Ferreira et al. 2011). *P. azorica* is found in coastal forests, where it is resistant to sea spray (Ferreira et al. 2012). Fossils of *Picconia* (*P. excelsa*) have been found in Pliocene deposits of central Europe indicating a serious loss of range into present times. If so, its possession of a torus would preadapt it to the environment of the refuge it now inhabits.

5.2 *Thymelaeaceae*

The Thymelaeaceae as a whole is a family of dry environments (Rendle 1925). For example, species of *Diarthron* and *Stellera* are found in the steppe regions of central and eastern Asia, where the plants are exposed to summer drought as well as to winter cold (Tan 1982). In those species that possess them, torus-bearing pit membranes are associated with narrow diameter tracheary elements (both vessel members and vascular tracheids) containing distinct helical thickenings (Dute et al. 1996, 2011). Here again, torus-bearing pit membranes are hypothesized to be part of a backup water-conducting system that has a low rate of flow but has a high safety factor.

5.3 *Rosaceae*

Cercocarpus is a genus of four species found in North America extending from Mexico north to Washington state and from the Great Plains of the United States west to the Pacific coast (McGregor et al. 1986; Hickman 1993; Cronquist et al. 1997; Chap. 5, this book). The plants are shrubs or small trees that prefer dry situations such as rocky slopes, cliffs, and chaparral (Bailey and Bailey 1976; McGregor et al. 1986; Hickman 1993; Cronquist et al. 1997). *Cercocarpus* is also the only known member of the Rosaceae to possess torus-bearing pit membranes in its wood (Jansen et al. 2007). The centrally located torus is found associated with both vessel members and tracheids with a tendency toward better development in cells of smaller diameter. Pit membrane pores could not be detected in the margo using SEM (Jansen et al. 2007). Measurements indicate both torus and aperture tend toward circular (circularity ratios of 0.88 and 0.8, respectively), and the average diameter was greater for torus than aperture indicating a complete seal during aspiration in most instances (Jansen et al. 2007). Interestingly, measurements of pits in *Cercocarpus montanus* indicate tight scalings between tori and pit dimensions similar to that of torus-bearing pit membranes in gymnosperms (Dute et al. 2010a). For example, in *Cercocarpus* the torus represents 46 % of pit membrane diameter compared to 48 % for three species of Pinaceae and to 45 % in enlarged tracheids of Douglas-fir (Domec et al. 2006; Hacke and Jansen 2009). Values for torus overlap in *C. montanus* are also similar to torus overlap in gymnosperms (Hacke et al. 2004; Hacke and Jansen 2009). The average torus overlap for *C. montanus* (0.3 for tracheids and 0.26 for vessel elements) agrees with Hacke et al. (2004) who hypothesized a maximum pit conductivity in gymnosperms for a torus overlap between 0.21 and 0.38.

Recent physiological data exist for water flux through tracheary elements of *Cercocarpus*. Wheeler et al. (2005) conducted a study of safety versus efficiency trade-off with regard to water stress-induced cavitation. The authors chose various species within the Rosaceae including *Cercocarpus ledifolius* and *C. montanus*. It was found that cavitation resistance and transport efficiency are controlled by total pit area per vessel. The greater the pit area, the lower the P_{50}, where this term represents the xylem pressure inducing 50 % loss of hydraulic conductivity.

By decreasing pit area per conduit, average size of the largest pore per vessel is reduced thus increasing the air-seeding pressure. On the debit side, greater safety leads to decreased water transport efficiency. It was not until 2007 that tori in *Cercocarpus* pit membranes were reported by Jansen et al. Thus, Wheeler et al. in 2005 assumed the pit membranes in *Cercocarpus* to be homogeneous. Experiments showed relatively high P_{50} values for *C. betuloides* (−7.5 MPa), *C. ledifolius* (−4.9 MPa), and *C. montanus* (−5.8 MPa) (Wheeler et al. 2005; Hacke et al. 2006). Rather than preventing rupture during aspiration or providing an impermeable seal, the torus in angiosperm pit membranes could function simply by reducing permeable area of the membrane surface (q.v. discussion in Dute et al. 2010a).

5.4 *Ulmaceae/Cannabaceae*

Wheeler (1983) was the first to observe tori in pit membranes associated with vascular tracheids and possibly narrow diameter vessel members in species of *Ulmus* and *Celtis*. When a pit membrane connected a vascular tracheid to a vessel element, a thickening was produced only on the pit membrane surface of the former cell. Wheeler suggested that the torus could prevent membrane tearing during displacement. Dute and Rushing (1990) were able to confirm torus thickenings involving both vascular tracheids and small diameter vessel members in *U. alata*, *C. laevigata*, and *C. occidentalis*. Jansen et al. (2004) investigated woods of 19 species of the Ulmaceae and observed tori in ring porous species from cold, temperate climates but not from diffuse porous species of (sub)tropical environments. They noted that torus-bearing pit membranes were restricted to narrow diameter tracheary elements, which also possessed distinct helically thickened walls. The authors hypothesized that although possessed of low conductivity, these narrow diameter tracheary elements provided greater resistance to freezing-induced cavitation. A later study regarding *Planera aquatica*, also a member of the Ulmaceae, located tori in narrow diameter tracheary elements of the wood (Dute et al. 2004). In this species as well, the cells in question possessed well-developed helical wall thickenings and were thought to provide a reserve water-conducting system. Thickened tori would either prevent membrane tearing during aspiration and/or provide a tight seal against an aperture (Dute et al. 2004). A point of interest is that *P. aquatica* (as the name suggests) is found growing in swamps in the southeastern United States (Radford et al. 1968).

Type of Imagining

Carbon/metal replicas viewed by transmission electron microscopy: 1B, 7A, 11A, B.
Resin-embedded, ultrathin sections stained with uranyl acetate and lead citrate and viewed by transmission electron microscopy: 2A, B, C, F, 4A, B, C, D, E, 5A, B, C, 6A, 8A, B, C, 9A, B, C, E, G, 10A, B, C, 12A, B, C, D.
Scanning electron microscopy of metal coated surfaces: 1C, 1D, 3B, C, 9D, F.
Atomic force microscopy: 2D, E, 6B, 7B, 11C.
Sectioned, stained, and viewed by light microscopy: 3A, D.

References

Arteaga MA, González G, Delgado JD, Arévalo JR, Fernández-Palacios JM (2006) Offspring spatial patterns in *Picconia excelsa* (Oleaceae) in the Canarian laurel forest. Flora 201:642–651

Baas P, Esser PM, van der Westen MET, Zandee M (1988) Wood anatomy of the Oleaceae. IAWA Bull New Ser 9:103–182

Bailey LH, Bailey EZ (1976) Hortus third, a concise dictionary of plants cultivated in the United States and Canada. Macmillan, New York

Barnett JR (1982) Plasmodesmata and pit development in secondary xylem elements. Planta 155:251–260

Barnett JR (1987) Changes in the distribution of plasmodesmata in developing fibre-tracheid pit membranes of *Sorbus aucuparia* L. Ann Bot 59:269–279

Barnett JR, Harris JM (1975) Early stages of bordered pit formation in radiata pine. Wood Sci Technol 9:233–241

Bauch J, Liese W, Schultze R (1972) The morphological variability of the bordered pit membranes in gymnosperms. Wood Sci Technol 6:165–184

Beck CB, Coy K, Schmid R (1982) Observations on the fine structure of *Callixylon* wood. Am J Bot 69:54–76

Bilderback TE (2007) Development, evaluation, and promotion of underutilized and novel ornamental plants. Hatch Project, accession no. 0139259

Coleman CM, Prather BL, Valente MJ, Dute RR, Miller ME (2004) Torus lignification in hardwoods. IAWA J 25:435–447

Cronquist A, Holmgren NH, Holmgren PK (1997) Intermountain flora: vascular plants of the Intermountain West, U.S.A. The New York Botanical Garden, Bronx

Czaninski Y (1979) Cytochimie ultrastucturelle des parois du xyléme secondaire. Biol. Cellulaire 35:97–102

Domec J-C, Lachenbruch B, Meinzer FC (2006) Bordered pit structure and function determine spatial patterns of air-seeding thresholds in xylem of Douglas-fir (*Pseudotsuga menziesii*; Pinaceae) trees. Am J Bot 93:1588–1600

Dute RR (1994) Pit membrane structure and development in *Ginkgo biloba*. IAWA J 15:75–90

Dute RR, Elder T (2011) Atomic force microscopy of torus-bearing pit membranes. IAWA J 32:4154430

Dute RR, Rushing AE (1987) Pit pairs with tori in the wood of *Osmanthus americanus* (Oleaceae). IAWA Bull New Ser 8:237–244

Dute RR, Rushing AE (1988) Notes on torus development in the wood of *Osmanthus americanus* (L.) Benth. & Hook. *ex* Gray (Oleaceae). IAWA Bull New Ser 9:41–51

Dute RR, Rushing AE (1990) Torus structure and development in the woods of *Ulmus alata* Michx., *Celtis laevigata* Willd., and *Celtis occidentalis* L. IAWA Bull New Ser 11:71–83

Dute RR, Rushing AE, Perry JW (1990) Torus structure and development in species of *Daphne*. IAWA Bull New Ser 11:401–412

Dute RR, Rushing AE, Freeman JD (1992) Survey of intervessel pit membrane structure in *Daphne* species. IAWA Bull New Ser 13:113–123

Dute RR, Freeman JD, Henning F, Barnard LD (1996) Intervascular pit membrane structure in *Daphne* and *Wikstroemia*—Systematic implications. IAWA J 17:161–181

Dute RR, Miller ME, Carollo RR (2001) Intervascular pit structure in selected species of Thymelaeaceae. J Ala Acad Sci 72:14–26

Dute RR, Martin AL, Jansen S (2004) Intervascular pit membranes with tori in wood of *Planera aquatica* J.F. Gmel. J Ala Acad Sci 75:7–21

Dute R, Hagler L, Black A (2008) Comparative development of intertracheary pit membranes in *Abies firma* and *Metasequoia glyptostroboides*. IAWA J 29:277–289

Dute R, Patel J, Jansen S (2010a) Torus-bearing pit membranes in *Cercocarpus*. IAWA J 31:53–66

Dute R, Rabaey D, Allison J, Jansen S (2010b) Torus-bearing pit membranes in species of *Osmanthus*. IAWA J 31:217–226
Dute R, Jandrlich MD, Thornton S, Callahan N, Hansen CJ (2011) Tori in species of *Diarthron*, *Stellera* and *Thymelaea* (Thymelaeaceae). IAWA J 32:54–66
Dute RR, Hubbard Z, Patel RV (2012a) Intervascular pit membranes in roots of two species of *Osmanthus* (Oleaceae). J Ala Acad Sci 83:8–19
Dute RR, Zwack PJ, Craig E, Baccus SM (2012b) Torus presence and distribution in leaves of *Osmanthus armatus*. IAWA J 33:257–268
Dute R, Hansen CJ, Baker M, Lambert E, Michael W (2013) Presence of torus-bearing pit membranes in reproductive organs of *Osmanthus americanus* (Oleaceae). J Ala Acad Sci 84:17–27
Dute RR, Bowen LA, Schier S, Vevon AG, Best TL, Auad M, Elder T, Bouche P, Jansen J (2014) Pit membranes of *Ephedra* resemble gymnosperms more than angiosperms. IAWA J 35(3):217–235
Esau K (1977) Anatomy of seed plants, 2nd edn. Wiley, New York
Fengel D (1972) Structure and function of the membrane in softwood bordered pits. Holzforschung 26:1–9
Ferreira RC, Piredda R, Bagnoli F, Bellarosa R, Attimonelli M, Fineschi S, Schirone B, Simeone MC (2011) Phylogeography and conservation perspectives of an endangered Macaronesian endemic: *Picconia azorica* (Tutin) Knobl. (Oleaceae). Eur J For Res 130:181–195
Ferreira RC, Lo Monaco A, Picchio R, Schirone A, Vessella F, Schirone B (2012) Wood anatomy and technological properties of an endangered species: *Picconia azorica* (Oleaceae). IAWA J 33:375–390
Fujikawa S, Ishida S (1972) Study on the pit of wood cells using scanning electron microscopy. III. Structural variation of bordered pit membrane on the radial wall between tracheids in Pinaceae species. J Jap Wood Res Soc 18:477–483
Gomes ABA (1998) Creating native displays at the Faial Botanic Garden, Azores, Portugal http://www.bgci.org/congress/congress_1998_cape/html/gomes.htm
Green PS (1958) A monographic revision of *Osmanthus* in Asia and America. Notes Roy Bot Gard Edinburgh 22:439–542
Hacke UG, Jansen S (2009) Embolism resistance of three boreal conifer species varies with pit structure. New Phytol 182:675–686
Hacke UG, Sperry JS, Pittermann J (2004) Analysis of circular bordered pit function II. Gymnosperm tracheids with torus-margo pit membranes. Am J Bot 91:386–400
Hacke UG, Sperry JS, Wheeler JK, Castro L (2006) Scaling of angiosperm xylem structure with safety and efficiency. Tree Physiol 26:689–701
He S-A, Yin G, Pang Z-J (1997) Resources and prospects of *Ginkgo biloba* in China. In: Hori T, Ridge RW, Tulecke W, Del Tredici P, Trémouillaux-Guiller J, Tobe H (eds) *Ginkgo biloba*—a global treasure. Springer, Tokyo, pp 373–382
Hickman JC (ed) (1993) The Jepson manual: higher plants of California. University of California Press, Berkeley
Humphries CJ (1979) Endemism and evolution in Macaronesia. In: Bramwell D (ed) Plants and islands. Academic, London, pp 171–191
Imamura Y, Harada H (1973) Electron microscopic study on the development of the bordered pit in coniferous tracheids. Wood Sci Technol 7:189–205
Imamura Y, Harada H, Saiki H (1974) Embedding substances of pit membranes in softwood tracheids and their degradation by enzymes. Wood Sci Technol 8:243–254
Jansen S, Choat B, Vinckier S, Lens F, Schols P, Smets E (2004) Intervascular pit membranes with a torus in the wood of *Ulmus* (Ulmacea) and related genera. New Phytol 163:51–59
Jansen S, Sano Y, Choat B, Rabaey D, Lens F, Dute RR (2007) Pit membranes in tracheary elements of Rosaceae and related families: new records of tori and pseudotori. Am J Bot 94:503–514
Jansen S, Choat B, Pletsers A (2009) Morphological variation of intervessel pit membranes and implications to xylem function in angiosperms. Am J Bot 96:409–419
Jansen S, Lamy J-B, Burlett R, Cochard H, Gasson P, Delzon S (2012) Plasmodesmatal pores in the torus of bordered pit membranes affect cavitation resistance of conifer xylem. Plant Cell Environ 35:1109–1120

Lancashire JR, Ennos AR (2002) Modelling the hydrodynamic resistance of bordered pits. J Exp Bot 53:1485–1493
McGregor RL, Barkley TM, Brooks RE, Schofield EK (1986) Flora of the Great Plains. University Press of Kansas, Lawrence
Mirov NT (1967) The genus *Pinus*. Ronald, New York
Murmanis L, Sachs IB (1969) Seasonal development of secondary xylem in *Pinus strobus* L. Wood Sci Technol 3:177–193
Nesom GL (2012) Synopsis of American *Cartrema* (Oleaceae). Phytoneuron 2012–96:1–11
Ohtani J (1983) SEM investigation on the micromorphology of vessel wall sculptures. Research Bulletin of the College Experiment Forests, College of Agriculture, Hokkaido University 40: 323–386
Ohtani J, Ishida S (1978) Pit membrane with torus in dicotyledonous woods. Mokuzai Gakkaishi 24:673–675
Parameswaran N, Gomes AV (1981) Fine structural aspects of helical thickenings and pits in vessels of *Ligustrum lucidum* Ait. (Oleaceae). IAWA Bull New Ser 2:179–185
Parameswaran N, Liese W (1973) Anomalous structures in the bordered pits of fiber-tracheids of *Ribes sanguineum*. Wood Fiber 5:76–79
Parameswaran N, Liese W (1981) Torus-like structures in interfibre pits of *Prunus* and *Pyrus*. IAWA Bull New Ser 2:89–93
Pittermann J, Sperry JS, Hacke UG, Wheeler JK, Sikkema EH (2005) Torus-margo pits help conifers compete with angiosperms. Science 310:1924
Plavcová L, Hacke UG (2011) Heterogeneous distribution of pectin epitopes and calcium in different pit types of four angiosperm species. New Phytol 192:885–897
Rabaey D, Lens F, Smets E, Jansen S (2006) The micromorphology of pit membranes in tracheary elements of Ericales: New records of tori and pseudo-tori. Ann Bot 98:943–951
Rabaey D, Huysmans S, Lens F, Smets E, Jansen S (2008a) Micromorphology and systematic distribution of pit membrane thickenings in Oleaceae: tori and pseudo-tori. IAWA J 29:409–424
Rabaey D, Lens F, Huysmans S, Smets E, Jansen S (2008b) A comparative ultrastructural study of pit membranes with plasmodesmata associated thickenings in four angiosperm species. Protoplasma 233:255–262
Radford AE, Ahles HE, Bell CR (1968) Manual of the vascular flora of the Carolinas. The University of North Carolina Press, Chapel Hill
Rendle AB (1925) The classification of flowering plants, vol II, Dictoyledons. Cambridge University Press, Cambridge
Roth-Nebelsick A, Voigt D, Gorb S (2010) Cryo-scanning electron microscopy studies of *Pinus wallichiana* and *Mallotus japonicas*. IAWA J 31:257–267
Rushforth K (1987) Conifers. Christopher Helm, London
Sano Y (2005) Inter- and intraspecific structural variations among intervascular pit membranes, as revealed by field-emission scanning electron microscopy. Am J Bot 92:1077–1084
Sano Y, Kawakami Y, Ohtani J (1999) Variation in the structure of intertracheary pit membranes in *Abies sachalinensis* as observed by field-emission scanning electron microscopy. IAWA J 20:375–388
Sano Y, Utsumi Y, Nakada R (2013) Homoplastic occurrence of perforated pit membranes and torus-bearing pit membranes in ancestral angiosperms as observed by field-emission scanning electron microscopy. J Wood Sci 59:95–103
Schmid R, Machado RD (1968) Pit membranes in hardwoods—fine structure and development. Protoplasma 66:185–204
Shen Z-H, Zhang X-S (2000) A study on the classification of the plant functional types based on the topographical pattern of plant distribution. Acta Bot Sin 42:1190–1196
Tan K (1982) Studies in the Thymelaeaceae III: the status of *Diarthron*, *Dendrostellera*, *Stelleropsis* and *Stellera*. Notes Roy Bot Gard Edinburgh 40:213–221
Thomas RJ (1968) The development and ultrastructure of the bordered pit membrane in the southern yellow pines. Holzforschung 22:38–44

Thomas RJ (1969) The ultrastructure of southern pine bordered pit membranes as revealed by specialized drying techniques. Wood Fiber 1:110–123

Thomas RJ (1970) Origin of bordered pit margo microfibrils. Wood Fiber 2:285–288

Thomas RJ (1972) The ultrastructure of differentiating and mature bordered pit membranes from cypress (*Taxodium distichum* L. Rich.). Wood Fiber 4:87–94

Timell TE (1979) Formation of compression wood in Balsam Fir (*Abies balsamea*) II. Ultrastructure of the differentiating xylem. Holzforschung 33:181–191

Wallander E, Albert VA (2000) Phylogeny and classification of Oleaceae based on *Rps 16* and *TRNL-F* sequence data. Am J Bot 87:1827–1841

Wheeler EA (1983) Intervascular pit membranes in *Ulmus* and *Celtis* native to the United States. IAWA Bull 4:79–88

Wheeler JK, Sperry JS, Hacke UG, Hoang N (2005) Inter-vessel pitting and cavitation in woody Rosaceae and other vesselled plants: a basis for a safety versus efficiency trade-off in xylem transport. Plant Cell Environ 28:800–812

Wright JG (1928) The pit-closing membrane in the wood of the lower gymnosperms. Trans R Soc Can 22:63–94

Wu Z-Y, Raven PH (1996) Osmanthus. Flora China 15:286–292

Xiang Q, Liu L, Wu X (2008) Studies on the cultivar classification of *Osmanthus*. Acta Horticult 799:61–65

Zimmermann MH (1983) Xylem structure and the ascent of sap. Springer, Berlin

Chapter 4
The Hydraulic Architecture of *Populus*

Uwe G. Hacke

1 *Populus* as a Model System for Studying Plant Hydraulics, Wood Development, and Other Aspects of Tree Biology

Tree species of the genus *Populus* (poplars, cottonwoods, aspens) have a wide natural distribution in the Northern Hemisphere. Trembling aspen (*P. tremuloides*) is the most widely distributed tree species in North America. *Populus* species are among the fastest-growing temperate trees, consistent with their role as vegetational pioneers (Eckenwalder 1996). Hybrid poplar and aspen genotypes are used as a source of fiber, fuel, and shelter (Fig. 4.1). Riparian cottonwoods provide stream-bank stabilization and wildlife habitat (Rood et al. 2003). Aspen and other *Populus* species are used in land reclamation (Pinno et al. 2012). *Populus* is also a model organism for the study of tree biology (Bradshaw et al. 2000; Mellerowicz et al. 2001; Cooke and Rood 2007). Many physiological, cell biological, molecular, and genomics tools are available to study wood development and other biological processes (Jansson and Douglas 2007; Groover et al. 2010).

From a water relations perspective, *P. trichocarpa* has been used as a model system to better understand the role of aquaporins in embolism repair (Secchi and Zwieniecki 2011). Many ongoing and completed studies have focused on hydraulic traits of *Populus* species (Hacke et al. 2012) and on the effects of elevated CO_2 concentration on water use (Uddling et al. 2008). While gene expression in the vascular cambium has been extensively studied (e.g., Goue et al. 2008; Zhong et al. 2011), a goal for future work will be to identify gene candidates that impact important xylem hydraulic and structural traits such as vessel diameter and length.

U.G. Hacke (✉)
Department of Renewable Resources, University of Alberta,
Edmonton, AB, T6G 2E3, Canada
e-mail: uwe.hacke@ualberta.ca

U. Hacke (ed.), *Functional and Ecological Xylem Anatomy*,
DOI 10.1007/978-3-319-15783-2_4

Fig. 4.1 An aspen plantation in Alberta, Canada at the beginning of the growing season. Leaves are still unfolding. Photo: U. Hacke

2 Wood Anatomy and Water Transport

2.1 General Characteristics of Poplar Xylem

The wood of most *Populus* species is diffuse porous and light in weight (Blake et al. 1996; Telewski et al. 1996). Poplar wood consists of about 50 % cellulose, 30 % hemicellulose, and 20 % lignin (Balatinecz et al. 2001). In terms of cellular composition, poplar xylem includes vessels with simple perforation plates for water transport and fibers for mechanical support of the plant body (Fig. 4.2). Vessel elements of hybrid poplar saplings had a length of 0.23 mm; mean vessel length was 3–5 cm (Plavcova et al. 2011; Plavcová et al. 2013). Fibers had a mean length of 0.58 mm (Plavcova et al. 2011), similar to what was reported for another hybrid poplar clone (Arend and Fromm 2007). For mature poplar wood, the average length of vessel elements is in the range of 0.58–0.67 mm; average fiber length is ~1.3 mm. Fiber width is ranging between 24 and 32 μm (Telewski et al. 1996; Balatinecz et al. 2001; Groover et al. 2010).

Fiber structure in a xylem cross section is rarely homogeneous due to differences in fiber diameter and the presence of tension wood. Tension wood is characterized by the presence of gelatinous fibers. These fibers have an additional gelatinous

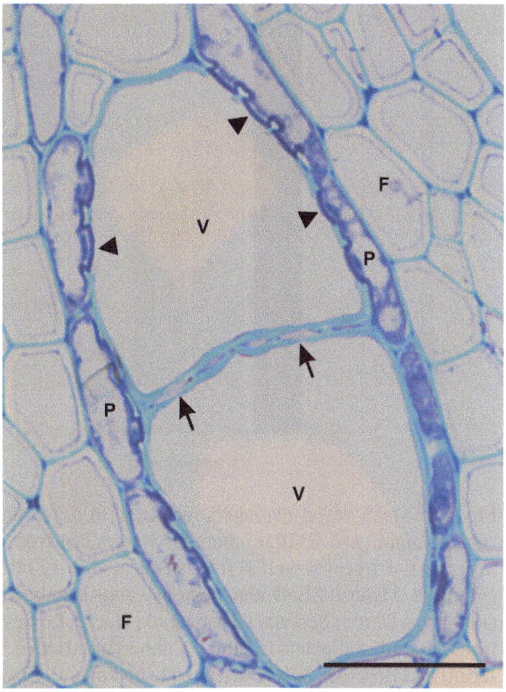

Fig. 4.2 Xylem cross section of hybrid poplar (clone H11-11) stained with toluidine blue. F: wood fibers; P: ray parenchyma cells; V: vessel elements. Arrows point to intervessel pits; arrow heads point to ray-vessel pits. Scale bar = 25 μm. Photo: L. Plavcová

layer, the *G*-layer, on the lumen side of the cell wall. The *G*-layer consists mainly of cellulose but may also include hemicellulose and other components (Arend 2008).

Vessel diameters vary widely, depending on the plant organ being studied, the age of the tree, and environmental factors (Hacke and Sauter 1996; Hajek et al. 2014). In two hybrid poplar clones, the mean vessel diameter increased from leaves to roots (Fig. 4.3), a pattern consistent with previous observations on free standing plants and proposed hormonal mechanisms of vascular differentiation (Aloni and Zimmermann 1983). By contrast, vessels of lianas tend to be wider in stems than in roots (see Table 6.1 in Chap. 6).

Populus species typically experience a substantial amount of native embolism during the growing season (Sperry et al. 1994; Tyree et al. 1994; Schreiber et al. 2013), which is consistent with reports that water transport occurs mainly in the outermost growth rings (Blake et al. 1996). This is also confirmed by dye perfusion experiments. Figure 4.4 shows a xylem cross section of a branch that was collected at a plantation in Alberta at the end of the growing season. Staining indicates that outer growth rings had mostly functional xylem while vessels in inner rings were mostly embolized. After vessels become embolized, they may be occluded by tyloses and phenolics (Fig. 4.5). Tyloses in *Populus* species are not frequently reported but have been observed as a response to wounding (Telewski et al. 1996) and have also been found in vessels of genetically modified low-lignin trees (Kitin et al. 2010).

Vessels are often arranged in radial clusters. Walls of adjacent vessels are connected via numerous intervessel pits (Fig. 4.2, arrows). Vessels and ray

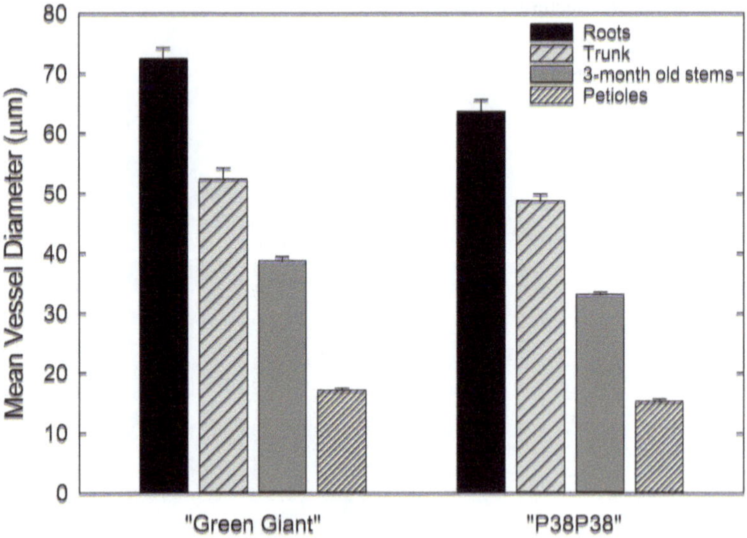

Fig. 4.3 Mean vessel diameters measured in different plant organs of the hybrid poplar clones "Green Giant" and P38P38. Samples were taken from roots and trunks of 6-year-old plantation trees ($n=6$–8 trees) as well as from stems ($n=12$–17 plants) and petioles ($n=4$ plants) of 3-month-old plants. Means + SE. Green Giant had consistently wider vessels than P38P38. Data from plantation trees is from Schreiber, Marceniuk, Hacke, Lowe & Thomas (unpublished manuscript). Data from 3-month-old plants is from Chamberland, Hacke, Schreiber & Thomas (unpublished manuscript) and Brocious & Hacke (unpublished manuscript)

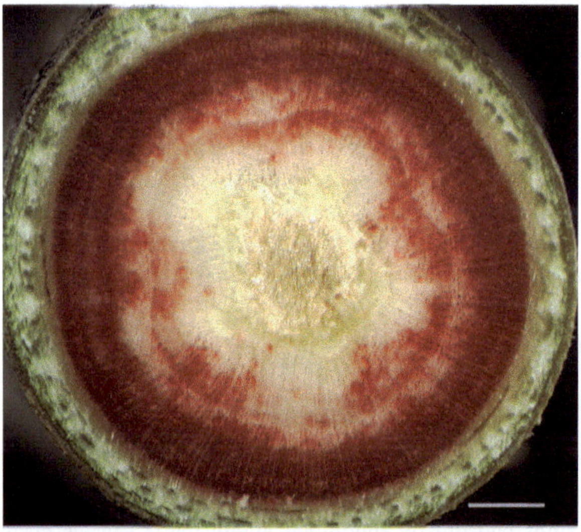

Fig. 4.4 Branch xylem of a hybrid poplar tree (clone DTAC 24) at the end of the growing season. The branch segment was cut under water and was subsequently perfused with safranin solution. The outer growth ring was stained red, indicating the presence of many functional vessels. By contrast, many vessels in inner growth rings were nonfunctional (not stained). Branch xylem of this clone had on average ~48 % embolism when the dye perfusion experiment was conducted. Scale bar = 1 mm. Photo: S. Schreiber

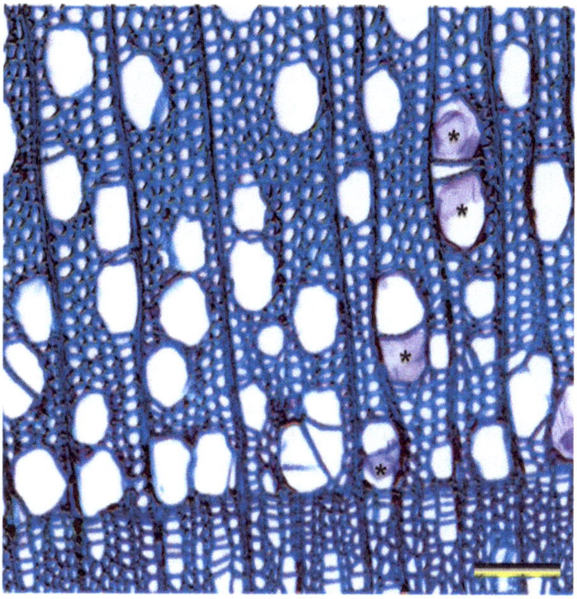

Fig. 4.5 Earlywood vessels of the hybrid poplar clone "Walker." The section was taken from the trunk. Asterisks mark vessel occlusions, which probably represent tyloses. The tree was growing in a plantation in Alberta, Canada. It had experienced several winters and many freeze–thaw cycles, which probably caused freezing-induced embolism. Scale bar = 100 μm. Photo: U. Hacke

parenchyma cells are also connected via pits (Fig. 4.2, arrow heads). Immunolabeling studies on poplar indicated that intervessel and vessel-ray pits have a different chemical composition. In contrast to intervessel pits, membranes of vessel-ray pits are rich in pectins (Plavcova and Hacke 2011; Kim and Daniel 2013).

The xylem contains a three-dimensional network of living cells, consisting of uniserate (only one cell wide) rays and axial parenchyma cells. Ray cells in *Populus* account for 11 % to 14 % of the total wood volume (Balatinecz et al. 2001). Parenchyma has multiple functions including storage and distribution of food materials, exchange of solutes with vessels, and the compartmentalization and sealing of wounds (Chap. 8; Sauter et al. 1973; Kitin et al. 2010). Ray parenchyma includes contact and isolation cells. Contact cells facilitate the exchange of solutes and water with vessels (see later) and exhibit high levels of respiratory and enzymatic activity (Telewski et al. 1996). Isolation cells lack contact with vessels and are adapted for radial translocation (Sauter and Kloth 1986).

2.2 *Transport Efficiency*

Transport efficiency can be defined as hydraulic conductance (flow rate per pressure difference for water at 20 °C). Normalizing by stem length yields hydraulic conductivity (K_h), which is the ratio of water flux through an excised stem segment and the pressure gradient causing the flow (Tyree and Ewers 1991; Sperry et al. 2006). Hydraulic conductivity can be measured with a conductivity apparatus (Sperry et al. 1988). When measuring K_h, it is useful to consider that water not only flows through the conduit lumen but also through the pits of the conduit end walls. End walls contribute on average more than half of the total conduit flow resistivity (Sperry et al. 2006). It follows that K_h should be measured on longer segments that include many vessel ends. Most conductivity measurements reported in this chapter were done on segments that were 14 cm (in some cases 27 cm) in length, and this is longer than the length of most vessels, at least in diffuse-porous branches (Jacobsen et al. 2012). This length has been used frequently because of the size of the centrifuge rotors used to measure vulnerability to cavitation curves (Hacke et al. 2006; Jacobsen et al. 2007).

2.2.1 Xylem Area-Specific Conductivity

When hydraulic conductivity is divided by the sapwood area, we obtain the xylem area-specific conductivity, K_S, which is a proxy of the porosity of the stem segment (Tyree and Ewers 1991). The K_S of well-watered greenhouse or growth chamber-grown hybrid poplar saplings typically ranges from ~5 to 8 kg s^{-1} m^{-1} MPa^{-1} (Voelker et al. 2011; Plavcova and Hacke 2012), which is high for diffuse-porous xylem (Fig. 4.6a). Flushed branch segments of hybrid poplar and trembling aspen trees growing at a boreal planting site exhibited specific conductivities between 1 and 2 kg s^{-1} m^{-1} MPa^{-1} (Schreiber et al. 2011).

Trends in K_S correspond with the mean vessel diameters measured in these plant groups. Branches of field-grown aspen and hybrid poplar genotypes exhibited mean vessel diameters between ~25 and 31 μm (Figs. 4.6b, and 4.7). These values are lower than those measured in hybrid poplar plants growing in the growth chamber. Growing conditions therefore had a strong effect on K_S and vessel diameters. Vessels were narrower under field conditions, probably because field-grown trees experienced more negative water potentials and were subjected to other stress factors that were absent in the growth chamber.

Xylem-specific conductivity was positively correlated with the maximum attainable height of a species and with leaf-to-xylem area ratios across Australian angiosperm species (Gleason et al. 2012). A positive correlation between the K_S of branches and growth rate was found in genetically distinct aspen populations growing in a common garden experiment in Germany (Hajek et al. 2014). Similarly, a study on *P. deltoides* × *P. nigra* genotypes revealed that vessel diameter and theoretical xylem-specific conductivity correlated positively with stomatal conductance and stem height (Fichot et al. 2009).

However, recent studies involving *Populus* genotypes found a negative correlation between the mean vessel diameter in branch xylem and growth, and this may be related to the occurrence of freezing-induced embolism. At a boreal planting site, trees with narrow xylem vessels exhibited reduced freezing-induced embolism and showed superior performance after 16 growing seasons (Schreiber et al. 2011; Schreiber et al. 2013) (Fig. 4.7a, b). This finding suggests that vessel diameter could impact plant function in boreal environments, and that narrow vessels can be advantageous when plants face drought or freezing stress (Schreiber, Hacke, Hamann, in press).

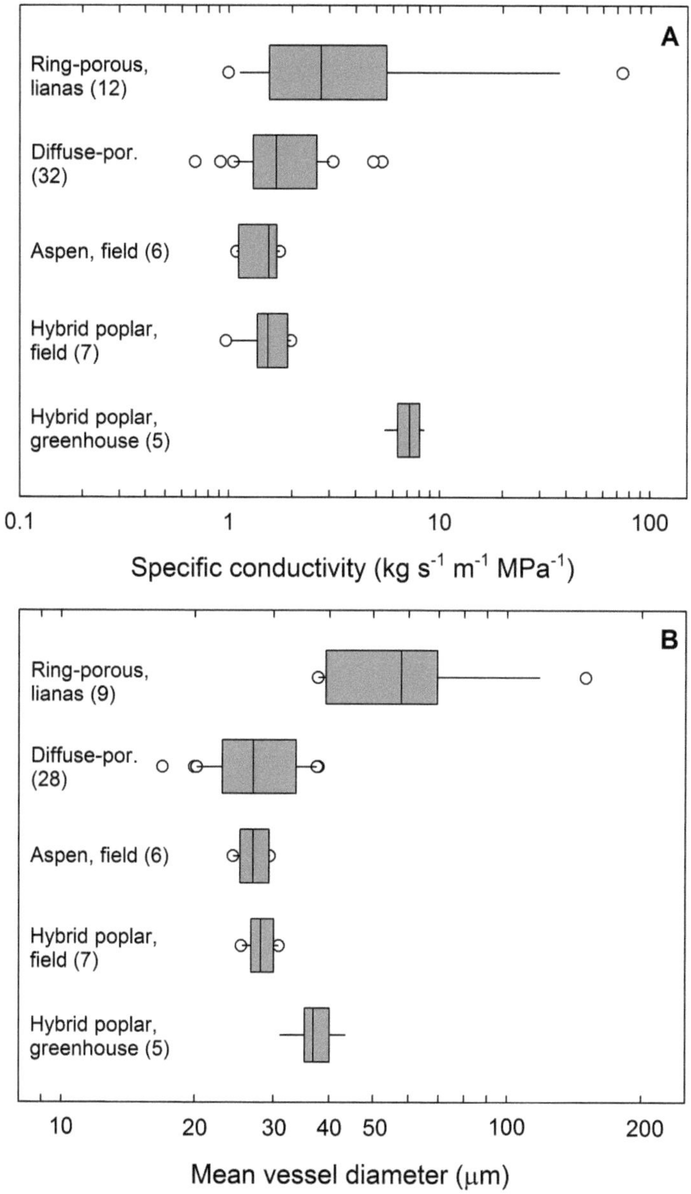

Fig. 4.6 Box plots showing xylem area-specific conductivity, K_S (**a**) and mean vessel diameters (**b**) for different plant groups. Vertical center lines represent the medians; box limits indicate the 25th and 75th percentiles; whiskers extend to the 10th and 90th percentiles as calculated by SigmaPlot 10. Outliers are shown as open circles. The numbers in parentheses refer to the sample size. Data for diffuse-porous, ring-porous, and liana species is from Hacke et al. (2006), Li et al. (2008), and Lens et al. (2010). K_S values refer to flushed segments that were typically ≥14 cm in length. Data for trembling aspen and hybrid poplar trees is for branches collected from field-grown plantation trees (Schreiber et al. 2011). Data for hybrid poplar saplings (Hybrid poplar, greenhouse) refers to well-watered plants growing in a growth chamber (Plavcova and Hacke 2012) or greenhouse (Harvey and van den Driessche 1997)

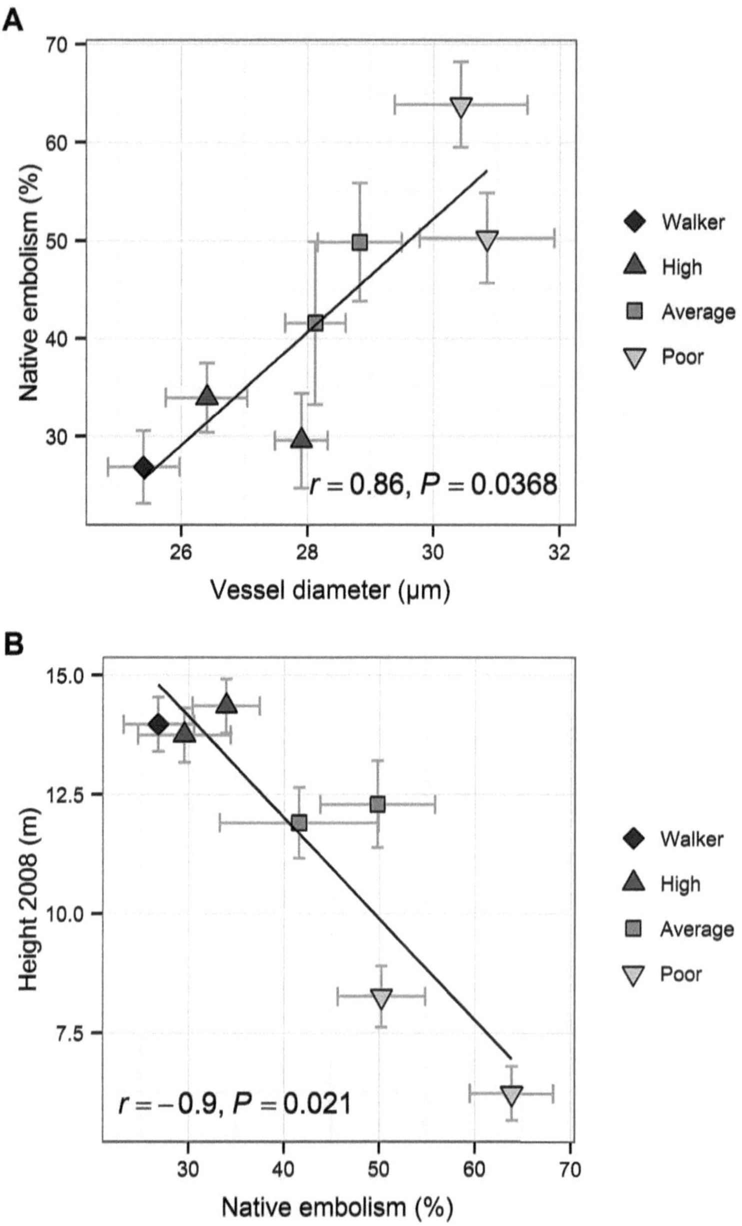

Fig. 4.7 Correlations between mean native embolism measured in May of 2011 and mean vessel diameter (**a**) as well as tree height (**b**) for seven hybrid poplar clones grouped by performance. The clone "Walker" is shown as a reference. Clones with poor performance exhibited wider vessels and had higher embolism levels than clones showing average or high growth. From Schreiber et al. (2013)

2.2.2 Leaf-Specific Conductivity

Leaf-specific conductivity (LSC) refers to the hydraulic conductivity of a stem segment divided by the leaf area distal to the segment and is a useful measure of the hydraulic "sufficiency" of the segment to supply water to leaves (Tyree and Ewers 1991). The higher LSC is, the lower the water potential gradient across the plant to maintain a particular transpiration rate (Chap. 6, Table 6.2). Via its potential impact on leaf water potential, LSC may be related to gas exchange and meristem growth (Brodribb et al. 2002; Tyree and Zimmermann 2002).

LSC for field-grown hybrid poplar and aspen trees was well within range of values reported for *Acer* and *Quercus* species (Woodrum et al. 2003; Maherali et al. 2006; Bhaskar et al. 2007). Hybrid poplar clones tended to have higher leaf-specific conductivities than aspen genotypes, consistent with a tendency for higher leaf-to-xylem area ratios and more negative leaf water potentials in aspen compared with hybrid poplars (Schreiber et al. 2011).

As mentioned, greenhouse-grown hybrid poplar saplings had wider vessels and more efficient xylem than field-grown trees. However, since plants grown in a controlled environment also exhibited much higher leaf areas per unit xylem area than field-grown trees, LSC values of greenhouse and field-grown hybrid poplar plants were similar. Extremely high leaf-to-xylem area ratios and growth rates were found in hybrid poplar saplings receiving high N fertilization (Hacke et al. 2010). Such high leaf-to-xylem area ratios developed because plants were kept well watered, shoots were structurally supported by stakes, and because of very high xylem-specific conductivities.

The picture emerging from the work on hybrid poplars is that K_S, vessel diameters, and leaf-to-xylem area ratios are quite variable depending on growing conditions, while variation in LSC is more constrained. Given the link between LSC and water potential gradients, this pattern is consistent with the need to maintain water potential homeostasis. Efficient control of the water potential range is required to minimize xylem cavitation (Hacke and Sauter 1995; Schreiber et al. 2011).

2.3 *Transport Safety*

Over a wide range of taxa, agreement exists between cavitation resistance and the water potential range plants experience in the field (Chap. 7, Fig. 7.2; Pockman and Sperry 2000; Jacobsen et al. 2005; Pratt et al. 2007; Choat et al. 2012). Plants that experience higher levels of water stress are typically more resistant to cavitation than plants whose water potential always remains favorable. Poplars generally prefer sites with an abundant and continuous supply of moisture (Heilman et al. 1996). Among woody plants, poplars are among the most drought susceptible and their productivity is associated with large water requirements (Silim et al. 2009). It is therefore not surprising that most *Populus* species are highly vulnerable to cavitation with 50 % loss of xylem conductivity occurring between−1 and−2.5 MPa (Fig. 4.8).

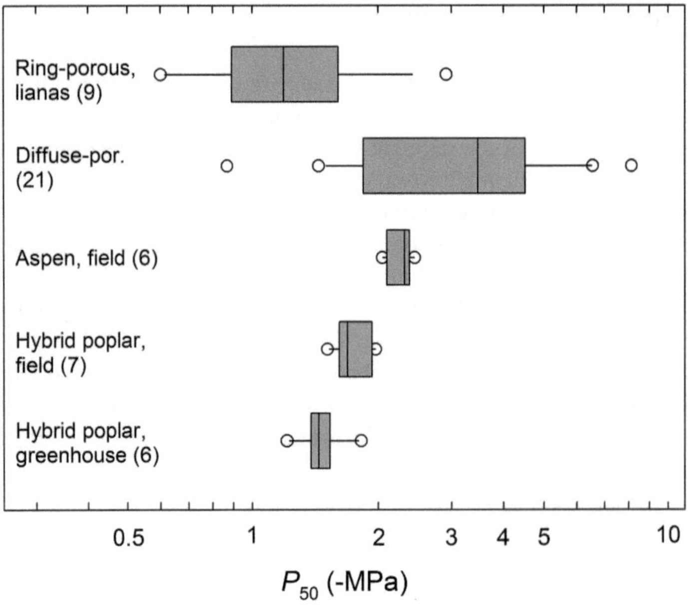

Fig. 4.8 Cavitation resistance of different plant groups. Cavitation resistance is expressed as the xylem pressure inducing 50 % loss of hydraulic conductivity (P_{50}). The vertical center lines in the box plot represent the medians; box limits indicate the 25th and 75th percentiles; whiskers extend to the 10th and 90th percentiles as calculated by SigmaPlot 10. Outliers are shown as open circles. The numbers in parentheses refer to the sample size. Values for diffuse-porous, ring-porous, and liana species is from Hacke et al. (2006). Data for trembling aspen and hybrid poplar trees is for branches collected from field-grown plantation trees (Schreiber et al. 2011). Data for hybrid poplar saplings (Hybrid poplar, greenhouse) refers to well-watered plants growing in a growth chamber (Plavcova and Hacke 2012) or greenhouse (Harvey and van den Driessche 1997)

While poplar xylem tends to be highly vulnerable to cavitation, differences between species and genotypes exist. A number of case studies shed light on how cavitation resistance impacts the way different *Populus* species cope with drought.

The cottonwood species *P. deltoides* and *P. fremontii* are adapted to riparian zones in semiarid regions. A common characteristic of these species is the patchy canopy architecture in which branch dieback is common (Fig. 4.9). The xylem of *P. deltoides* and *P. fremontii* is exceptionally vulnerable to xylem cavitation (Tyree et al. 1994), consistent with trees being restricted to stream-side zones along perennial streams (Rood et al. 2000). The extreme vulnerability to cavitation can be interpreted as adaptive: Cavitation in small branches may precede branch dieback. This would reduce transpirational water loss, enabling maintenance of favorable water potentials in surviving plant parts (Rood et al. 2000). This proposed model assumes a nonuniform pattern of cavitation across branches.

In contrast to riparian cottonwoods, aspen can grow in seasonally drier habitats and typically exhibits more negative water potentials than other *Populus* species (Schreiber et al. 2011; Galvez et al. 2013). Blake et al. (1996) noted that *P. tremuloides* stems had more resistant xylem than *P. deltoides*, *P. angustifolia*, *P. balsamifera*, and

Fig. 4.9 A cottonwood tree (*Populus deltoides*) growing along the South Saskatchewan River, Alberta. The photograph shows the extensive branch and crown dieback that is typical of this species. From Rood et al. (2000)

other poplar species. The greater cavitation resistance of aspen xylem relative to other *Populus* species and their hybrids has been confirmed in subsequent studies (Hacke et al. 2001b; Cai and Tyree 2010; Arango-Velez et al. 2011; Schreiber et al. 2011; Anderegg et al. 2013) and is illustrated in Fig. 4.8. Working with aspen saplings, Cai and Tyree (2010) found a tight correlation between vessel diameter and vulnerability to drought-induced cavitation. Narrow vessels tended to be more resistant than wide conduits, allowing the authors to predict variations in vulnerability curves from vessel diameter data. While this may be possible for stems of a single species when plants grow under uniform conditions, such a correlation cannot always be expected in field-grown trees and when different species are compared (Hacke et al. 2006; Lens et al. 2010).

Despite the high tolerance of aspen to climatic stress (Lieffers et al. 2001), aspen forests have recently experienced unusual dieback and mortality (Worrall et al. 2013). Mortality occurred in the southwestern United States (Worrall et al. 2010) and in the aspen parkland region of Alberta and Saskatchewan (Michaelian et al. 2011). The aspen parkland varies in climatic suitability for aspen, but suitability in nearly all of the parkland area decreased greatly in recent times (Worrall et al. 2013). The analysis of Worrall et al. (2013) indicates that exceptional drought episodes were a major cause of the decline episodes, and that aspen is sensitive to drought in much of its range. Sudden aspen decline (SAD) is not primarily induced by pathogens or changes in nonstructural carbohydrate reserves but is associated with high embolism levels of branches and roots (Anderegg et al. 2012).

However, there can be lags between drought stress and mortality rates. Extended multiyear tree mortality following drought suggests that hydraulic failure could be a gradual, longer term process as opposed to a sudden, catastrophic event. Anderegg et al. (2013) documented a major role of accumulated hydraulic changes. Although SAD-affected and healthy ramets experienced similar water potentials, SAD branches exhibited increased vulnerability as a consequence of previous cavitation damage, a phenomenon known as "cavitation fatigue" (Hacke et al. 2001b). Therefore, even if aspen trees refill embolized vessels, these vessels will remain more vulnerable to future drought episodes than the vessels of healthy trees.

2.4 Stomatal Control of Water Loss

Stomatal behavior plays an important role in avoiding excessive levels of xylem embolism (Tyree and Sperry 1988; Jones and Sutherland 1991). Based on the analysis of Blake et al. (1996), many poplar species maintain a very small margin of safety from runaway cavitation in their native habitat. The responsiveness of stomata to changes in leaf water potential therefore plays a critical role in the water use strategies of *Populus* genotypes.

Interestingly, poplar species and their hybrids vary tremendously in their stomatal sensitivity to water deficits (Schulte et al. 1987; Ceulemans et al. 1988; Silim et al. 2009; Almeida-Rodriguez et al. 2010). Some species have been classified as isohydric while others were anisohydric, i.e., their leaf water potential exhibited a strong decrease with evaporative demand and/or soil drought. Sparks and Black (1999) examined four populations of the riparian species *P. trichocarpa* growing across a gradient in humidity and temperature. Populations from drier environments tended to exhibit higher resistance to drought-induced xylem cavitation than populations from moist environments. Trees from moist sites maintained higher stomatal conductances at leaf water potentials <−1.5 MPa than trees from dry sites. Some trees did not close stomata sufficiently to avoid cavitation. Sparks and Black (1999) concluded that individuals with nonresponsive stomata and vulnerable xylem can only persist in moist environments.

By contrast, *P. balsamifera* trees in Germany exhibited isohydric behavior. On days when evaporative demand was very high, stomata closed to maintain water potentials near the threshold at which xylem cavitation started to occur (Hacke and Sauter 1995). Isohydric behavior was also reported for *P. euramericana* (Tardieu and Simonneau 1998) and *P. tremuloides* (Hogg et al. 2000; Lieffers et al. 2001; Galvez et al. 2011).

Studying stomatal responses of nine poplar clones, Silim et al. (2009) found that drought-tolerant clones had lower stomatal conductance and closed their stomata more gradually than sensitive clones. Another study examined stomatal responses to drought and vulnerability to cavitation in hybrid poplar and balsam poplar clones (Arango-Velez et al. 2011). Based on its stomatal responses and relatively low vulnerability to cavitation, balsam poplar was considered the most drought tolerant of the examined clones. Other hybrid poplar clones exhibited poor stomatal control despite being relatively sensitive to cavitation.

In addition to genetic differences, stomatal sensitivity can be affected by growing conditions. A period of water stress preconditioning resulted in modified stomatal closure in *P. trichocarpa* leaves (Schulte and Hinckley 1987). Leaves previously subjected to water stress were capable of at least partial closure of stomata when turgor was lost from guard cells. By contrast, stomatal aperture remained nearly constant over a wide range of water potentials in leaves that had not previously experienced water stress. In a hybrid poplar clone, leaves growing at high relative humidity had larger stomatal length and aperture than leaves of plants growing at lower relative humidity (Laur and Hacke 2013). After application of abscisic acid to leaves, the pore apertures of high humidity plants remained larger than those of control plants indicating that high humidity plants exhibited incomplete closure (Fig. 4.10a–d).

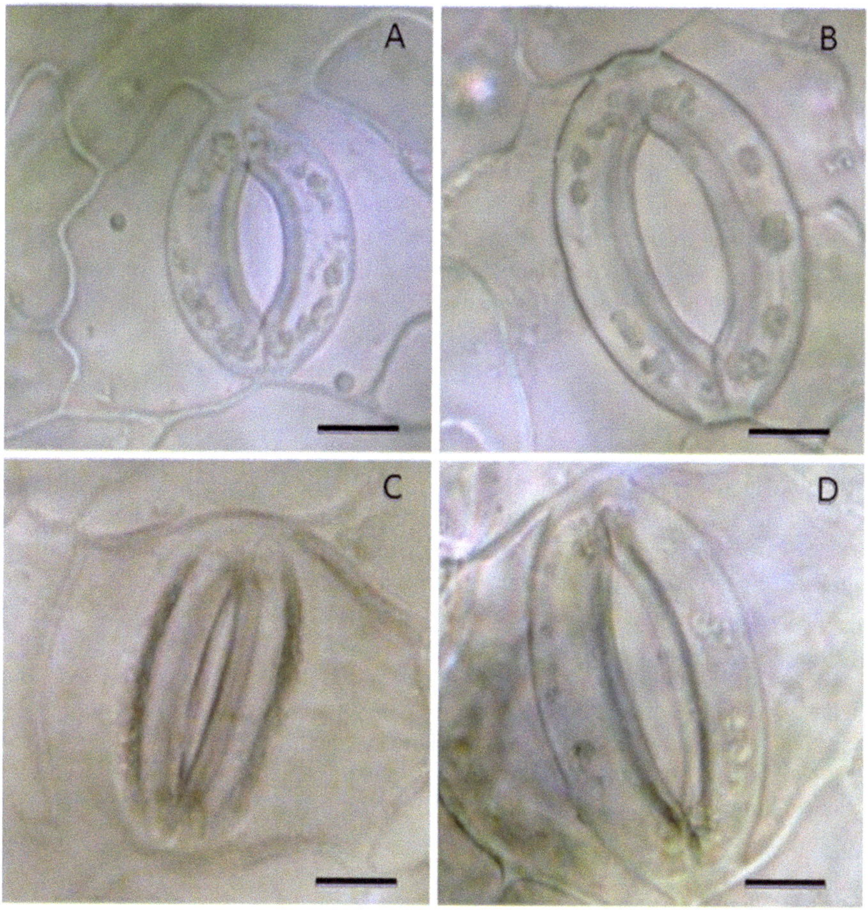

Fig. 4.10 Light microscope images of stomata from poplar leaves growing in moderate (**a**) and high relative humidity (**b**). The images were taken from the abaxial side of the leaves. Leaves that developed under high humidity had larger stomatal length and aperture. While application of 100 μM abscisic acid (ABA) triggered stomatal closure in plants growing at moderate humidity (**c**), the large stomata of high humidity grown plants failed to close fully (**d**). Bars = 10 μm. From Laur and Hacke (2013)

2.5 *Phenotypic Plasticity*

While we have begun to identify key genes involved in the regulation of secondary vascular growth (Groover et al. 2010; Spicer and Groover 2010; Zhong and Ye 2013), much remains to be learned about environmental influences on wood formation (e.g., Lautner 2013). With regards to xylem transport physiology, Telewski et al. (1996) posed the question if vessel structure could be altered in such a way that the xylem would become less vulnerable to cavitation. Indeed, we know that cavitation resistance of *Populus* species can be impacted by environmental conditions (Harvey and van den Driessche 1997; Awad et al. 2010; Fichot et al. 2010; Hacke et al. 2010). In recent years, we have learned much about phenotypic plasticity of cavitation resistance and related structural traits. A series of experiments was conducted with hybrid poplar plants growing at different levels of water availability, light, and nitrogen fertilization (Plavcova and Hacke 2012). Stem segments were sampled from two different positions along the plant's main stem to assess developmental plasticity. One set of segments was taken from a basal region of the stem. Another set of segments was sampled at a position closer to the apex of a plant ("distal" segments). Across all treatments, the amount of leaf area supported by stem segments scaled linearly with stem native hydraulic conductivity, suggesting that the total leaf area was constrained by the xylem transport capacity. The P_{50} ranged from −1.71 MPa to −0.15 MPa depending on treatment and segment position. Across all treatments, P_{50} was tightly correlated with wood density (Plavcova and Hacke 2012). The following is a summary of what we learned from these and many other experiments.

2.5.1 Water Availability

Basal segments taken from plants grown under drought conditions had narrower vessels and lower K_S than well-watered control plants (Plavcova and Hacke 2012). The P_{50} was not significantly different. By contrast, distal segments of drought-stressed plants were more resistant to cavitation than distal segments of control plants. Others reported an increase in cavitation resistance in response to drought in some, but not all clones that were tested (Awad et al. 2010; Fichot et al. 2010).

Water availability can affect wood growth through effects on cell division and cell expansion and via effects on photosynthesis and photosynthate allocation to the cambium. Since the expansion of vessel elements and fibers is driven by turgor pressure, vessel and fiber diameters in newly formed xylem should be sensitive to the water status of the plant during periods of high cambial activity. This was confirmed by several studies (Arend and Fromm 2007; Bogeat-Triboulot et al. 2007). Arend and Fromm (2007) also observed deformation of vessel elements in the newly formed xylem of drought-treated trees.

Given that narrow vessels can be more resistant to drought-induced cavitation than wide vessels within a stem (Harvey and van den Driessche 1997; Cai and Tyree 2010), it is not surprising that drought sometimes results in greater cavitation resistance. Since narrower (and fewer) vessels also tend to be less conductive than wide conduits, gains in cavitation resistance may be accompanied by reduced transport efficiency.

2.5.2 Nitrogen (N) Availability

The influence of N on xylem traits has been extensively studied and some key findings have emerged. Plants receiving high levels of N (high N plants) exhibited enhanced growth, larger leaves, as well as wider vessels and fibers (Harvey and van den Driessche 1999; Plavcová et al. 2013). Pitre et al. (2007) reported that high N plants produced fibers that were wider and shorter compared to the adequate N treatment. There was also a slight tendency for longer vessels (as well as vessel elements) in high N plants (Plavcová et al. 2013, their Fig. S1). These anatomical differences were paralleled by higher K_S in high N plants compared with plants receiving only adequate N fertilization (adequate N plants) (Hacke et al. 2010; Plavcová et al. 2013). Further, high N plants had lower wood densities and were more vulnerable to cavitation than adequate N plants. Distal segments of high N plants were extremely vulnerable and exhibited an exponentially shaped curve (Fig. 4.11a, open circles). For each treatment, vulnerability curves generated by the standard centrifuge method (Hacke et al. 2015) were in good agreement with the native embolism values plotted against the native xylem pressure (Fig. 4.11b, squares).

Differences in xylem phenotype between high and adequate N plants were underpinned by changes in transcription of hundreds of genes in the developing xylem region (Plavcová et al. 2013). Gene candidates were identified that may affect vessel dimensions and cell wall thickness, although detailed functional characterization of these genes in poplar will be required to corroborate the proposed functions.

The form of N taken up by plants (NO_3^- versus NH_4^+) was recently found to impact many physiological and morphological parameters (Cirelli 2014). Growth was reduced in NH_4^+-fed plants compared with NO_3^--fed plants. Large differences were also found in the vessel diameters of roots and in root diameter.

When interpreting some of the findings described above, it is important to consider that most of the work described here has been done in the greenhouse or growth chamber while plants were kept well watered. Concern has been expressed that fertilization may increase the susceptibility of hybrid poplars to drought stress (Harvey and van den Driessche 1999; DesRochers et al. 2007). More work is needed to study the potentially complex interactions between N fertilization and water availability.

2.5.3 Light Availability

The effect of shading on xylem structure and function was studied in 3-month-old hybrid poplar saplings (Plavcova et al. 2011). Shading resulted in more vulnerable and less efficient xylem. Increased vulnerability of shaded plants was associated with thinner intervessel pit membranes. Vessels of shaded plants were narrower and longer than in control stems. Shaded plants also had much lower LSC than control plants. Wood density was also lower in shaded plants. The xylem of shaded plants therefore appears less optimized from a hydraulic standpoint (Hacke 2014).

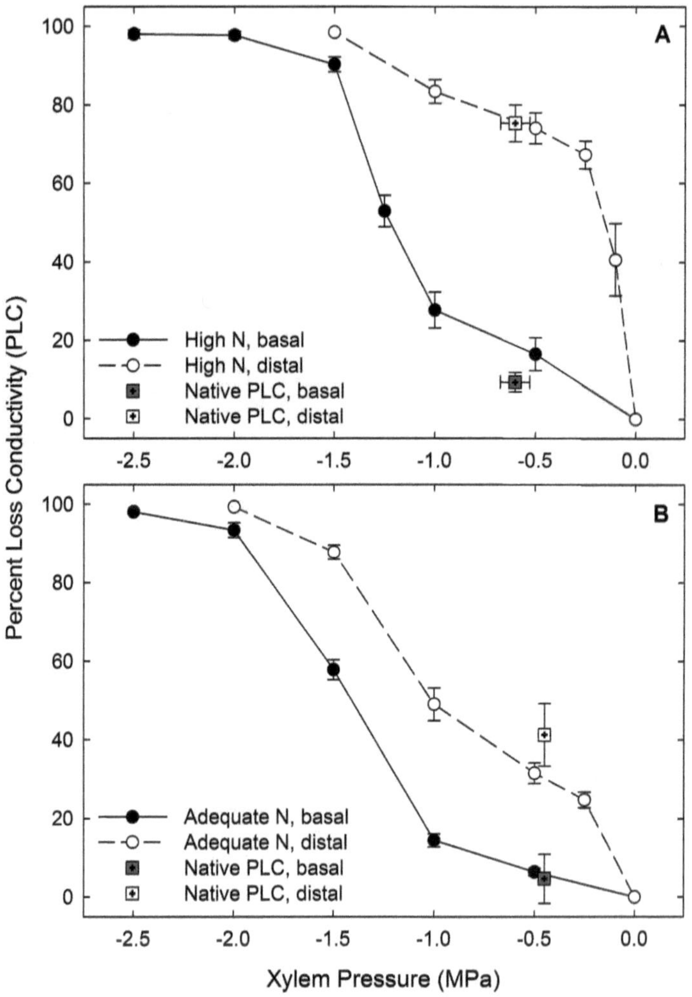

Fig. 4.11 Vulnerability to cavitation curves (*circles*) and native values of percentage loss of conductivity (PLC) plotted against the native xylem pressure (*squares*) for basal (*filled symbols*) and distal (*open symbols*) stem segments in hybrid poplar saplings grown under (**a**) high N and (**b**) adequate N fertilization. Note the profoundly different shape of the vulnerability curves in distal segments (*open circles*), ranging from sigmoidal to exponential. Centrifuge-generated curves agreed well with native PLC data. Means ± SE are shown. Replotted from Plavcova and Hacke (2012)

2.5.4 Salt Stress

After two weeks of salt stress, leaves of hybrid poplar saplings exhibited brown lesions and turned yellow. Salt-stressed plants developed fewer radial cell layers in the cambial zone and narrower vessels than control trees (Escalante-Pérez et al. 2009). Junghans et al. (2006) reported that high salt concentrations reduced cambial activity and vessel diameters of *P.* × *canescens* plants.

Auxin is one of the factors affecting vessel diameter (Junghans et al. 2004). The wider vessel diameters of nonstressed *P.* × *canescens* plants may be associated with higher indole-3-acetic acid (IAA) concentrations in the xylem of control plants (Junghans et al. 2006). Salinity may also cause increased ABA synthesis and accumulation, which in turn may modulate various physiological processes related to plant hydraulics (Escalante-Pérez et al. 2009; Shatil-Cohen et al. 2011).

2.5.5 CO_2 Concentration

The increases in atmospheric CO_2 concentrations observed in recent decades in conjunction with changing temperature and precipitation patterns will likely cause significant alterations in plant water relations (for a detailed analysis, see Chap. 9). Free-air CO_2 enrichment (FACE) experiments with trembling aspen and *Populus* × *euramericana* showed that long-term exposure to elevated [CO_2] resulted in greater whole-plant transpiration (Uddling et al. 2008; Tricker et al. 2009). Higher [CO_2] led to lower stomatal conductance on the leaf level while total leaf area increased. Interestingly, elevated [CO_2] may also increase fine root biomass, especially at deeper soil depths (Norby and Zak 2011, and literature cited therein). Four-year-old *P. deltoides* trees growing at elevated [CO_2] had greater sap flux per stem area and higher leaf area/sapwood area ratios than plants growing at ambient [CO_2] (Bobich et al. 2010). High CO_2 trees had lower wood densities than trees growing at ambient CO_2. Vessel diameters were not different.

At a European FACE site, trees growing at elevated [CO_2] had smaller percentages of cell wall area and increased fiber diameters compared with plants at ambient [CO_2] (Luo et al. 2005). This finding agrees with the lower wood densities reported by Bobich et al. (2010). Furthermore, elevated [CO_2] was associated with increased percentages of ray parenchyma areas compared with ambient [CO_2] (Luo et al. 2005). There was a tendency for larger vessel diameters at elevated [CO_2], suggesting that trees growing at elevated [CO_2] were more sensitive to drought. Trembling aspen trees growing under elevated [CO_2] for 11 years showed increased radial growth and vessel diameters (Kostiainen et al. 2014). However, these responses to CO_2 were not consistent throughout the experiment and were more often seen early in the experiment. Future work will likely explore the complex interactions between [CO_2] and various abiotic and biotic factors, including temperature, drought, nutrients, ozone, and insects.

2.6 Effect of Low Lignin Content on Wood Structure and Function

Considerable attention has been given to the potential to manipulate the lignin biosynthesis pathway of poplars to increase pulping efficiencies (Pilate et al. 2002; Mansfield et al. 2012). The success of this pursuit will likely depend on getting a better understanding of the role of lignin on the multiple functions of xylem within living trees (Koehler et al. 2006). Lignin impacts both the mechanical and the hydraulic properties of wood. In the context of water transport, an important function of lignin is to support the xylem pipeline against implosion by negative pressure (Sperry 2003). Large bending stresses can arise in the double wall between water- and gas-filled (embolized) conduits (Hacke et al. 2001a). Across species, a greater degree of mechanical strength is typically correlated with greater cavitation resistance (Jacobsen et al. 2005; Pratt et al. 2007).

Decreased lignification can lead to premature cavitation and collapse of vessels (Coleman et al. 2008). Voelker et al. (2011) reported that reduced lignification was associated with increased vulnerability to cavitation and lower K_S. Lower hydraulic conductivities were caused by collapsed vessels as well as noncollapsed vessels filled with phenolics and/or tyloses (Kitin et al. 2010). Low lignin trees also showed more frequent dieback and more frequent mortality than controls, probably as a result of hydraulic failure (Voelker et al. 2011). It therefore seems that trees with lower lignin content will be at risk for reduced growth and higher mortality, especially if grown in environments with water limitations. Long-term field trials of low lignin trees will be required to fully evaluate the effects of genetic modifications (Pilate et al. 2002).

3 Putative Functions of Aquaporins

Aquaporins (AQPs) are membrane transport proteins and their primary function is to facilitate water movement across cell membranes; some AQPs also transport other molecules (Maurel et al. 2008). As such, AQPs are interesting candidates to investigate for their role in xylem development and xylem physiology. AQPs fall into several homology subgroups. The plasma membrane intrinsic proteins (PIPs) and the tonoplast intrinsic proteins (TIPs) can be distinguished based on sequence similarity and subcellular localization. PIPs and TIPs represent the most abundant AQPs in the plasma membrane and in the tonoplast, respectively.

Thirteen mammalian AQPs have been identified, and most of these have been shown to transport water (Papadopoulos and Verkman 2013). AQP4, for instance, is expressed at the borders between the brain and major water-containing compartments, suggesting that AQP4 facilitates water flow into and out of the brain (Papadopoulos and Verkman 2013).

Plants have a large number of AQP genes (Gomes et al. 2009). More than fifty full-length aquaporin sequences have been identified in the *Populus trichocarpa* genome (Gupta and Sankararamakrishnan 2009; Almeida-Rodriguez et al. 2010; Lopez et al. 2012). Similar to human AQPs (Verkman 2012), water channels in plants are involved in a wide range of physiological functions. This is particularly evident in leaves where AQPs may impact both water and gas transport. In leaves of *P. trichocarpa*, PIPs belonging to the PIP1 and PIP2 groups are present in vascular tissue (Fig. 4.12b). However, PIPs also appear to be present in chloroplast membranes of palisade parenchyma (Fig. 4.12a), suggesting that these channels conduct CO_2 in addition to water (Herrera and Garvin 2011; Secchi and Zwieniecki 2013).

From a water relations perspective, TIP genes appear to play an important role in reversing drought-induced reductions in leaf hydraulic conductance (Laur and Hacke 2014). An obvious function of water channels in a stem cross

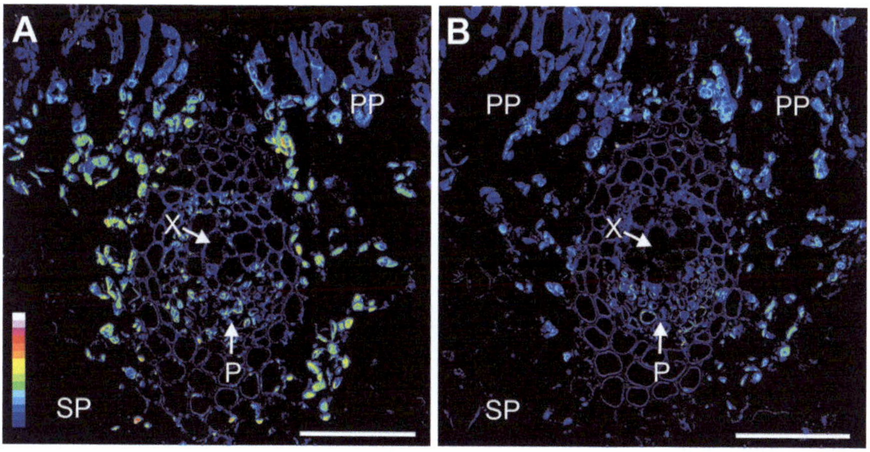

Fig. 4.12 Confocal laser scanning micrographs showing the localization of PIP1 (**a**) and PIP2 proteins (**b**) in *Populus trichocarpa* leaf sections. PIP1 labelling is apparent in phloem and palisade parenchyma cells. PIP2 labeling in phloem cells is mainly confined to the plasma membrane. Images were taken at an identical setting and were color-coded with an intensity look-up table (LUT; displayed in (**a**)), in which *black* was used to encode background, and *blue*, *green*, *yellow*, *red*, and *white* to encode increasing signal intensities. *P* phloem, *PP* palisade parenchyma, *SP* spongy parenchyma; *X* xylem. Scale bars = 50 µm. Photo: J. Laur

section would be to facilitate the radial exchange of water between phloem, xylem, and pith. The following sections discuss some specific putative functions of AQPs in the context of xylem development and water transport.

3.1 Wood Formation

Genes encoding PIPs and TIPs are frequently found to be expressed in the cambial region of poplar species (Schrader et al. 2004; Goue et al. 2008; Berta et al. 2010; Nilsson et al. 2010; Song et al. 2011; Plavcová et al. 2013) and in bulk xylem tissue (Hacke et al. 2010). RNA in situ hybridization revealed that *AQP*s were expressed in developing xylem and phloem cells (including living fibers), and in rays (Almeida-Rodriguez and Hacke 2012) (Fig. 4.13a). What is the biological function of water channel proteins in these cells and tissues? Returning to some of the N fertilization experiments mentioned above may provide some answers. As outlined above, one of the effects of N fertilization on xylem structure is increased vessel diameter. This suggests that the expansion of vessel elements occurs more rapidly in high N plants than in adequate N plants. Cell expansion is driven by water uptake, which in turn can be facilitated by AQPs.

Hacke et al. (2010) found that five *PIP2*s and two *TIP*s showed higher transcript abundance under conditions of high N. All five of these PIP2s have been identified as water transporters by a swelling analysis in oocytes (Secchi et al. 2009; Almeida-Rodriguez et al. 2010). In another N fertilization experiment, three *TIP*s were

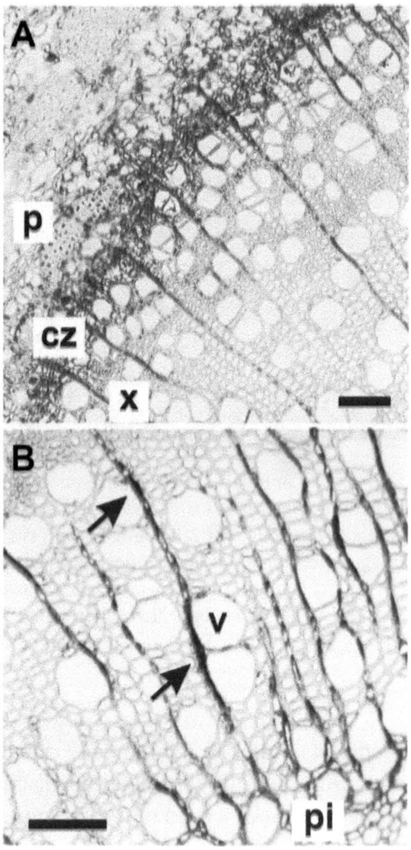

Fig. 4.13 In situ mRNA hybridization of an aquaporin gene (*PtPIP2;3*) in stem cross sections of *Populus* plants. Regions of aquaporin expression are indicated by dark-purple staining. High expression occurred in the cambial region (**a**) and in ray cells. Ray cells adjacent to vessels showed particularly strong labeling (*arrows* in (**b**)). Expression extended into the cell expansion zone. *cz* cambial zone, *p* phloem, *pi* pith, *v* vessel, *x* xylem. Bars = 100 μm. From Almeida-Rodriguez and Hacke (2012)

up-regulated in the cambial region of high N plants (Plavcová et al. 2013). Some or all of these AQPs may facilitate water influx into newly formed cells during their expansion and elongation phase. Further, when expressed in rays, these aquaporins could facilitate radial transfer of water from functional vessels to the region of rapid cell expansion (Groover et al. 2010; Hacke et al. 2010; Almeida-Rodriguez and Hacke 2012). If this hypothesis were confirmed, *AQP* expression in the xylem could be functionally linked with the development of wider vessels, which in turn could have implications for transport efficiency and safety.

3.2 *Vessel Refilling*

While *AQP*s were expressed in a symplastic continuum extending from the periderm to the pith, expression was particularly high in contact cells (Almeida-Rodriguez and Hacke 2012) (Fig. 4.13b, arrows). This suggests that AQPs in these cells increase water exchange between apoplast and symplast, which necessarily involves the crossing of cell membranes. Water could move from apoplast (vessel) to symplast (contact cell) or vice versa, depending on the driving force.

A specific role of water channels in contact cells could be to facilitate the refilling of embolized vessels (Sakr et al. 2003; Nardini et al. 2011). Expression levels of several AQP genes were shown to increase during the refilling process (Secchi and Zwieniecki 2010, 2011). While the refilling mechanism is still not completely understood, it seems likely that the efflux of water out of contact cells (into vessels) during refilling will be facilitated by up-regulation of water channels from the PIP and TIP subfamilies (Secchi et al. 2011).

3.3 Root Water Uptake

In isohydric plants, stomatal control of water loss maintains leaf water potential relatively constant during periods of water stress. However, plants can also modulate water *uptake* in a dynamic fashion (Siemens and Zwiazek 2004). AQPs can modify root hydraulic conductance in response to changing environmental conditions. Hybrid poplar plants were grown under contrasting irradiance and relative humidity conditions, and were subsequently exposed to a sudden increase in transpirational demand (Laur and Hacke 2013). Only 4 h after a sudden decrease in relative humidity, there was a 75 % increase in the cumulative transcript copy numbers of six *PIP* genes in roots. This increase in transcripts was mainly due to a twofold increase in the transcript copy numbers of the three *PIP1* genes that were studied.

In roots of control plants, PIP1 protein was present in epidermis and cortex cells as well as in the endodermis and in vascular tissue (Fig. 4.14a). Weak labelling was observed in roots of shaded plants (Fig. 4.14b). In contrast, root sections taken after the increase in light level exhibited strong immunolabeling of the epidermis, endodermis, and of cells adjacent to the endodermis. Labelling was particularly abundant after 28 h when a continuous fluorescence signal occurred in the epidermis (Fig. 4.14c, d). A similar trend was observed in plants that were exposed to decreasing humidity (Fig. 4.14e–g), although strong signals were already detected after 4 h.

4 Root System

Many aspects of the water use strategy and whole-plant physiology of *Populus* species are linked to the structure and function of the root system; yet the root systems of *Populus* remain the most poorly understood portion of the plant (Pregitzer and Friend 1996). An important variable for water uptake is the spatial distribution of roots. Horizontal poplar roots can be found several tree lengths away from the stem. Vertical "sinker" roots appear to be common and may allow water uptake during

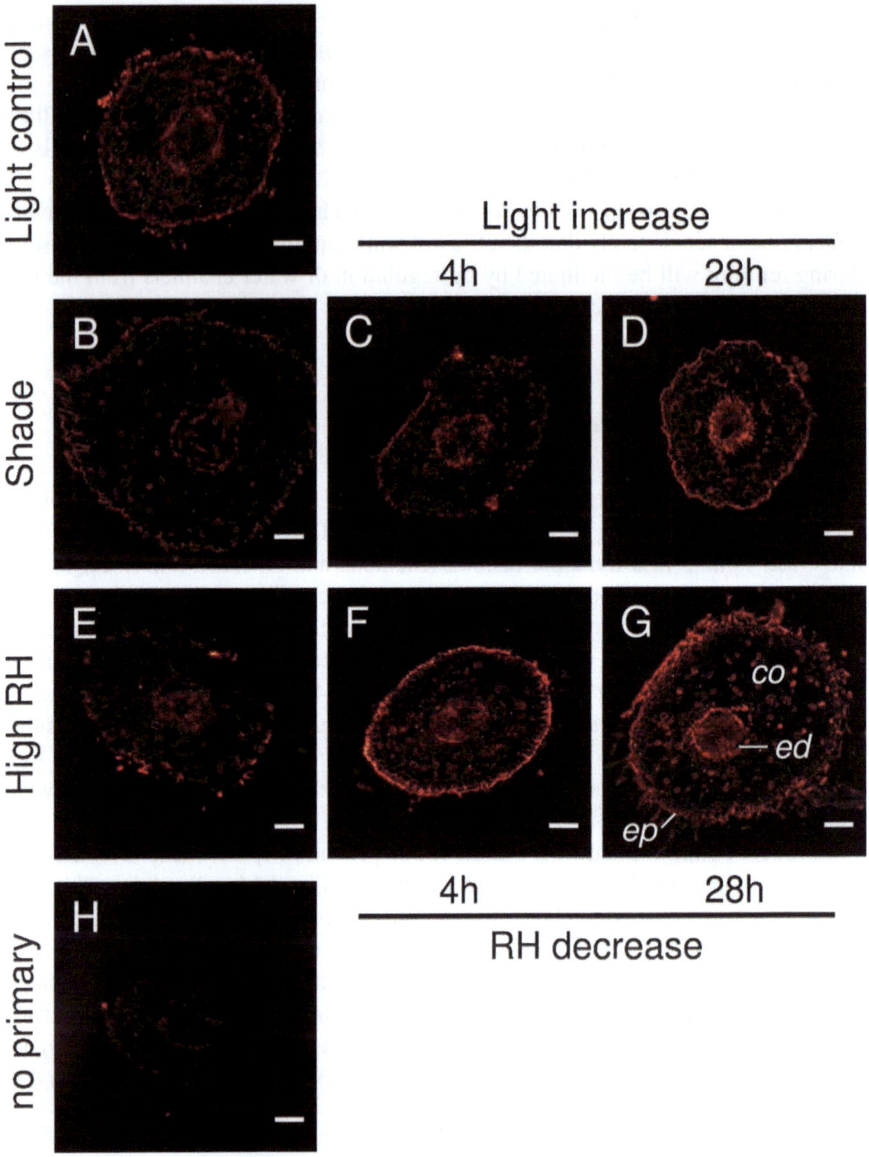

Fig. 4.14 Immunolocalization of PIP1 protein in root cross-sections of a hybrid poplar plant. Transverse sections were taken at 25–30 mm from the root tip. The intensity of the red color is equivalent to the abundance of PIP1 protein. (**a**) Roots of control plants growing at full light in the growth chamber. (**b**)–(**d**) Roots of shaded plants before (**b**) and after a step change in light level (**c**, **d**). (**e**)–(**g**) Roots of plants growing at high relative humidity before (**e**) and after a step change in humidity (**f**, **g**). (**h**) Control with no primary antibody indicates minimal background autofluorescence. *co* cortex, *ed* endodermis, *ep* epidermis. Bars = 100 μm. From Laur and Hacke (2013)

periods when the soil surface contains little extractable water (Pregitzer and Friend 1996). Rooting depth of >2 m were reported for phreatophytic trees such as *Populus fremontii* (Stromberg 2013).

In aspen, intra- and inter-clonal root connections occur, which can play a role in the sharing of water and other resources. When the aboveground trees die following a stand-replacing disturbance, shallow roots are the main source of suckers that reestablish the next stand (Lieffers et al. 2001). Regeneration from an established root system seems to be a more reliable means of reproduction in water-stressed environments than from seed.

In the boreal forest, aspen is often confined to the warmest positions on the landscape. There is virtually no root growth of aspen at soil temperatures lower than 6 °C (Lieffers et al. 2001). Growth inhibition of aspen in cold soils appears to be caused by reduced root water flow (Wan et al. 1999). Membrane permeability and water viscosity reduce the ability of roots to deliver water to the foliage at low root temperatures.

Acknowledgements Work in my laboratory was supported by grants from the Natural Sciences and Engineering Research Council of Canada, the Canada Foundation for Innovation, and by the Canada Research Chair program. Thanks to current and previous members of the Hacke lab who contributed data to this manuscript.

References

Almeida-Rodriguez AM, Cooke JEK, Yeh F, Zwiazek JJ (2010) Functional characterization of drought-responsive aquaporins in *Populus balsamifera* and *Populus simonii* × *balsamifera* clones with different drought resistance strategies. Physiol Plant 140:321–333

Almeida-Rodriguez AM, Hacke UG (2012) Cellular localization of aquaporin mRNA in hybrid poplar stems. Am J Bot 99:1249–1254

Aloni R, Zimmermann MH (1983) The control of vessel size and density along the plant axis. Differentiation 24:203–208

Anderegg WRL, Berry JA, Smith DD, Sperry JS, Anderegg LDL, Field CB (2012) The roles of hydraulic and carbon stress in a widespread climate-induced forest die-off. Proc Natl Acad Sci U S A 109:233–237

Anderegg WRL, Plavcová L, Anderegg LDL, Hacke UG, Berry JA, Field CB (2013) Drought's legacy: multi-year hydraulic deterioration underlies widespread aspen forest die-off and portends increased future risk. Glob Chang Biol 19:1188–1196

Arango-Velez A, Zwiazek JJ, Thomas BR, Tyree MT (2011) Stomatal factors and vulnerability of stem xylem to cavitation in poplars. Physiol Plant 143:154–165

Arend M (2008) Immunolocalization of (1, 4)-β-galactan in tension wood fibers of poplar. Tree Physiol 28:1263–1267

Arend M, Fromm J (2007) Seasonal change in the drought response of wood cell development in poplar. Tree Physiol 27:985–992

Awad H, Barigah T, Badel E, Cochard H, Herbette S (2010) Poplar vulnerability to xylem cavitation acclimates to drier soil conditions. Physiol Plant 139:280–288

Balatinecz JJ, Kretschmann DE, Leclercq A (2001) Achievements in the utilization of poplar wood-guideposts for the future. Forestry Chronicle 77:265–269

Berta M, Giovannelli A, Sebastiani F, Camussi A, Racchi ML (2010) Transcriptome changes in the cambial region of poplar (*Populus alba* L.) in response to water deficit. Plant Biol 12:341–354

Bhaskar R, Valiente-Banuet A, Ackerly DD (2007) Evolution of hydraulic traits in closely related species pairs from mediterranean and nonmediterranean environments of North America. New Phytol 176:718–726

Blake TJ, Sperry JS, Tschaplinski TJ, Wang SS (1996) Water relations. In: Stettler RF, Bradshaw HD, Heilman PE, Hinckley TM (eds) Biology of *Populus* and its implications for management and conservation. NRC, Ottawa, pp 401–442

Bobich EG, Barron-Gafford GA, Rascher KG, Murthy R (2010) Effects of drought and changes in vapour pressure deficit on water relations of *Populus deltoides* growing in ambient and elevated CO_2. Tree Physiol 30:866–875

Bogeat-Triboulot M-B, Brosché M, Renaut J, Jouve L, Le Thiec D, Fayyaz P, Vinocur B, Witters E, Laukens K, Teichmann T (2007) Gradual soil water depletion results in reversible changes of gene expression, protein profiles, ecophysiology, and growth performance in *Populus euphratica*, a poplar growing in arid regions. Plant Physiol 143:876–892

Bradshaw HD, Ceulemans R, Davis J, Stettler R (2000) Emerging model systems in plant biology: poplar (*Populus*) as a model forest tree. J Plant Growth Regul 19:306–313

Brodribb TJ, Holbrook NM, Gutierrez MV (2002) Hydraulic and photosynthetic co-ordination in seasonally dry tropical forest trees. Plant Cell Environ 25:1435–1444

Cai J, Tyree MT (2010) The impact of vessel size on vulnerability curves: data and models for within-species variability in saplings of aspen, *Populus tremuloides* Michx. Plant Cell Environ 33:1059–1069

Ceulemans R, Impens I, Imler R (1988) Stomatal conductance and stomatal behavior in *Populus* clones and hybrids. Can J Bot 66:1404–1414

Choat B, Jansen S, Brodribb TJ, Cochard H, Delzon S, Bhaskar R, Bucci SJ, Feild TS, Gleason SM, Hacke UG, Jacobsen AL, Lens F, Maherali H, Martinez-Vilalta J, Mayr S, Mencuccini M, Mitchell PJ, Nardini A, Pittermann J, Pratt RB, Sperry JS, Westoby M, Wright IJ, Zanne AE (2012) Global convergence in the vulnerability of forests to drought. Nature 491:752–755

Cirelli D (2014) Whole-plant transpiration in *Populus* sp.: its determination, nocturnal effects and influence by form of nitrogen. University of Alberta, Edmonton

Coleman HD, Samuels AL, Guy RD, Mansfield SD (2008) Perturbed lignification impacts tree growth in hybrid poplar—a function of sink strength, vascular integrity, and photosynthetic assimilation. Plant Physiol 148:1229–1237

Cooke JEK, Rood SB (2007) Trees of the people: the growing science of poplars in Canada and worldwide. Can J Bot 85:1103–1110

DesRochers A, van den Driessche R, Thomas BR (2007) The interaction between nitrogen source, soil pH, and drought in the growth and physiology of three poplar clones. Can J Bot 85:1046–1057

Eckenwalder JE (1996) Systematics and evolution of *Populus*. In: Stettler RF, Bradshaw JR, Heilman PE, Hinckley TM (eds) Biology of *Populus* and its implications for management and conservation. Part I, Chapter 1. NRC, Ottawa, pp 7–32

Escalante-Pérez M, Lautner S, Nehls U, Selle A, Teuber M, Schnitzler J-P, Teichmann T, Fayyaz P, Hartung W, Polle A (2009) Salt stress affects xylem differentiation of grey poplar (*Populus* × *canescens*). Planta 229:299–309

Fichot R, Barigah TS, Chamaillard S, Le Thiec D, Laurans F, Cochard H, Brignolas F (2010) Common trade-offs between xylem resistance to cavitation and other physiological traits do not hold among unrelated *Populus deltoides* × *Populus nigra* hybrids. Plant Cell Environ 33:1553–1568

Fichot R, Laurans F, Monclus R, Moreau A, Pilate G, Brignolas F (2009) Xylem anatomy correlates with gas exchange, water-use efficiency and growth performance under contrasting water regimes: evidence from *Populus deltoides* × *Populus nigra* hybrids. Tree Physiol 29:1537–1549

Galvez DA, Landhäusser S, Tyree M (2013) Low root reserve accumulation during drought may lead to winter mortality in poplar seedlings. New Phytol 198:139–148

Galvez DA, Landhäusser SM, Tyree MT (2011) Root carbon reserve dynamics in aspen seedlings: does simulated drought induce reserve limitation? Tree Physiol 31:250–257

Gleason SM, Butler DW, Ziemińska K, Waryszak P, Westoby M (2012) Stem xylem conductivity is key to plant water balance across Australian angiosperm species. Functional Ecol 26:343–352

Gomes D, Agasse A, Thiébaud P, Delrot S, Gerós H, Chaumont F (2009) Aquaporins are multifunctional water and solute transporters highly divergent in living organisms. Biochim Biophys Acta 1788:1213–1228

Goue N, Lesage-Descauses MC, Mellerowicz EJ, Magel E, Label P, Sundberg B (2008) Microgenomic analysis reveals cell type-specific gene expression patterns between ray and fusiform initials within the cambial meristem of *Populus*. New Phytol 180:45–56

Groover AT, Nieminen K, Helariutta Y, Mansfield SD (2010) Wood Formation in *Populus*. In: Jansson S, Bhalerao RP, Groover AT (eds) Genetics and genomics of *Populus*, vol 8, Plant genetics and genomics: crops and models. Springer, New York, pp 201–224. doi:10.1007/978-1-4419-1541-2_10

Gupta AB, Sankararamakrishnan R (2009) Genome-wide analysis of major intrinsic proteins in the tree plant *Populus trichocarpa*: characterization of XIP subfamily of aquaporins from evolutionary perspective. BMC Plant Biol 9:134

Hacke U, Sauter JJ (1995) Vulnerability of xylem to embolism in relation to leaf water potential and stomatal conductance in *Fagus sylvatica* f. *purpurea* and *Populus balsamifera*. J Exp Bot 46:1177–1183

Hacke U, Sauter JJ (1996) Drought-induced xylem dysfunction in petioles, branches, and roots of *Populus balsamifera* L. and *Alnus glutinosa* (L.) Gaertn. Plant Physiol 111:413–417

Hacke UG (2014) Irradiance-induced changes in hydraulic architecture. Botany 92:437–442

Hacke UG, Jacobsen AL, Brandon Pratt R, Maurel C, Lachenbruch B, Zwiazek J (2012) New research on plant–water relations examines the molecular, structural, and physiological mechanisms of plant responses to their environment. New Phytol 196:345–348

Hacke UG, Plavcová L, Almeida-Rodriguez A, King-Jones S, Zhou W, Cooke JEK (2010) Influence of nitrogen fertilization on xylem traits and aquaporin expression in stems of hybrid poplar. Tree Physiol 30:1016–1025

Hacke UG, Sperry JS, Pockman WT, Davis SD, McCulloh KA (2001a) Trends in wood density and structure are linked to prevention of xylem implosion by negative pressure. Oecologia 126:457–461

Hacke UG, Sperry JS, Wheeler JK, Castro L (2006) Scaling of angiosperm xylem structure with safety and efficiency. Tree Physiol 26:689–701

Hacke UG, Stiller V, Sperry JS, Pittermann J, McCulloh KA (2001b) Cavitation fatigue. Embolism and refilling cycles can weaken the cavitation resistance of xylem. Plant Physiol 125:779–786

Hacke UG, Venturas MD, MacKinnon ED, Jacobsen AL, Sperry JS, Pratt RB (2015) The standard centrifuge method accurately measures vulnerability curves of long-vesselled olive stems. New Phytol 205:116–127

Hajek P, Leuschner C, Hertel D, Delzon S, Schuldt B (2014) Trade-offs between xylem hydraulic properties, wood anatomy and yield in *Populus*. Tree Physiol 34:744–756

Harvey HP, van den Driessche R (1997) Nutrition, xylem cavitation and drought resistance in hybrid poplar. Tree Physiol 17:647–654

Harvey HP, van den Driessche R (1999) Nitrogen and potassium effects on xylem cavitation and water-use efficiency in poplars. Tree Physiol 19:943–950

Heilman PE, Hinckley TM, Roberts DA, Ceulemans R (1996) Production physiology. In: Stettler RF, Bradshaw HD, Heilman PE, Hinckley TM (eds) Biology of *Populus* and its implications for management and conservation. NRC, Ottawa, pp 459–489

Herrera M, Garvin JL (2011) Aquaporins as gas channels. Pflugers Arch 462:623–630

Hogg EH, Saugier B, Pontailler J-Y, Black T, Chen W, Hurdle P, Wu A (2000) Responses of trembling aspen and hazelnut to vapor pressure deficit in a boreal deciduous forest. Tree Physiol 20:725–734

Jacobsen AL, Ewers FW, Pratt RB, Paddock WA, Davis SD (2005) Do xylem fibers affect vessel cavitation resistance? Plant Physiol 139:546–556

Jacobsen AL, Pratt RB, Davis SD, Ewers FW (2007) Cavitation resistance and seasonal hydraulics differ among three arid Californian plant communities. Plant Cell Environ 30:1599–1609
Jacobsen AL, Pratt RB, Tobin MF, Hacke UG, Ewers FW (2012) A global analysis of xylem vessel length in woody plants. Am J Bot 99:1583–1591
Jansson S, Douglas CJ (2007) *Populus*: a model system for plant biology. Annu Rev Plant Biol 58:435–458
Jones HG, Sutherland R (1991) Stomatal control of xylem embolism. Plant Cell Environ 14:607–612
Junghans U, Langenfeld-Heyser R, Polle A, Teichmann T (2004) Effect of auxin transport inhibitors and ethylene on the wood anatomy of poplar. Plant Biol 6:22–29
Junghans U, Polle A, Duchting P, Weiler E, Kuhlman B, Gruber F, Teichmann T (2006) Adaptation to high salinity in poplar involves changes in xylem anatomy and auxin physiology. Plant Cell Environ 29:1519–1531
Kim JS, Daniel G (2013) Developmental localization of homogalacturonan and xyloglucan epitopes in pit membranes varies between pit types in two poplar species. Iawa J 34:245–262
Kitin P, Voelker SL, Meinzer FC, Beeckman H, Strauss SH, Lachenbruch B (2010) Tyloses and phenolic deposits in xylem vessels impede water transport in low-lignin transgenic poplars: a study by cryo-fluorescence microscopy. Plant Physiol 154:887–898
Koehler L, Ewers FW, Telewski FW (2006) Optimizing for multiple functions: mechanical and structural contributions of cellulose microfibrils and lignin in strengthening tissues. In: Stokke DD, Groom LH (eds) Characterization of the cellulosic cell wall. Blackwell, Ames, pp 20–29
Kostiainen K, Saranpaa P, Lundqvist S-O, Kubiske ME, Vapaavuori E (2014) Wood properties of *Populus* and *Betula* in long-term exposure to elevated CO_2 and O_3. Plant Cell Environ 37:1452–1463
Laur J, Hacke UG (2013) Transpirational demand affects aquaporin expression in poplar roots. J Exp Bot 64:2283–2293
Laur J, Hacke UG (2014) The role of water channel proteins in facilitating recovery of leaf hydraulic conductance from water stress in *Populus trichocarpa*. PLoS One 9:e111751
Lautner S (2013) Wood formation under drought stress and salinity. In: Fromm J (ed) Cellular aspects of wood formation. Springer, Berlin, pp 187–202
Lens F, Sperry JS, Christman MA, Choat B, Rabaey D, Jansen S (2010) Testing hypotheses that link wood anatomy to cavitation resistance and hydraulic conductivity in the genus *Acer*. New Phytol 190:709–723
Li YY, Sperry JS, Taneda H, Bush SE, Hacke UG (2008) Evaluation of centrifugal methods for measuring xylem cavitation in conifers, diffuse- and ring-porous angiosperms. New Phytologist 177:558–568
Lieffers VJ, Landhausser SM, Hogg EH (2001) Is the wide distribution of aspen a result of its stress tolerance? In: Sustaining Aspen in Western landscapes: symposium proceedings, pp. 311–324
Lopez D, Bronner G, Brunel N, Auguin D, Bourgerie S, Brignolas F, Carpin S, Tournaire-Roux C, Maurel C, Fumanal B, Martin F, Sakr S, Label P, Julien JL, Gousset-Dupont A, Venisse JS (2012) Insights into *Populus* XIP aquaporins: evolutionary expansion, protein functionality, and environmental regulation. J Exp Bot 63:2217–2230
Luo ZB, Langenfeld-Heyser R, Calfapietra C, Polle A (2005) Influence of free air CO_2 enrichment (EUROFACE) and nitrogen fertilisation on the anatomy of juvenile wood of three poplar species after coppicing. Trees-Struct Funct 19:109–118
Maherali H, Moura CF, Caldeira MC, Willson CJ, Jackson RB (2006) Functional coordination between leaf gas exchange and vulnerability to xylem cavitation in temperate forest trees. Plant Cell Environ 29:571–583
Mansfield SD, Kang KY, Chapple C (2012) Designed for deconstruction—poplar trees altered in cell wall lignification improve the efficacy of bioethanol production. New Phytol 194:91–101
Maurel C, Verdoucq L, Luu DT, Santoni V (2008) Plant aquaporins: membrane channels with multiple integrated functions. Annu Rev Plant Biol 59:595–624
Mellerowicz EJ, Baucher M, Sundberg B, Boerjan W (2001) Unravelling cell wall formation in the woody dicot stem. Plant Mol Biol 47:239–274

Michaelian M, Hogg EH, Hall RJ, Arsenault E (2011) Massive mortality of aspen following severe drought along the southern edge of the Canadian boreal forest. Global Change Biol 17: 2084–2094

Nardini A, Lo Gullo MA, Salleo S (2011) Refilling embolized xylem conduits: is it a matter of phloem unloading? Plant Sci 180:604–611

Nilsson R, Bernfur K, Gustavsson N, Bygdell J, Wingsle G, Larsson C (2010) Proteomics of plasma membranes from poplar trees reveals tissue distribution of transporters, receptors, and proteins in cell wall formation. Mol Cell Proteom 9:368–387

Norby RJ, Zak DR (2011) Ecological lessons from free-air CO_2 enrichment (FACE) experiments. Annu Rev Ecol Evol Syst 42:181

Papadopoulos MC, Verkman AS (2013) Aquaporin water channels in the nervous system. Nat Rev Neurosci 14:265–277

Pilate G, Guiney E, Holt K, Petit-Conil M, Lapierre C, Leplé J-C, Pollet B, Mila I, Webster EA, Marstorp HG (2002) Field and pulping performances of transgenic trees with altered lignification. Nat Biotechnol 20:607–612

Pinno BD, Landhäusser SM, MacKenzie MD, Quideau SA, Chow PS (2012) Trembling aspen seedling establishment, growth and response to fertilization on contrasting soils used in oil sands reclamation. Can J Soil Sci 92:143–151

Pitre FE, Cooke JEK, Mackay JJ (2007) Short-term effects of nitrogen availability on wood formation and fibre properties in hybrid poplar. Trees-Struct Funct 21:249–259

Plavcová L, Hacke UG (2011) Heterogeneous distribution of pectin epitopes and calcium in different pit types of four angiosperm species. New Phytol 192:885–897

Plavcová L, Hacke UG (2012) Phenotypic and developmental plasticity of xylem in hybrid poplar saplings subjected to experimental drought, nitrogen fertilization, and shading. J Exp Bot 63:6481–6491

Plavcová L, Hacke UG, Almeida-Rodriguez AM, Li E, Douglas CJ (2013) Gene expression patterns underlying changes in xylem structure and function in response to increased nitrogen availability in hybrid poplar. Plant Cell Environ 36:186–199

Plavcová L, Hacke UG, Sperry JS (2011) Linking irradiance-induced changes in pit membrane ultrastructure with xylem vulnerability to cavitation. Plant Cell Environ 34:501–513

Pockman WT, Sperry JS (2000) Vulnerability to xylem cavitation and the distribution of Sonoran desert vegetation. Am J Bot 87:1287–1299

Pratt RB, Jacobsen AL, Ewers FW, Davis SD (2007) Relationships among xylem transport, biomechanics and storage in stems and roots of nine Rhamnaceae species of the California chaparral. New Phytol 174:787–798

Pregitzer KS, Friend AL (1996) The structure and function of *Populus* root systems. Biology of *Populus* and its implications for management and conservation. NRC, Ottawa, pp 331–354

Rood SB, Braatne JH, Hughes FMR (2003) Ecophysiology of riparian cottonwoods: stream flow dependency, water relations and restoration. Tree Physiol 23:1113–1124

Rood SB, Patino S, Coombs K, Tyree MT (2000) Branch sacrifice: cavitation-associated drought adaptation of riparian cottonwoods. Trees 14:248–257

Sakr S, Alves G, Morillon RL, Maurel K, Decourteix M, Guilliot A, Fleurat-Lessard P, Julien JL, Chrispeels MJ (2003) Plasma membrane aquaporins are involved in winter embolism recovery in walnut tree. Plant Physiol 133:630–641

Sauter JJ, Iten W, Zimmermann MH (1973) Studies on the release of sugar into the vessels of sugar maple (*Acer saccharum*). Can J Bot 51:1–8

Sauter JJ, Kloth S (1986) Plasmodesmatal frequency and radial translocation rates in ray cells of poplar (*Populus* × *canadensis* Moench 'robusta'). Planta 168:377–380

Schrader J, Nilsson J, Mellerowicz E, Berglund A, Nilsson P, Hertzberg M, Sandberg G (2004) A high-resolution transcript profile across the wood-forming meristem of poplar identifies potential regulators of cambial stem cell identity. Plant Cell 16:2278–2292

Schreiber S, Hacke UG, Hamann A (in press) Variation of xylem vessel diameters across a climate gradient: insight from a reciprocal transplant experiment with a widespread boreal tree. Funct Ecol. doi:10.1111/1365-2435.12455
Schreiber SG, Hacke UG, Hamann A, Thomas BR (2011) Genetic variation of hydraulic and wood anatomical traits in hybrid poplar and trembling aspen. New Phytol 190:150–160
Schreiber SG, Hamann A, Hacke UG, Thomas BR (2013) Sixteen years of winter stress: an assessment of cold hardiness, growth performance and survival of hybrid poplar clones at a boreal planting site. Plant Cell Environ 36:419–428
Schulte P, Hinckley T (1987) The relationship between guard cell water potential and the aperture of stomata in *Populus*. Plant Cell Environ 10:313–318
Schulte P, Hinckley T, Stettler R (1987) Stomatal responses of *Populus* to leaf water potential. Can J Bot 65:255–260
Secchi F, Gilbert ME, Zwieniecki MA (2011) Transcriptome response to embolism formation in stems of *Populus trichocarpa* provides insight into signaling and the biology of refilling. Plant Physiol 157:1419–1429
Secchi F, Maciver B, Zeidel ML, Zwieniecki MA (2009) Functional analysis of putative genes encoding the PIP2 water channel subfamily in *Populus trichocarpa*. Tree Physiol 29:1467–1477
Secchi F, Zwieniecki MA (2010) Patterns of PIP gene expression in *Populus trichocarpa* during recovery from xylem embolism suggest a major role for the PIP1 aquaporin subfamily as moderators of refilling process. Plant Cell Environ 33:1285–1297
Secchi F, Zwieniecki MA (2011) Sensing embolism in xylem vessels: the role of sucrose as a trigger for refilling. Plant Cell Environ 34:514–524
Secchi F, Zwieniecki MA (2013) The physiological response of *Populus tremula* × *alba* leaves to the down-regulation of *PIP1* aquaporin gene expression under no water stress. Front Plant Sci 4:507
Shatil-Cohen A, Attia Z, Moshelion M (2011) Bundle-sheath cell regulation of xylem-mesophyll water transport via aquaporins under drought stress: a target of xylem-borne ABA? Plant J 67:72–80
Siemens JA, Zwiazek JJ (2004) Changes in root water flow properties of solution culture-grown trembling aspen (*Populus tremuloides*) seedlings under different intensities of water-deficit stress. Physiol Planta 121:44–49
Silim S, Nash R, Reynard D, White B, Schroeder W (2009) Leaf gas exchange and water potential responses to drought in nine poplar (*Populus* spp.) clones with contrasting drought tolerance. Trees-Struct Funct 23:959–969
Song DL, Xi W, Shen JH, Bi T, Li LG (2011) Characterization of the plasma membrane proteins and receptor-like kinases associated with secondary vascular differentiation in poplar. Plant Mol Biol 76:97–115
Sparks JP, Black RA (1999) Regulation of water loss in populations of *Populus trichocarpa*: the role of stomatal control in preventing xylem cavitation. Tree Physiol 19:453–459
Sperry JS (2003) Evolution of water transport and xylem structure. Int J Plant Sci 164:S115–S127
Sperry JS, Donnelly JR, Tyree MT (1988) A method for measuring hydraulic conductivity and embolism in xylem. Plant Cell Environ 11:35–40
Sperry JS, Hacke UG, Pittermann J (2006) Size and function in conifer tracheids and angiosperm vessels. Am J Bot 93:1490–1500
Sperry JS, Nichols KL, Sullivan JEM, Eastlack SE (1994) Xylem embolism in ring-porous, diffuse-porous, and coniferous trees of northern Utah and interior Alaska. Ecology 75:1736–1752
Spicer R, Groover A (2010) Evolution of development of vascular cambia and secondary growth. New Phytol 186:577–592
Stromberg J (2013) Root patterns and hydrogeomorphic niches of riparian plants in the American Southwest. J Arid Environ 94:1–9

Tardieu F, Simonneau T (1998) Variability among species of stomatal control under fluctuating soil water status and evaporative demand: modelling isohydric and anisohydric behaviours. J Exp Bot 49:419–432

Telewski F, Aloni R, Sauter J (1996) Physiology of secondary tissues of *Populus*. In: Stettler RF, Bradshaw HD, Heilman PE, Hinckley TM (eds) Biology of *Populus* and its implications for management and conservation. NRC, Ottawa, pp 301–329

Tricker PJ, Pecchiari M, Bunn SM, Vaccari FP, Peressotti A, Miglietta F, Taylor G (2009) Water use of a bioenergy plantation increases in a future high CO_2 world. Biomass Bioenergy 33:200–208

Tyree MT, Ewers FW (1991) Tansley review no. 34: the hydraulic architecture of trees and other woody plants. New Phytol 119:345–360

Tyree MT, Kolb KJ, Rood SB, Patino S (1994) Vulnerability to drought-induced cavitation of riparian cottonwoods in Alberta: a possible factor in the decline of the ecosystem? Tree Physiol 14:455–466

Tyree MT, Sperry JS (1988) Do woody plants operate near the point of catastrophic xylem dysfunction caused by dynamic water stress? Answers from a model. Plant Physiol 88:574–580

Tyree MT, Zimmermann MH (2002) Xylem structure and the ascent of sap, 2nd edn. Springer, Berlin

Uddling J, Teclaw RM, Kubiske ME, Pregitzer KS, Ellsworth DS (2008) Sap flux in pure aspen and mixed aspen–birch forests exposed to elevated concentrations of carbon dioxide and ozone. Tree Physiol 28:1231–1243

Verkman A (2012) Aquaporins in clinical medicine. Annu Rev Med 63:303

Voelker SL, Lachenbruch B, Meinzer FC, Kitin P, Strauss SH (2011) Transgenic poplars with reduced lignin show impaired xylem conductivity, growth efficiency and survival. Plant Cell Environ 34:655–668

Wan XC, Landhausser SM, Zwiazek JJ, Lieffers VJ (1999) Root water flow and growth of aspen (*Populus tremuloides*) at low root temperatures. Tree Physiol 19:879–884

Woodrum CL, Ewers FW, Telewski FW (2003) Hydraulic, biomechanical, and anatomical interactions of xylem from five species of *Acer* (Aceraceae). Am J Bot 90:693–699

Worrall JJ, Marchetti SB, Egeland L, Mask RA, Eager T, Howell B (2010) Effects and etiology of sudden aspen decline in southwestern Colorado, USA. Forest Ecol Manag 260:638–648

Worrall JJ, Rehfeldt GE, Hamann A, Hogg EH, Marchetti SB, Michaelian M, Gray LK (2013) Recent declines of *Populus tremuloides* in North America linked to climate. Forest Ecol Manag 299:35–51

Zhong R, Ye Z-H (2013) Transcriptional regulation of wood formation in tree species. In: Fromm J (ed) Cellular aspects of wood formation. Springer, Heidelberg, pp 141–158

Zhong RQ, McCarthy RL, Lee C, Ye ZH (2011) Dissection of the transcriptional program regulating secondary wall biosynthesis during wood formation in poplar. Plant Physiol 157:1452–1468

Chapter 5
Grapevine Xylem Development, Architecture, and Function

Anna L. Jacobsen, F. Daniela Rodriguez-Zaccaro, Tamani F. Lee, Jessica Valdovinos, Hayden S. Toschi, Justin A. Martinez, and R. Brandon Pratt

1 *Vitis* as a Model Xylem System

The genus *Vitis* (Vitaceae) has been thoroughly studied, in part because of commercial and agricultural interests in grapevine. This has led to myriad studies on many aspects of grapevine biology, including efforts to understand grapevine molecular, physiological, structural, and development biology.

Early efforts to sequence entire genomes of model plant species such as *Arabidopsis*, poplar, and rice also included efforts to sequence *Vitis*. This resulted in the early sequencing of the entire genome of *V. vinifera* "Pinot Noir" (Jaillon et al. 2007; Velasco et al. 2007). Additional molecular research has also been conducted on the *Vitis* chloroplast genome (Jansen et al. 2006) and on the use of other molecular techniques to compare across the genus *Vitis* (Myles et al. 2010).

In addition to ongoing molecular and genetic studies, many physiological studies have examined grapevine functional biology, including water relations and hydraulics (Zufferey et al. 2000; Schultz 2003; Domec and Johnson 2012). These studies have included xylem responses to pathogens and infection (Chatelet et al. 2006; Thorne et al. 2006; Sun et al. 2013) and to drought (Schultz and Matthews 1988; Lovisolo and Schubert 1998; Lovisolo and Schubert 1998; Lovisolo et al. 2010). Additionally, active processes that occur within the xylem have been examined, including the formation of tyloses and gels following infection, wounding, and embolism (Sun et al. 2007, 2008; Pérez-Donoso et al. 2007; Jacobsen and Pratt 2012).

A.L. Jacobsen (✉) • F.D. Rodriguez-Zaccaro • T.F. Lee • J. Valdovinos • H.S. Toschi
J.A. Martinez • R.B. Pratt
Department of Biology, California State University, 9001 Stockdale Hwy, 61 SCI, Bakersfield, CA 93311, USA
e-mail: ajacobsen@csub.edu; dann_zaccaro@hotmail.com; tamani_g@hotmail.com; jvaldovinos3@yahoo.com; htoschi@yahoo.com; jmartin8@poets.whittier.edu; rpratt@csub.edu

U. Hacke (ed.), *Functional and Ecological Xylem Anatomy*,
DOI 10.1007/978-3-319-15783-2_5

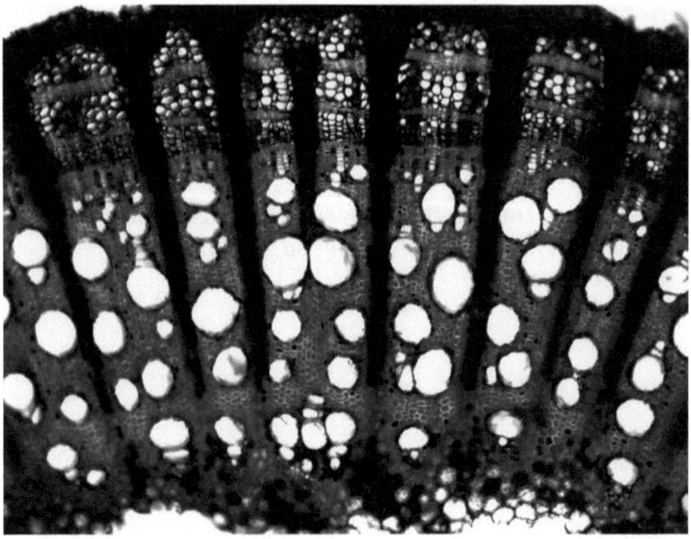

Fig. 5.1 Representative xylem micrograph from a 1-year-old stem of *Vitis vinifera*. The starch within the section has been stained dark to highlight the location of living cells with the xylem, particularly as they occur in wide xylem rays and in paratracheal axial parenchyma cells. This section is from a stem harvested toward the end of the growing season as is evidenced by the presence of many very narrow vessels near the cambium containing secondary cell walls and the absence of large expanding vessel elements lacking secondary cell walls. The cells of the phloem are also clearly visible

The hydraulics of grapevine stems has also been examined (e.g., Sperry et al. 1987; Tibbetts and Ewers 2000; Choat et al. 2010; Jacobsen and Pratt 2012).

Finally, grapevine hydraulic structure has been the subject of many recent studies. Grapevine has several unique features of its xylem and phloem structure, many of which are associated with its liana habit (Fig. 5.1; Chap. 6). Grapevine stems have wide diameter vessels, which are relatively isolated from other large vessels within the secondary xylem tissue (Fig. 5.1). This makes these vessels relatively easy to resolve using newly developing images technologies (Holbrook et al. 2001; Brodersen et al. 2010, 2011).

2 Secondary Xylem Vessel Development

2.1 General Patterns of Woody Plant Vessel Development

Within flowering plants, water is transported throughout the plant body primarily through xylem vessel elements. Many individual vessel elements connect at axially located perforation plates to form long continuous tubes through xylem tissue that are termed vessels. Vessels begin and end at terminal vessel elements, which contain one end wall that is not open (i.e., a perforation plate is only present on one side of the vessel element) (Handley 1936). Individual vessels may extend several meters in length, but most vessels are quite short (Zimmermann and Jeje 1981; Zimmermann and Potter 1982; Ewers et al. 1990; Jacobsen et al. 2012).

Xylem vessel elements are formed from meristematic cells in either the procambium or the vascular cambium and each vessel element matures individually (Esau 1953). Following differentiation, vessel elements expand to what will become their mature volume before starting to form secondary wall layers along their sides (Esau and Charvat 1978). At this stage of vessel element development, the protoplasm is highly vacuolated (Esau 1936; Esau and Charvat 1978). During early vessel element development, the end walls, located basally and apically in the cell where perforation plates will later occur, become thickened (Esau and Hewitt 1940; Esau and Charvat 1978; Benayoun et al. 1981). The term perforation partition has also been used to refer to the areas of the cell wall that will later open to form perforation plates (Meylan and Butterfield 1981).

Maturation of vessel elements into hydraulically functional tubes is a multistage process that may occur over several weeks in the secondary xylem. Vessel element differentiation and expansion are followed by the deposition of lateral secondary walls. Deposition does not occur in the areas that will become perforation plates (usually at the end walls, but they may occur on the lateral walls in some species; Esau 1953) or in the areas along lateral walls that will become pits (Barnett 1982; Chaffey et al. 1997; see Chap. 3 for information on pit formation in conifers). Following the formation of lateral secondary walls within vessel elements, the end walls begin to thin (Esau 1936). Other cellular changes also occur, including the rupture of the tonoplast, the degradation of other organelles within the cell, and the deposition of lignin (reviewed in Fukuda 1997 and Bollhöner et al. 2012). At this stage, each individual vessel element maintains end walls and intact cell membranes filled with dilute protoplasm and may remain in this stage for up to several weeks (Esau and Hewitt 1940; Halis et al. 2012; Bollhöner et al. 2012).

Death of vessel elements, the opening of end walls, and the lysing of the remaining cellular contents occur during the final stages of vessel element development and signal their transition to becoming hydraulically active. Vessel element end walls open at the same time that the cell membrane disintegrates (Esau and Hewitt 1940; Murmanis 1978; Benayoun et al. 1981). Within a single vessel, individual vessel elements die at different times, although usually starting first basally and progressing in the apical direction. Both open basal and closed apical vessel elements may be found simultaneously within a single vessel (Esau 1936; Eames and MacDaniels 1947; Halis et al. 2012). Thus, individual vessel elements may become hydraulically active before the entire vessel activates (Halis et al. 2012) and the end walls of still living vessels form the barrier between the apoplast and symplast during this time. Once all of the vessel elements within a vessel undergo autolysis, the vessel becomes fully hydraulically functional, with water able to move from element to element through open perforation plates along the entire length of a vessel or through pits into lateral vessels.

The differentiation, development, and maturation of vessels may extend over several weeks in woody species. It has been reported that vessel element development took about five weeks in the earlywood vessel elements of the bole and branches of mature oak trees (Zasada and Zahner 1969). Others have reported that the final stages of vessel element maturation, including the disintegration of the end wall was quite gradual (Esau 1953; Murmanis 1978). Cell death of vessel elements and the onset of hydraulic function may also be delayed an additional several weeks after the full expansion of cells and formation of mature secondary walls. This has been described previously in grapevine (Halis et al. 2012). Indeed, vessel elements as well as fibers may remain alive (i.e., retain an intact cell membrane) at locations distant from the vascular cambium in some species (Zasada and Zahner 1969; Almeida-Rodriguez and Hacke 2012; Bollhöner et al. 2012).

The timing of these processes in woody plants differs from cell culture-, primary xylem-, or *Arabidopsis*-based studies which have reported that vessel elements differentiate, mature, and undergo cell death rather rapidly (Benayoun et al. 1981; Fukuda 1997; Groover and Jones 1999; Turner et al. 2007; Bollhöner et al. 2012), although metaxylem vessel elements have been described as remaining alive for longer periods and for up to 30 cm from growing tips in some nonwoody species as well (St. Aubin et al. 1986). Rapid vessel element maturation and subsequent cell death suggest that nonwoody systems may have limitations as models for understanding development of vessels within the secondary xylem of woody plants.

2.2 Vessel Differentiation, Expansion, and Maturation in Grapevine

Secondary xylem vessel element development in grapevine follows a similar pattern as described for other woody angiosperms. Vessel elements within the secondary xylem tissue are initiated early in the growing season, following initial shoot expansion. Throughout most of the growing season, new vessels are continuously formed. Thus, when xylem is examined in cross section, secondary vessels that are still differentiating and expanding, that are still living and have complete secondary walls, and hydraulically functional vessels are all visible simultaneously (Fig. 5.2a–d). This pattern has been described in *Vitis vinifera* "Cabernet" (Halis et al. 2012) and is also apparent in samples from both *V. vinifera* "Glenora" and *V. vinifera* "Chardonnay" (Fig. 5.2a–d), and thus may be general to *V. vinifera*. Interestingly, these vessels are not easily apparent unless active xylem staining is employed (compare Fig. 5.1 to Fig. 5.2a–d; see below for additional method details).

We examined vessel lifespan in 1-year-old shoots of field-grown *V. vinifera* L. "Glenora," a table grape that grows well in southern California and produces pale-red medium-sized seedless berries (see Jacobsen and Pratt 2012 for additional information on this variety) and from which samples were readily available and in close proximity to the laboratory. Additionally, a limited number of observations were also conducted on 1-year-old shoots of field-grown *V. vinifera* L. "Chardonnay." All samples were collected from well-watered plants on or near the campus of California State University, Bakersfield during the 2012 and 2013 growing seasons.

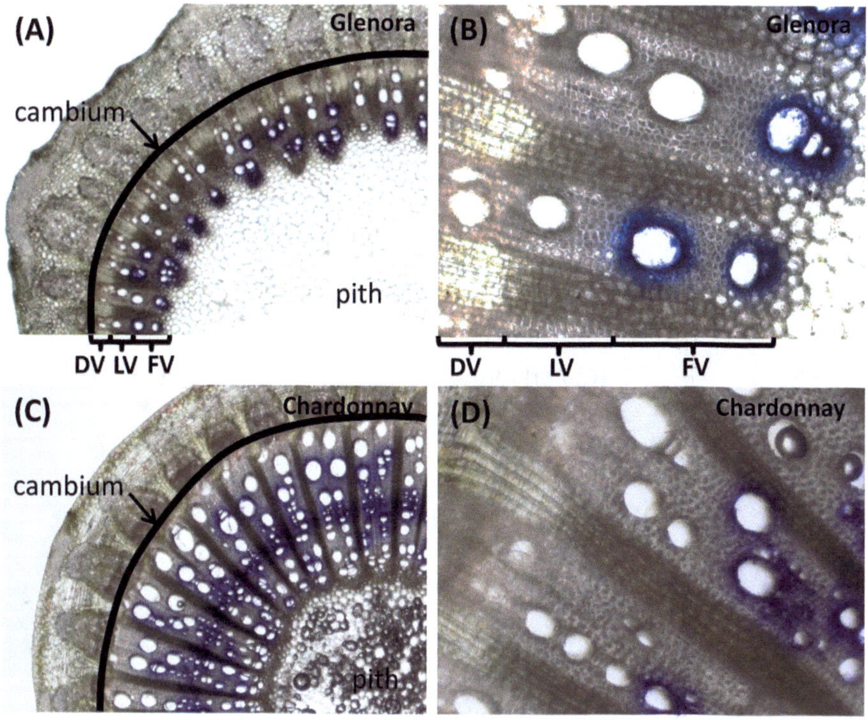

Fig. 5.2 Xylem cross sections collected during the middle of the growing season from 1-year-old stems of *Vitis vinifera* "Glenora" (**a**, **b**) and *V. vinifera* "Chardonnay" (**c**, **d**) that have been stained for active xylem vessels using crystal violet and are shown at both low (**a**, **c**; 25×) and higher (**b**, **d**; 100×) magnification. Only vessels which have already undergone autolysis and are hydraulically functional stain purple (i.e., they are functional vessels, FV). The remaining, nonstained vessels, are shown in two different stages of vessel development, those that do not yet have a complete secondary cell wall and may still be expanding (differentiating/developing vessels, DV) and those that have a complete secondary cell wall but with vessel elements that are still alive and filled with dilute protoplasm (living vessels, LV). Both examined varieties have many nonhydraulically functional vessels and contain a relatively narrow band of vessels that would be contributing significantly to long distance water transport within the stem. The vascular cambium is identified in panels A and C to highlight that vessels only become hydraulically functional once they are relatively distant from the cambium and that the large vessels nearest the cambium are not hydraulically active. This pattern is apparent throughout most of the growing season as long as new vessels are continuing to form (see Figs. 5.3 and 5.4)

Samples collected in 2012 were used to hone the staining technique described below to ensure that flushing, staining, and sample preparation did not disrupt living vessel elements (see St. Aubin et al. 1986 for discussion of the challenges of preparing samples without disruption of end walls, particularly when fixation was used).

In 2013, 1-year-old shoots were removed from plants at the site of initial bud emergence so that the entire portion of 1-year-old growth was removed from the point of its connection to older tissue. *Vitis vinifera* "Glenora," samples were collected 15 April, 23 April, 3 May, 7 June, 27 June, 17 July, and 26 October.

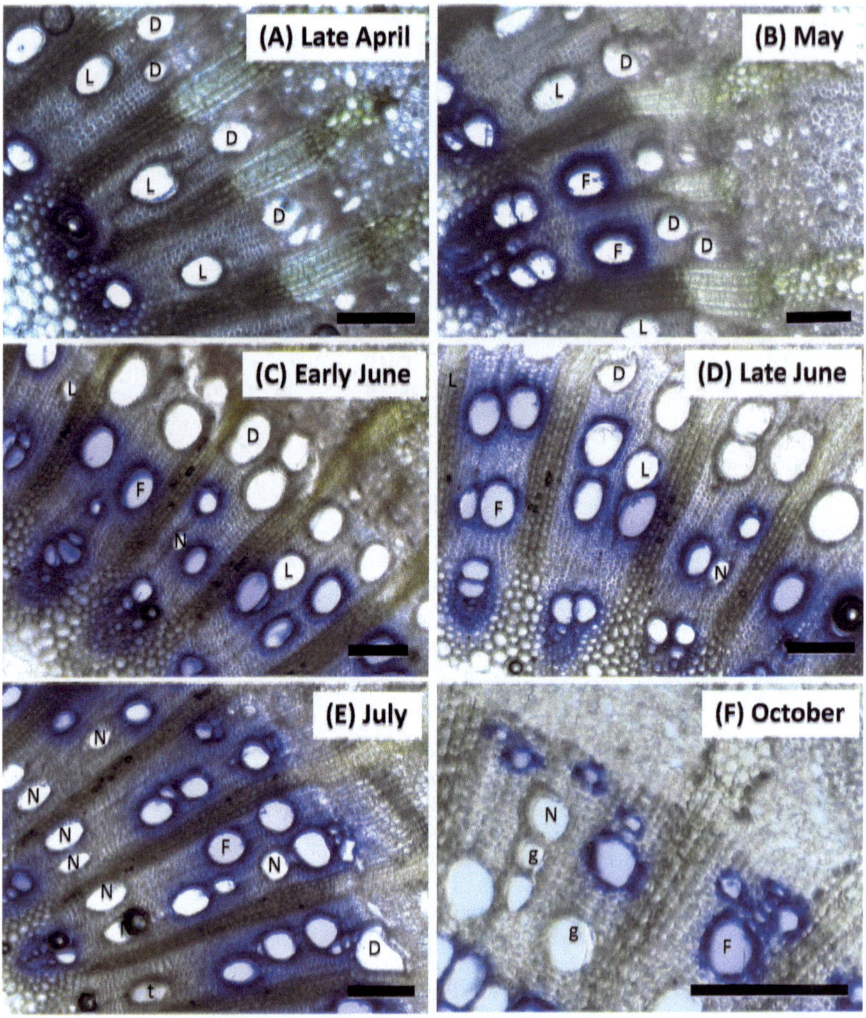

Fig. 5.3 Micrographs showing representative staining patterns from sampling times (scale bars indicate 200 μm) selected to illustrate patterns in vessel development throughout the growing season. Vessels were identified as falling within one of the following vessel lifespan classes: developing (D), living (L), potentially hydraulically functional (F), or nonhydraulically functional (N). Several vessels in each of these classes are identified in the panels above to illustrate their appearance. Potentially hydraulically active vessels were stained with crystal violet and appear purple in micrographs. New vessel formation only ceased at the end of the growing season, at the sampling period coinciding with the onset of leaf abscission (October). Vessel occlusions, including tyloses (t) and gels (g) are visible in some sections

These collection times corresponded to early shoot growth (April), blooming (May), fruit development (June), fruit ripening (July), and the initiation of leaf abscission (October). Samples were collected from *V. vinifera* "Chardonnay" on 18 June and 17 July.

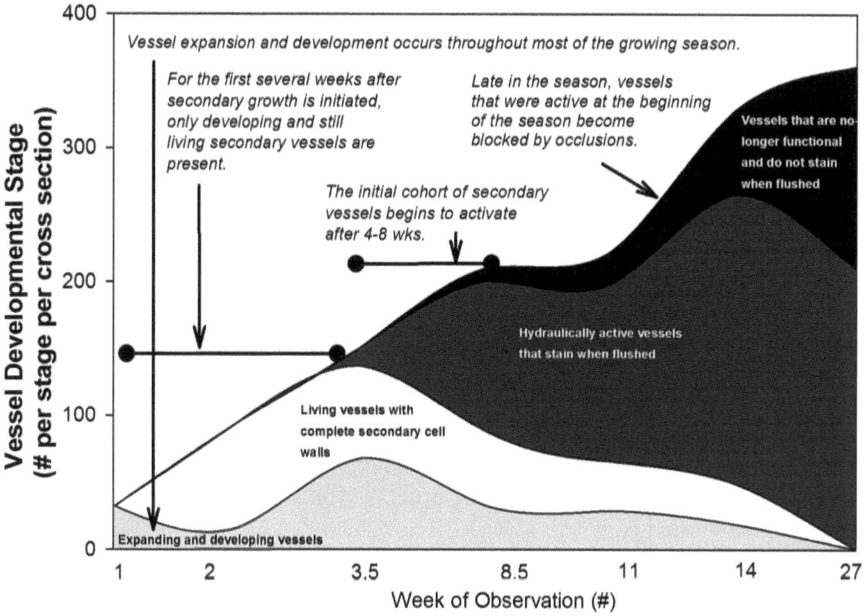

Fig. 5.4 The number of vessels occurring in different vessel lifespan classes as determined throughout the 2013 growing season, beginning in April and ending in October, and based on observations from 25 stem sections of current year growth from well-watered *V. vinifera*. Gray-scale fills indicate different vessels classes and are based on mean vessel numbers. Important transitions in vessel development are indicated in italics within the figure. When analyzed across the growing season, potentially hydraulically active vessels accounted for approximately 35 % of vessels in cross sections. Vessels that are not able to transport sap represented the majority of vessels within the secondary xylem tissue, including developing and living vessels (54 %) and vessels that were no longer hydraulically functional (11 %)

At each sampling date four large shoots were collected at predawn, double bagged in large plastic bags with a moist paper towel, and transported to the laboratory (<10 min transport time). In October, only a single large shoot was sampled due to limited ability to find shoots that still maintained green nonabscising leaves from among the shoots that had been tagged and followed since bud growth initiation earlier in the season. Immediately upon arrival in the laboratory, a 0.20 m stem segment was trimmed underwater from each large shoot at 1 m distance from the proximal end. All samples were collected at this same distance so that samples were of progressively older stems throughout the growing season from shoots initiated at a similar time. This meant that samples became progressively wider in diameter at each sampling time due to continuing secondary growth.

The ends of shoot segments were shaved with fresh razor blades and samples were inserted into a tubing apparatus and flushed for 1 h at 100 kPa using a degassed 20 mM KCl solution filtered to 0.1 μm (inline filter, GE Water and Process Technologies, Trevose, PA, USA). A 0.10 m segment was then cut underwater 0.06 m from the proximal end of the longer flushed segments. This shorter segment was stained using a 0.1 % (mass/volume) crystal violet dye solution, adjusted to a pH of 2 using HCl, and ultra-filtered (0.1 μm, MAGNA, nylon supported filter, Osmonics, Inc., Minnetonka, MN, USA). Stain was pulled up through stem samples at approximately 1–2 kPa for 20–30 min following the methods of Jacobsen et al. (2007). At the end of this time period, dye could be seen emerging from the distal ends of stem samples. This stain was selected because it easily passes through pit membranes, readily stains secondary cell walls (also termed gentian violet, Chamberlain 1901; Ruzin 1999) with a preference for lignified walls, and has been used in many studies as an indicator of active xylem vessels (Kolb and Davis 1994; Tibbetts and Ewers 2000; Jaquish and Ewers 2001; Jacobsen et al. 2007; Jacobsen and Pratt 2012).

After staining, segments were briefly (1–3 min) flushed at 1–2 kPa with the same solution used to flush samples prior to staining. This removes dye from many of the hydraulically active vessel lumens, but does not remove dye from already stained cell walls. This reduced the spread of dye into adjacent tissue prior to imaging and also reduced the amount of dye transferred by sectioning. Dyed stems were sectioned in the central portion of segments, in a region of the stem that was not submerged beneath dye during the staining process. This ensured that dye was limited to the region around conductive vessels into which dye had been pulled. Because dye slowly spreads following staining, each stem was sectioned and immediately photographed following staining.

All secondary xylem vessels were quantified into one of the four following classes: (1) Developing (not stained purple, near the cambium, without a thick secondary wall), (2) living (not stained purple, close to the cambium, thick secondary wall), (3) potentially hydraulically functional (stained purple), and (4) nonhydraulically functional (not stained purple, in a region of the stem where other vessels are or have been hydraulically functional; may contain gels or tyloses). Vessels located within the primary xylem tissue were also observed, but they were classified only as being either conductive or nonconductive. To confirm the presence of gels in older nonconductive secondary xylem vessels, some fresh sections were analyzed under a dissecting scope with no coverslip or mounting material. This allowed for easy identification of gel-filled vessels because gels do not recede from cut vessels and were often visible actively extruding from cut vessels (see Jacobsen and Pratt 2012, their Fig. 4; see also Fig. 5.11a–d in the present chapter).

Vessels located within primary xylem tissue, located in vascular bundles, were already hydraulically active during the first sampling period in early April (not shown). Interestingly, the vast majority of these primary xylem vessels remained potentially hydraulically active throughout the growing season. More than 93 % of the primary xylem vessels in all of the examined sections were hydraulically active from early April through late June. The percentage of potentially hydraulically active primary xylem vessels began to decline in July (mean of 76 % potentially hydraulically active) and had further declined by the onset of leaf abscission in October (mean of 27 % potentially hydraulically active vessels).

Secondary xylem production was initiated in a few of the observed samples prior to early April, but at the time of the first observations of samples there were no secondary xylem vessel elements yet displaying fully formed secondary cell walls. The first secondary xylem vessel elements to fully expand and start to form secondary cell walls were observed in late April (Fig. 5.3a). These vessel elements were still living and not yet potentially hydraulically active. The first potentially hydraulically active secondary xylem vessels were observed in May (Fig. 5.3b), more than 3 weeks after initial observations of vessel differentiation. The entire initial cohort of secondary xylem vessels were not observed to be potentially hydraulically active until early June, nearly 8 weeks after initial observations of vessel differentiation. Thus, the observed amount of time from the initial differentiation of vessel elements to the appearance of hydraulically active vessels was approximately 4–8 weeks (Figs. 5.3a–f and 5.4). This included at least 2 weeks to differentiate, expand, and form secondary walls and an additional 2–4 weeks before autolysis and opening of perforation plates. The same patterns of vessel element development were also apparent in samples from Chardonnay (data not shown; see Fig. 5.2d).

The production of secondary xylem and secondary xylem vessel development continued throughout most of the growing season (Figs. 5.3a–f and 5.4). A band of differentiating and developing vessels could be found near the vascular cambium at every sampling period, except for October (Figs. 5.3a–f and 5.4). Beginning at the time of fruit harvest (July) and continuing to the period of leaf abscission (October), vessel development declined and a cohort of narrow diameter vessels became hydraulically active immediately adjacent to the vascular cambium (Fig. 5.3f). Similarly, living vessels were present throughout most of the growing season and were visible as fully expanded vessel elements, usually removed from the vascular cambium by at least one or two developing vessels, that were not yet potentially hydraulically active (Figs. 5.2a–d and 5.3a–f).

Individual vessels did not remain potentially hydraulically active throughout the growing season. Rather, there was a steady change in the suite of vessels that were potentially hydraulically active and vessels that were active early in the season were mostly nonhydraulically functional later in the season (Fig. 5.3e, f). In general, none of the secondary xylem vessels that would have been hydraulically active in May and June were still active late in the season. This was indicated by the failure of early season vessels to stain with active xylem staining later in the season.

When analyzed across the growing season, hydraulically active vessels represented only a small proportion of total vessels (Fig. 5.4). The region of xylem containing vessels that were hydraulically functional was small and vessels are only active for a limited portion of time (Fig. 5.3a–f). Staining of samples that have been flushed to remove emboli indicated that potentially hydraulically active vessels account for only 35 % of vessels visible within cross sections of 1-year-old stems in *V. vinifera* and, due to embolism formation, the in situ number of hydraulically active vessels would likely have been even lower. Vessels not able to transport sap represented the majority of vessels. The majority of these vessels, particularly in the early half of the season, were vessels that were still developing or living (~54 % of visible secondary xylem vessels). This included a large band of vessels located near the cambium.

Later in the season, the proportion of vessels that were not able to transport sap also included an expanding band of inactive vessels occurring near the pith. This band slowly expanded out toward the vascular cambium throughout the late period of the growing season. These inactive vessels did not become conductive following flushing of stems under positive pressure, indicating that this loss was semipermanent to permanent. This was most likely due to the formation of gels or tyloses (Sun et al. 2007, 2008; Jacobsen and Pratt 2012; Sun et al. 2013). The high proportion of nonhydraulically functional vessels in grapevine may pose a particularly challenging problem in ongoing research examining grapevine structure and function (see Sect. 5 of this chapter).

3 Grapevine Stem Hydraulic Architecture

Grapevine stem xylem vessels are relatively unexceptional in their dimensions when standardized by sample diameter, as is required for cross-species comparisons (Jacobsen et al. 2012). Secondary vessels in 1-year-old grapevine stems, the type commonly used in hydraulic studies, contain vessels that are much smaller in diameter, on average, than the reported global mean (compare standardized values in Table 5.1 to a global mean

Table 5.1 Mean and maximum vessel diameters from stems of several *Vitis* species and varieties compiled from several different sources

Species	Variety (if reported)	Mean vessel diameter (μm)	Maximum vessel diameter (μm)	Source
Not standardized by sample diameter (ranges indicate that the reported mean varied with stem diameter, node, or treatment):				
V. labrusca	–	Approx. 300	–	Zimmermann and Jeje (1981)
V. rotundifolia	–	36–245.5	–	Ewers and Fisher (1989), Ewers et al. (1990)
V. vinifera	Freisa	74–101	>140	Lovisolo and Schubert (1998)
V. vinifera	Grenache	–	108.7	Lovisolo et al. (2008)
V. vinifera	Chardonnay	60.5–67.6	–	Sun et al. (2006)
V. vinifera	Glenora	28.6–57.9	68.6–158.1	Current chapter, Fig. 5.10
V. vinifera	–	69.18	–	Sperry et al. (2005)
V. vinifera	Nebbiolo	61–107	>140	Schubert et al. (1999)
Standardized samples (n = 6–23 per variety; 4–8 mm sampled stem diameters from 1-year-old segments):				
V. labrusca	Catawba	29.1 ± 2.7	64.7 ± 3.1 (75.5)	Jacobsen, unpublished data
V. labrusca	Concord	24.2 ± 1.17	49.8 ± 4.7 (99.3)	Jacobsen, unpublished data
V. vinifera	Cabernet Savignon	18.4 ± 0.8	42.5 ± 2.6 (55.6)	Jacobsen, unpublished data
V. vinifera	Chardonnay	25.0 ± 1.0	57.3 ± 1.6 (74.7)	Jacobsen, unpublished data
V. vinifera	Flame	29.6 ± 2.0	55.3 ± 2.3 (66.2)	Jacobsen, unpublished data
V. vinifera	Glenora	25.6 ± 1.1	59.8 ± 1.4 (73.4)	Jacobsen, unpublished data

Data are divided by those reported in the literature from a wide range of stem diameters, ages, growing locations, and treatments and those that have been standardized by stem diameter and age specifically so that values are comparable between varieties and comparable to the global values reported in Jacobsen et al. (2012). Samples that have been standardized by diameter are all from 1-year-old stem samples collected in summer 2013 from varieties growing on or near campus at California State University, Bakersfield, USA and mean ± 1 SE are reported. Parenthetical values reported after maximum vessel diameters represent the largest diameter vessel recorded within samples from that variety

Table 5.2 Mean and maximum vessel length from stems of several *Vitis* species and varieties compiled from several different sources

Species	Variety (if reported)	Mean vessel length (m)	Maximum vessel length (m)	Source
*V. labrusca/riparia**	–	0.416*	1.4*	Sperry et al. (1987)
*V. labrusca**	–	1.435*	7.5*	Zimmermann and Jeje (1981)
V. labrusca	Catawba	–	0.78±0.13	Jacobsen, unpublished data
V. labrusca	Concord	–	0.63±0.10	Jacobsen, unpublished data
*V. rotundifolia**	–	0.322*	1.73*	Ewers and Fisher (1989), Ewers et al. (1990)
V. vinifera	Cabernet Savignon	–	0.54±0.09	Jacobsen, unpublished data
V. vinifera	Chardonnay	–	0.67±0.09	Jacobsen, unpublished data
V. vinifera	Chardonnay	–	0.5 to 0.9	Choat et al. (2010)
V. vinifera	Flame	–	0.73 to 0.6	Jacobsen, unpublished data
V. vinifera	Glenora	0.116	0.85	Jacobsen and Pratt (2012)
V. vinifera	–	0.207	–	Sperry et al. (2005)
V. vinifera	–	0.128	–	Wheeler et al. (2005)

Samples that are followed by an asterisk (*) are from wider diameter samples and may be from stems that are multiple years of age, whereas all vessel length values that are not followed by an asterisk are from 1-year-old stem segments. All samples reported from Jacobsen (unpublished data) were sampled in July 2014 from plants growing on or near campus at California State University, Bakersfield, USA and mean ± 1 SE are reported (n=4–6 per variety). Mean vessel lengths were determined using different methods that have been shown to produce similar results (reviewed in Jacobsen et al. (2012)) and maximum vessel lengths were determined using air injection at 100 kPa (Greenidge 1952)

diameter of 67.89 μm for 3,005 woody angiosperms; Zanne et al. 2010). Similarly, mean and maximum vessel lengths are relatively short for lianas and are similar to the global mean across woody tree, shrub, and liana species (compare values in Table 5.2 to a global mean vessel length of 0.13 m and a global mean maximum vessel length of 0.93 m for 148 species; Jacobsen et al. 2012). However, when both vessel length and width are considered, grapevine has a combination of vessel length and width that results in the average size of grapevine conduits being larger than average when compared against the data available globally to date (Jacobsen and Pratt 2012).

Most species show an increase in both vessel length and width with increasing stem diameter (reviewed in Jacobsen et al. 2012) and, consistent with this general trend, grapevine vessel dimensions vary considerably with stem diameter and sample age (discussed below and shown in Fig. 5.10; Lovisolo and Schubert 1998; Schubert et al. 1999). Thus, especially in older samples or in wide diameter stems, grapevine xylem may contain very wide (approximately 300 μm diameter) and long (~8 m) vessels (Zimmermann and Jeje 1981).

The vessel network within grapevine secondary xylem has some unique features (Figs. 5.5a, b, 5.6a, b, and 5.7a, b). Large diameter vessels in grapevine are generally not directly connected laterally to other large vessels. Examination of serial sections of grapevine xylem suggests that large diameter vessels tend to terminate

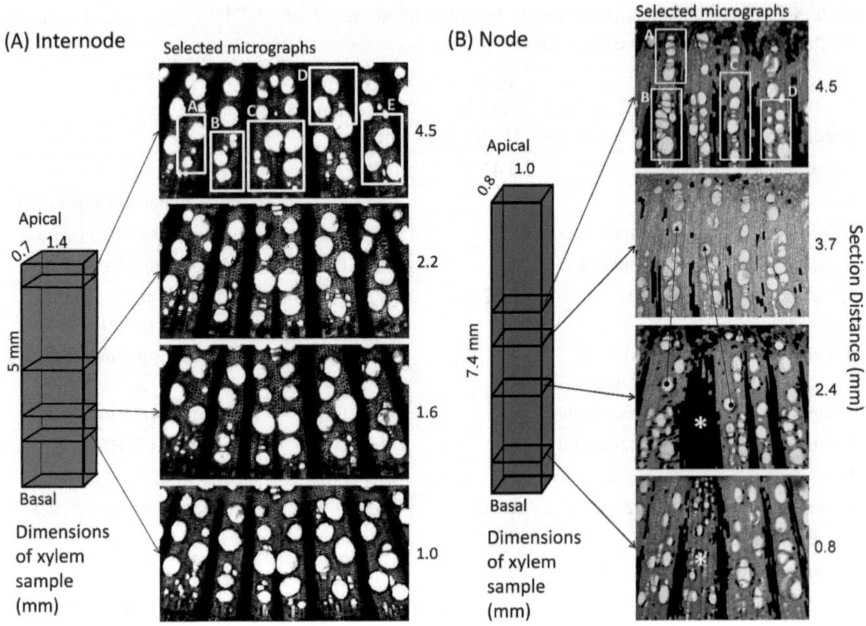

Fig. 5.5 Serial sections (at 100 μm intervals) were made through the xylem of internode (**a**) and node (**b**) segments of 1-year-old grapevine stems and were examined and used to construct three-dimensional models of the vessel network. Representative micrographs from the examined xylem samples are shown above along with a model illustrating the location of the sections. Inset boxes within the uppermost micrograph identify specific vessel groups that were selected for more detailed analysis of vessel network structure and vessel connectivity (see Figs. 5.6 and 5.7). For the node segment (**b**) the *asterisks* indicate a xylem section that exits the stem xylem to enter a leaf and the two-sided arrows that extend across panels identify the same vessel in each panel in their new location following the rearrangement of vessels around the exit of the leaf trace

abruptly, with one large vessel connecting to the next predominately at the vessel termini. The region of overlap between these large vessels occurs over just a few hundred microns and with only one or two vessel elements overlap (Jacobsen, unpublished data). Rather, the vast majority of connections between large diameter vessels occurs through the connection of large diameter vessels by relays of many shorter and narrower vessels (Fig. 5.6a; Brodersen et al. 2013). This is particularly evident within internode regions (Fig. 5.6a), but also occurred within nodes (Fig. 5.6b) and may be an important, but thus far unexplored, feature of grapevine hydraulic function.

Vessel network architecture differs between node and internode regions (Figs. 5.5a, b, 5.6a, b, and 5.7a, b). Xylem node regions contain a vessel network that is more disordered than the network found within internodes (Fig. 5.7a, b). Numerous vessels terminate within nodes, which represents a higher proportion of

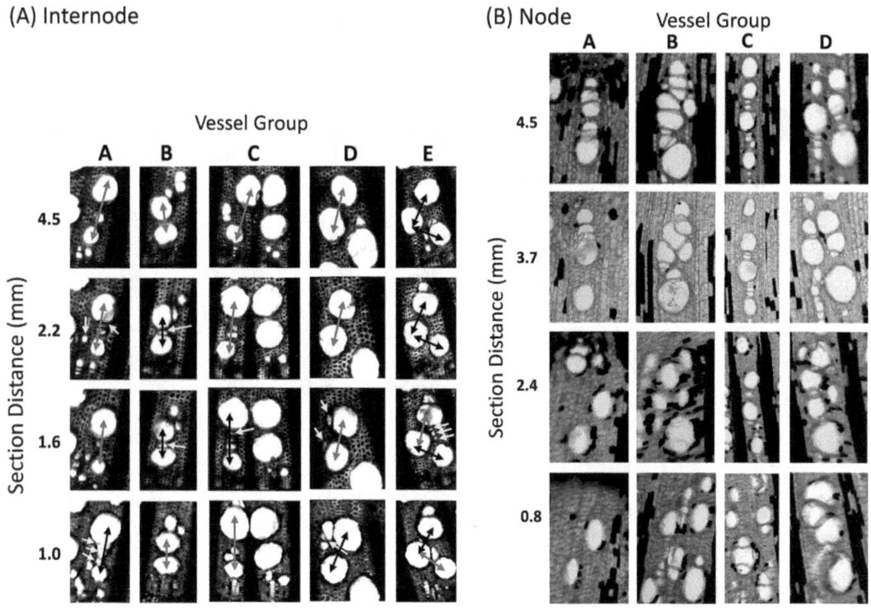

Fig. 5.6 Selected micrographs from serial sections of grapevine stem xylem located in an internode (**a**) and node (**b**) are shown, with specific vessel groups highlighted to show patterns of vessel connectivity through the xylem (see Fig. 5.5 for the location of vessel groups relative to one another). In Panel A, large vessels that are connected along part of their axial pathway are indicated with gray arrows, but they are not connected within the panels containing the gray arrows. Black arrows indicate the specific image(s) in which large vessels are visible as connected through vessel relays within that particular image. Large vessels rarely come into direct contact with one another and instead connect through the development of relays of narrower and shorter vessels (indicated with small white arrows). The connection of large vessels by shorter and narrower vessel relays is particularly evident in internode section (**a**), but is also visible within the node (**b**). Vessel connections are not indicated in panel B because the more disorganized vessel connections of the node make these connections much more complex; however, the connectivity of the node section shown above (**b**) can been seen in three-dimensional modeling for the entire sampled segment in Fig. 5.7b

vessels terminating per unit length than in internode sections (Figs. 5.6a, b and 5.7a, b; Jacobsen, unpublished data). This is similar to previous reports for *V. vinifera* as well as for other species, which also report that vessels are more likely to terminate within nodes (Salleo et al. 1984). Within the node, rays of parenchyma tissue were also more heavily rearranged, in part because of the loss of xylem segments into leaves, tendrils, and side branches at nodes, but also due to the termination and creation of additional xylem segments (Fig. 5.5a, b). This represents a significant nodal rearrangement in the connectivity of xylem conduits within the stem. This rearrangement is also visible in the much more tortured pathways of vessels through the xylem within nodes (Fig. 5.7a, b).

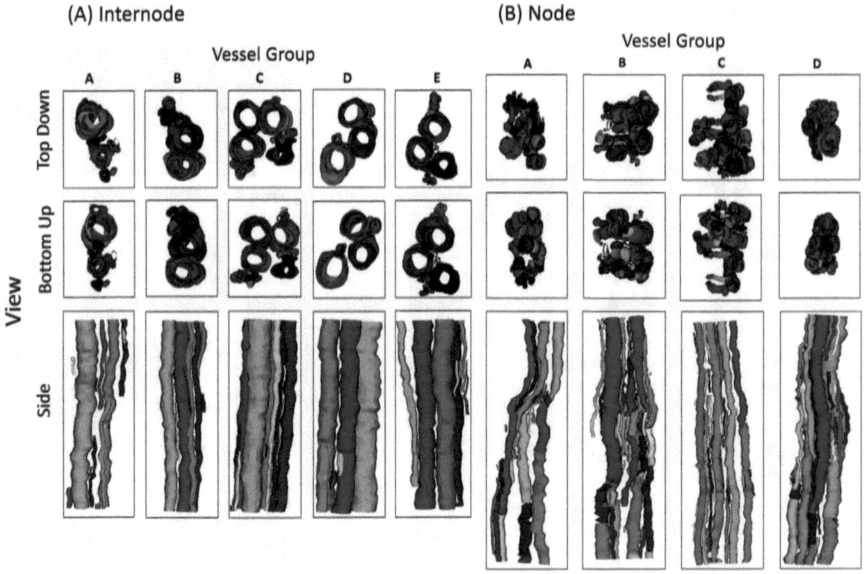

Fig. 5.7 Three-dimensional models of the vessels within a grapevine stem xylem internode (**a**) and node (**b**) xylem section based on combining serial sections (at 100 μm intervals) that were made through the same xylem segments presented in Figs. 5.5 and 5.6. Serial sections were aligned using Reconstruct (v. 1.1.0.1) and three-dimensional models were constructed using Image J (v. 1.44p) analysis of the stacked images as described in Yahya et al. (2011). Vessel terminations are modeled as being "flat" although they would likely be pyramidal within the examined specimens; this is largely due to the limitations of having images created from the stacking of 100 μm thick sections. The vessel network is relatively ordered within the internode (**a**) and open vessel lumens can be seen in both the bottom and top views of the xylem, illustrating that vessel pathways are relatively linear. The vessel network becomes much more tortured in the node (**b**) where numerous vessel groupings are rearranged following the exit of a leaf trace from the stem (Fig. 5.5) and the twisting, rearrangement, and tortured pathway of vessels means that lumens are not visible as being "open" through the "bottom up" and "top down" views through the node. Along the length of a single vessel, the diameter of the vessel changes considerably, as described previously in Akachuku (1987) and Ewers and Fisher (1989)

4 Hydraulic Function

4.1 Resistance to Water Stress-Induced Cavitation

Grapevine varieties are generally highly susceptible to water-stress induced cavitation as indicated by rapid declines in hydraulic conductivity with small declines in water potential (Fig. 5.8). Indeed, across many different varieties and several different studies, grapevine species and varieties have consistently been shown to be highly susceptible to cavitation and water stress (Table 5.3). Dehydrated grapevine shoots from two different grapevine varieties have also been reported to display a majority of acoustic emissions (AEs) between −0.2 and −1.5 MPa with early peaks

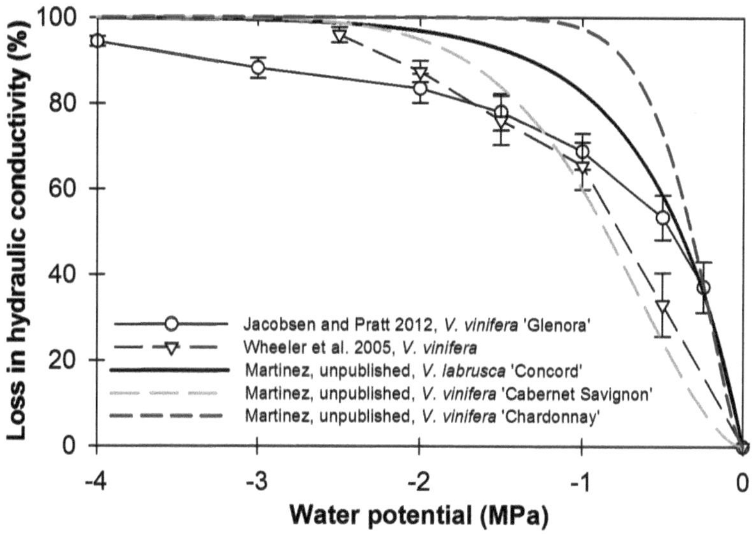

Fig. 5.8 Vulnerability to cavitation curves for five selected varieties of grapevine, showing the percentage loss in hydraulic conductivity with declining water potential. All curves were measured on 1-year-old samples using the centrifuge method (Alder et al. 1997)

in AEs at approximately –0.5 MPa (Schultz 2003). Acoustic emissions are indicative of cavitation events within the xylem. The production of many acoustic emissions (AEs) at relatively high water potentials as grapevine shoots are dehydrated represents an independent (of hydraulic methods) and in vivo confirmation that grapevine xylem is highly susceptible to water stress. Most recently, single vessel injection experiments conducted on 1-year-old stems of *V. vinifera* "Chardonnay" have confirmed the presence of a large proportion of highly susceptible vessels within the xylem (M. Venturas unpublished data).

High vulnerability to cavitation in the xylem of 1-year-old shoots of grapevine is consistent with grapevine physiology. Many grapevine varieties have been identified as being near isohydric and even varieties that display slightly anisohydric behavior do not typically experience water potentials below –2 MPa even when severely stressed. Indeed, the water potentials of "stressed" dehydrated plants have been reported to range from –0.6 to –2.0 MPa in many studies (Loveys and Kriedemann 1973; Matthews and Anderson 1988; Schultz and Matthews 1988; Lovisolo and Schubert 1998; Flexas et al. 1999; Choné et al. 2001; Schultz 2003; Lovisolo et al. 2008; Lovisolo et al. 2010). Whole plant leaf specific hydraulic conductance (k_l) has been shown to rapidly decline with minor declines in water potential (see whole shoot dehydration P_{50} reported in Table 5.3; Schultz 2003; Zufferey et al. 2011) and whole plant hydraulic resistance increased rapidly when water potentials declined from –0.6 to –1.2 MPa (Schultz and Matthews 1988). This decline in whole plant conductance is due to a combination of stomatal closure, cavitation, and other modifications of xylary and extraxylary pathways that lead to a decrease in conductance in roots, stems, branches, and leaves.

Table 5.3 Water potential at 50 % loss in hydraulic conductivity (P_{50}) calculated from centrifuge-based vulnerability curves (centrifuge), dehydration-based curves where the percentage loss in hydraulic conductivity was calculated relative to hydrated K_s (dehydration decline relative to hydrated plant K_s), and water potential at 50 % loss in whole plant leaf specific hydraulic conductance (k_l) as calculated for whole shoot dehydration experiments (dehydration decline in k_l) for two *Vitis* species and several different varieties

Vulnerability curve method[a]	Plant material	P_{50} (MPa)[b]	*n*	Sample length (m)	Source
Centrifuge	*V. labrusca* L. "Catawba"	−0.20±0.07	6	0.14	Martinez, unpublished data
Centrifuge	*V. labrusca* L. "Concord"	−0.37±0.07	6	0.14	Martinez, unpublished data
Centrifuge	*V. vinifera* L. "Cabernet Savignon"	−0.76±0.10	6	0.14	Martinez, unpublished data
Dehydration decline in k_l	*V. vinifera* L. "Chasselas"	−0.41	12	whole shoot	Zufferey et al. (2011)
Centrifuge	*V. vinifera* L. "Chardonnay"	−0.7	6	0.145	Drayton (2009), Choat et al. (2010)
Centrifuge	*V. vinifera* L. "Chardonnay"	−0.28±0.05	11	0.14	Martinez, unpublished data
Centrifuge	*V. vinifera* L. "Glenora"	−0.16±0.03	4	0.14	Jacobsen and Pratt (2012)
Centrifuge	*V. vinifera* L. "Glenora"	−0.31±0.11	4	0.27	Jacobsen and Pratt (2012)
Centrifuge	*V. vinifera* L. "Glenora"	−0.34±0.06	12	0.14	Jacobsen and Pratt (2012)
Centrifuge	*V. vinifera* L. "Glenora"	−0.55±0.12	12	0.27	Jacobsen and Pratt (2012)
Dehydration decline relative to hydrated plant K_s	*V. vinifera* L. "Glenora"	−0.51	31	0.14; K_s	Jacobsen and Pratt (2012)
Dehydration decline in k_l	*V. vinifera* L. "Grenache"	−0.32	19	whole shoot	Schultz (2003)
Dehydration decline in k_l	*V. vinifera* L. "Syrah"	−0.51	15	whole shoot	Schultz (2003)
Centrifuge	*V. vinifera* L. (variety not reported)	−0.76	≥6	0.142	Wheeler et al. (2005)

[a]Dehydration-based curves where percentage loss in hydraulic conductivity is calculated relative to a flushed maximum value have not been included due to methodological problems with this technique (see text for details)

[b]All P_{50} were calculated using a Weibull fit

For each reported P_{50} value, the sample size (n) and the length of measured samples are reported. Data were compiled from several different sources as indicated

4.2 Seasonal Changes in Hydraulics

Changes in xylem hydraulic function seasonally have been described previously for *Vitis* (Sperry et al. 1987; Tibbetts and Ewers 2000; Choat et al. 2010; Jacobsen and Pratt 2012) and have also been found in the *V. vinifera* and *V. labrusca* varieties reported on in the present chapter (Fig. 5.9; Jacobsen et al., unpublished data). In general, grapevine xylem-specific hydraulic conductivity (K_s) has been shown to increase throughout the growing season. This change occurs concurrent with changes in xylem cavitation resistance, with stems grown later in the growing season showing increased resistance to cavitation and reduced levels of conductivity loss at more negative water potentials (Choat et al. 2010; their Fig. 2).

The studies included above have examined changes in hydraulics at different times during the season, but seasonal changes in hydraulics have also been examined through the examination of changes occurring along a single 1-year-old shoot. Grapevines exhibit rapid growth and an individual stem may grow several meters throughout the growing season. Along this shoot, basal segments are initiated early in the season and the earlywood of these segments represents xylem formed during this period. These segments produce more secondary growth compared to those initiated later in the season and they are wider in diameter by the end of the growing season. Distal segments of 1-year-old stems represent xylem that was initiated later in the growing season and was produced during drier and hotter conditions. These stems have less time for secondary growth to occur and are narrower at the end of the growing season.

Examination of xylem structural and functional traits as they vary along 1-year-old shoots provides the opportunity to evaluate some of the changes that have occurred throughout the growing season. Within 1-year-old shoots, stem segments that are narrower in diameter (i.e., produced later in the season) are up to 10 times more resistant to cavitation (i.e., they have more negative P_{50}) than segments that are

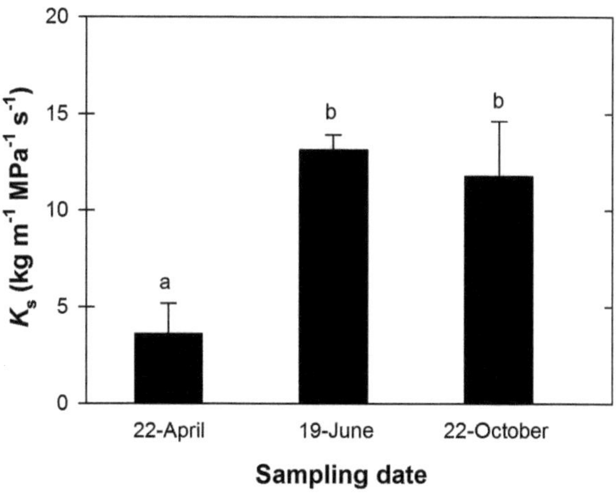

Fig. 5.9 Xylem-specific hydraulic conductivity (K_s) measured at three sample times in 2013 ($n=6$ per sample time; mean ± 1 SE) on *V. vinifera* "Glenora". Different letters indicate significant differences in K_s. Early season stem segments exhibited decreased xylem-specific conductivity (K_s) compared to samples measured later in the growing season ($P=0.017$, $F=5.71$, ANOVA)

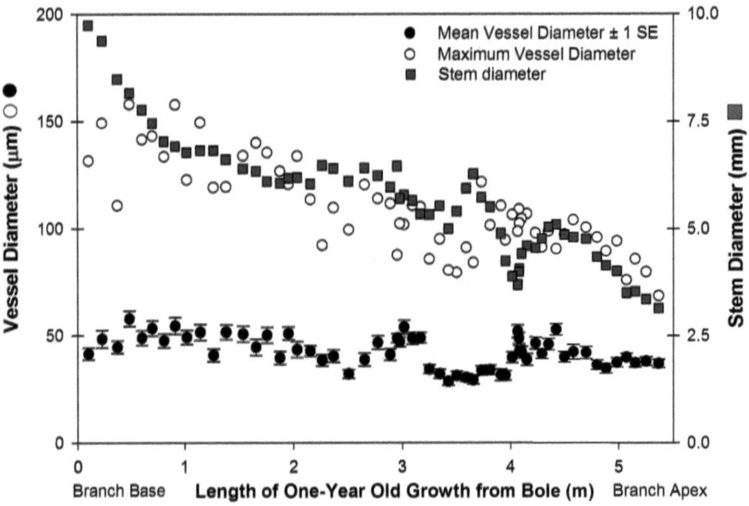

Fig. 5.10 Mean (±1 SE) and maximum vessel diameter as measured along the length of stem grown within a single growing season in grapevine. Each point represents a value from a single internode section beginning at the point of shoot initiation (internode 1) and continuing to the branch tip (internode 60). Stem diameter decreased from the branch base to the branch tip (*gray squares*) and vessel traits varied with both stem diameter and internode number. Maximum vessel diameter (*open circles*), in particular, was particularly sensitive to changes in internode number and stem diameter

wider in diameter (i.e., produced earlier in the season with a longer time period for secondary growth to occur) (Jacobsen and Pratt 2012; their Fig. 2). They also are higher in their stem hydraulic conductivity (Lovisolo and Schubert 1998; Schubert et al. 1999). These changes are consistent with the patterns in grapevine vessel development described above, whereby vessels take several weeks to become functional and different cohorts of vessels are active during different times of the season. These changes are also consistent with changes in vessel anatomy.

Vessel structure changes throughout the growing season (Fig. 5.10). Stems grown earlier in the growing season contain vessels that are wider in diameter compared to those produced later in the growing season (Fig. 5.10; Lovisolo and Schubert 1998; Schubert et al. 1999). This pattern is particularly evident when only the widest vessels within a cross section are examined (Fig. 5.10). This may be due, at least in part, to a longer period of secondary growth in stem segments that are initiated early, as evidenced by their wider diameter, but may also relate to other factors.

Thus, plant hydraulic structural and functional traits, particularly in grapevine, appear to be quite variable intra-annually and this may be due to different vessels becoming hydraulically active throughout the season and/or the formation of structurally different vessels within stems that are initiated during different times during the season. This is a functionally interesting pattern and suggests that xylem may be quite dynamic in a way that has not been widely appreciated (for instance, see Fig. 4.11 in Uwe's Poplar Chapter). Plants may have considerable control over the types of vessels that are grown or activated and further research on the development of vessels and their onset of functionality (as well as loss in functionality) throughout the growing season may be particularly interesting.

5 Methodological Considerations in Study of Grapevine Hydraulic and Function

Although much of the data on grapevine hydraulics has been consistent and suggests high susceptibility to water stress, a few recent studies have called some of these findings into question (e.g., Choat et al. 2010; McElrone et al. 2012). Some of this confusion is likely linked to issues related to imaging of grapevine stems containing living or gel-filled vessels, but may also be linked to other complications. It is, therefore, worthwhile to consider how some of the unique aspects of grapevine xylem may impact the methodologies used to examine it and may impact the interpretation of some studies and results.

5.1 Challenges Related to In Vivo Imaging

The high proportion of nonhydraulically functional vessels in grapevine may be a particularly challenging problem in ongoing research examining grapevine structure and function using current in vivo imaging technologies. These nonhydraulically active vessels, including cells with completed secondary walls and degraded organelles that are still alive and also inactivated vessels that are gel filled (Fig. 5.11a–d), appear to be difficult to identify when the latest imaging technologies such as magnetic resonance imaging (MRI), nuclear magnetic resonance (NMR), and high resolution computed tomography (HRCT) are utilized. The problem arises because living or developing vessels and gel-filled vessels that are not conductive appear similar in images to cells that are mature and hydraulically active, as demonstrated by the inability of MRI images to differentiate glass tubes filled with water, saline solution, or pectin gel (Pérez-Donoso et al. 2007).

This limitation of imaging techniques is, in some cases, at the heart of a debate regarding the accuracy of methods to measure resistance to cavitation of xylem. Some authors have taken the data generated from imaging as a "gold standard" by which to judge other methods and interpreted disagreements with imaging as an indication that the nonimaging methods are flawed (Choat et al. 2010; Cochard et al. 2014). However, it appears that an alternative explanation is likely. To date, imaging data have yet to be shown to accurately estimate hydraulic conductivity, percentage loss in hydraulic conductivity, or cavitation resistance when compared against standard techniques. The difficulty in properly analyzing and interpreting grapevine images generated using these techniques is perhaps best illustrated through the repeated inability of these imaging techniques (e.g., Choat et al. 2010; McElrone et al. 2012) to generate data

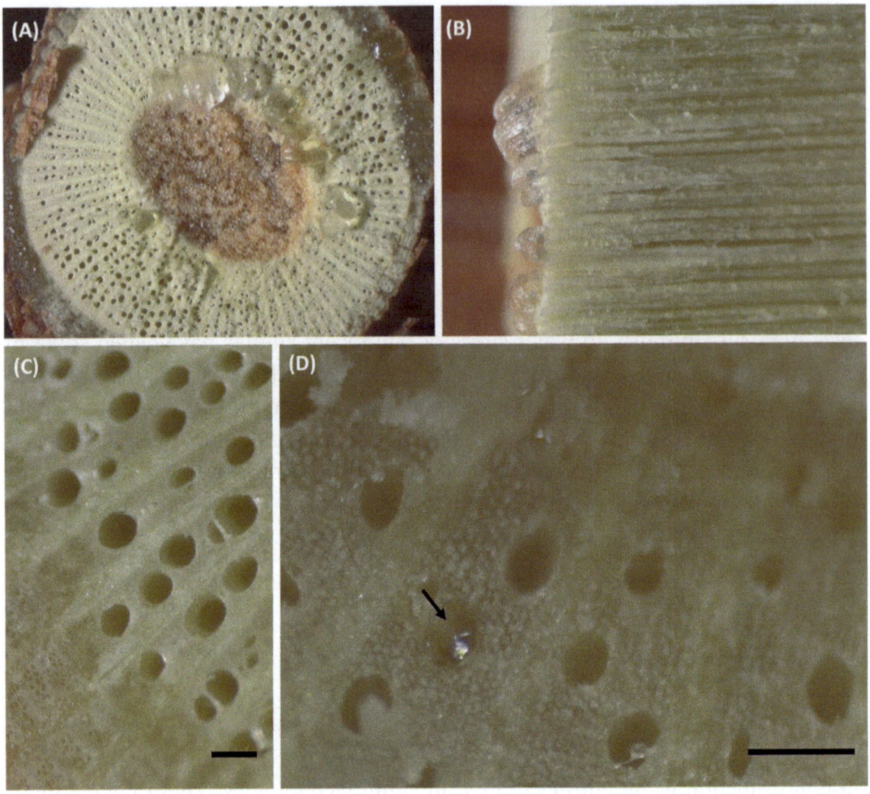

Fig. 5.11 Grapevine stems produce gels in response to both drought and wounding as seen in both cross (**a**) and longitudinal (**b**) sections of 1-year-old stems, where gels are visible exuding out of vessels. Gels appear to be produced in vessels during dehydration, but are not usually apparent when hydrated samples are sectioned ((**c**); note the absence of gels). When gels are present in vessels, they are easily seen when samples are freshly sectioned because gels do not recede from vessel ends ((**d**); gel-filled vessel is indicated by the *black arrow*). Scale bars in panels (**c**) and (**d**) indicate 200 μm

that agree with other standard and thoroughly verified hydraulic methods (i.e., the centrifuge method and dehydration methods as presented in Jacobsen and Pratt 2012 using the methods verified by Sperry et al. 2012; Tobin et al. 2013, and Hacke et al. 2015). This suggests that image analysis is not yet a verified tool in the examination of hydraulic function, at least in the case of grapevine xylem. Difficulties with in vivo imaging may also extend to imaging analyses conducted in other species where imaging data appear inconsistent with other physiological and standard hydraulic methods (e.g., Cochard et al. 2014), particularly for species such as oak and poplar where living vessel elements have been identified occurring well away from the vascular cambium (see Sect. 2.1 of the current chapter).

5.2 Challenges Posed by Dynamic Xylem Processes

Not all techniques or analyses for examination of resistance to water stress-induced cavitation may be appropriate for grapevine. The use of dehydration vulnerability curves, particularly those conducted over several weeks or longer (Alsina et al. 2007; Choat et al. 2010), may be heavily impacted by the formation of gels within vessels (Jacobsen and Pratt 2012) or the activation of new and functionally different xylem vessels. Of particular concern with dehydration measures is the formation of occlusions within the xylem during dehydration which results in error in measurements of the maximum xylem-specific hydraulic conductivity of samples (K_{smax}). This issue arises due to the analysis of these curves using percentage loss in hydraulic conductivity (PLC) as calculated relative to a flushed maximum value for each sample.

The K_{smax} error produced by the formation of gels within vessels and when flushed or vacuumed samples are affected by vessel clogging is not consistent across samples or sampling periods and it is difficult to identify and may be impossible to correct for. This variability and its potential impact on PLC estimates was illustrated in the data reported by Drayton (2009), the original source for much of the data published in Choat et al. (2010). Drayton (2009) reported that some, but not all samples, exhibited declines in conductivity following flushing and the declines were not consistent across seasonal measures. Declines were most notable in *V. vinifera* "Cabernet Sauvignon" samples collected in May, which exhibited a mean of approximately–20 % loss in conductivity following flushing (Drayton 2009). A decline in conductivity following flushing indicates the blockage of xylem conduits and is most likely linked to the exudation of gels in grapevine.

Although K_{smax} error may be dealt with by analyzing K_s of unflushed stems and possibly the calculation of PLC relative to hydrated and nonflushed samples, K_s is rarely reported for vulnerability curves. For this reason, dehydration vulnerability curves that are based on PLC calculated using a flushed maximum have not been included in the data reported within the present chapter. Additionally, centrifuge-based curves, which may be generated over a short time period, may be preferable when examining grapevine.

Changes in xylem seasonally may also be linked to some ongoing methods issues. For instance, this may explain, in part, the difficulty of interpreting hydraulic data collected over the growing season (Choat et al. 2010) and lends support to the suggestion that study design should carefully control for season (Jacobsen et al. 2007; Jacobsen and Pratt 2012; Hacke et al. 2015). These issues may be further complicated by seasonal variation in the formation of gels within the xylem (Sun et al. 2008). Additionally, few studies have examined plant hydraulic traits seasonally (however see Kolb and Sperry 1999; Jacobsen et al. 2007, and Jacobsen et al. 2014) and it is likely that measurements of plant hydraulics conducted at a single sampling time may not, in some cases, represent the function of a plant throughout the growing season. This may complicate the experimental design of hydraulic studies, but may be necessary to understand dynamic changes in the cohort of active vessels and the ability for xylem function to change.

5.3 *Active Xylem Staining Confirms Xylem Vulnerability to Water Stress*

Traditional staining protocols may provide the means to study xylem structure and function in a manner that avoids or corrects for most of the methodological issues described above. Staining allows for the examination of the distribution of hydraulically functional vessels and the identification of vessels that remain conductive during exposure to water stress treatments.

We examined the use of staining in grapevine as a means to examine which vessels remain conductive as plants were dehydrated. We collected 12 large branches (>3 m) from a single hydrated field-grown plant to reduce experimental variability and we divided these branches between two different water stress treatments: centrifugation or benchtop dehydration (six large shoots per treatment with three treated and three control). We included both benchtop dehydrated and centrifuged samples as an additional comparison of dehydration versus centrifuge methods for studying cavitation resistance in grapevine. All treatment samples were exposed to approximately −0.5 MPa, which is near the pressure at 50 % loss in hydraulic conductivity for the grapevine variety used for this experiment (Jacobsen and Pratt 2012). Control samples were collected from matched large branches that were kept hydrated, but were allowed to sit for the same amount of time as treated branches between collection and staining.

Following a water stress treatment through centrifugation of a 0.14 m stem segment or benchtop dehydration of a >3 m large shoot, one 0.10 m sample was excised and stained for active xylem vessels from each treated large branch and from a matched control branch. The 0.10 m sample was sampled from the center of centrifuged segments and from >1 m from the base of dehydrated shoots. Samples were taken from >1 m away from the cut end to avoid measurement of introduced emboli as well as potential wounding effects from sampling near the cut end (Sun et al. 2006). Sections were stained using crystal violet dye (1 % w/v) that was pulled up through segments using a mild tension (−2 kPa). Only vessels that were not embolized or otherwise blocked (i.e., by gels, tyloses, or closed end walls) were able to transport this dye, so that hydraulically active vessels could be easily identified as having purple stained secondary walls.

With this comparison, using traditional staining, we found that approximately 33 % of the vessels in hydrated control samples were not hydraulically active and were presumably developing and living or filled with gels or tyloses (Fig. 5.12a, b; 32.6 ± 1.8 % of vessels did not stain purple, $n = 6$). This is consistent with the pattern of vessel element development described above and the proportion of living vessels that we have previously found.

The percentage of active vessels was determined for each treated sample, using the control sections as a baseline for determining which vessels would have been hydraulically active prior to exposure to water stress so as not to overestimate the loss in conductivity. The estimated percentage loss in active vessels was not different between dehydrated (Fig. 5.12c, d) and centrifuged (Fig. 5.12e, f) stem samples

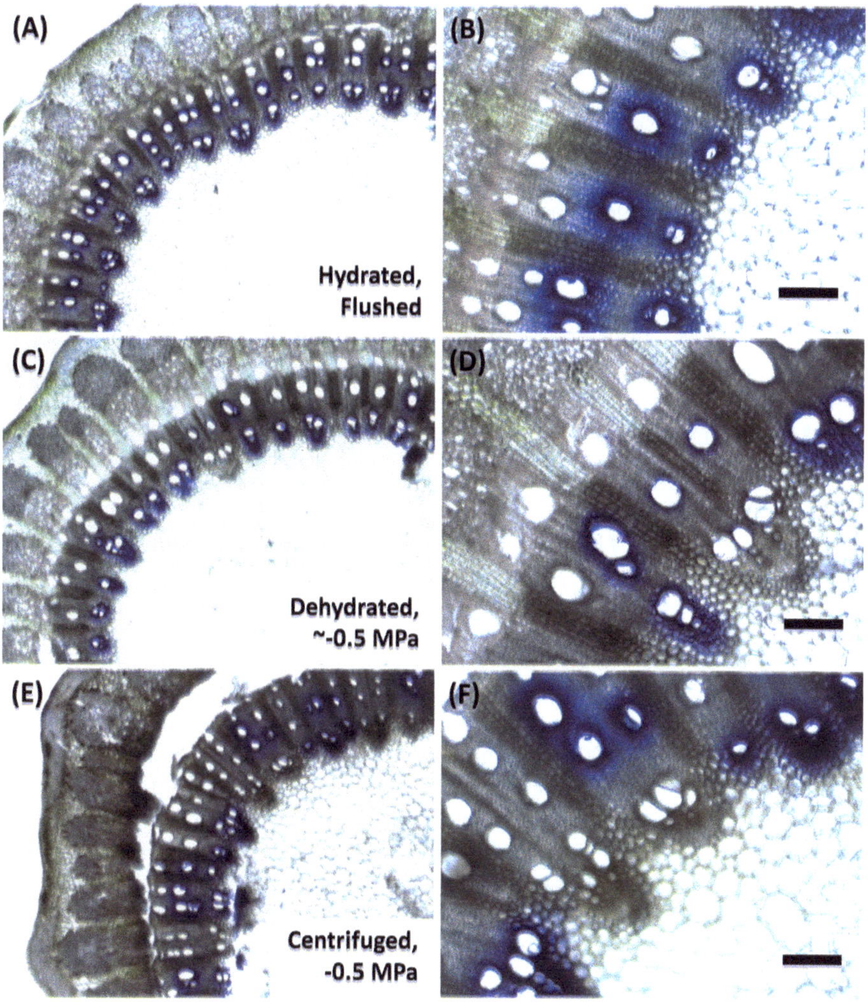

Fig. 5.12 Representative segments from approximately 8 mm diameter stems of grapevine that have been stained for active xylem vessels using crystal violet (purple) shown at both low ((**a**, **c**, **e**); 25×) and higher ((**b**, **d**, **f**); 100×; scale bars indicate 200 μm) magnification. Large branches (>3 m) were collected from a hydrated field-grown plant. Control segments were excised from hydrated branches, flushed, and then stained to determine which vessels were potentially active (**a**, **b**). Approximately 33 % of the vessels in hydrated control samples were not hydraulically active and were presumably developing or living ((**a**, **b**); 32.6 ± 1.8 % of vessels did not stain purple, $n = 6$). These vessels are visible predominantly near the cambium and are not stained purple (**a**, **b**). Additional segments were exposed to two treatments prior to staining, dehydration (**c**, **d**) of large branches or centrifugation (**e**, **f**), to a water potential of approximately −0.5 MPa ($n = 3$ per treatment). Segments from treated branches were then stained to determine which vessels remained functional (**c**, **d**, **e**, **f**). The percentage of active vessels was determined for each treated sample, assuming that if any vessels within a ring of vessels were active in control segments, then all of the vessels within that ring had previously been active. The estimated percentage loss of active vessels, when living vessels were excluded, was similar between dehydrated and centrifuged stem samples and this pattern was consistent among the three samples of each that were examined (54.2 ± 2.2 % and 60.9 ± 3.2 % of previously hydraulically active vessels were no longer active for dehydrated and centrifuged samples, respectively, $n = 3$ per treatment). Stem segments were sampled from a field grown plant of *V. vinifera* "Glenora" in April–May 2012 and sampling time was carefully matched across treatments and controls

(54.2 ± 2.2 % and 60.9 ± 3.2 % of previously hydraulically active vessels were no longer active for dehydrated and centrifuged samples, respectively, mean ± 1 SE; $P = 0.164$, $t_4 = -1.704$, $n = 3$ per treatment).

The agreement of centrifuged and dehydrated samples when stained using this standard technique suggests that: (1) grapevine is indeed highly susceptible to water stress, (2) centrifuge and dehydration treatments result in the same impact on hydraulically conductive vessels when measures are conducted in a way that avoids the errors that may result from PLC-based estimates (Choat et al. 2010) or from artifacts associated with the interpretation of images (McElrone et al. 2012), and (3) many vessels within grapevine xylem are not potentially hydraulically active at any given time. This also agrees with the findings of Sperry et al. (2012), Tobin et al. (2013), and Hacke et al. (2015) that centrifuge and dehydration measures generally agree.

5.4 *Open Vessels Do Not Impact the Reliability of Standard Hydraulic Measures*

The presence of open vessels (i.e., vessels that do not contain a terminal vessel element along their length within a measured sample) and how they may impact hydraulic measures has been an additional issue of concern to some (McElrone et al. 2004; Choat et al. 2010; Cochard et al. 2010; McElrone et al. 2012). While some vulnerability curve methods, such as the cavitron, appear to be prone to vessel-length associated measurement artifacts (Cochard et al. 2010; Wang et al. 2014), the standard centrifuge method has been rigorously tested for short- and long-vesselled samples and these studies have found no evidence for an open-vessel artifact when the standard, Alder et al. (1997), rotor design was used (Jacobsen and Pratt 2012; Sperry et al. 2012; Tobin et al. 2013; Jacobsen et al. 2014; Hacke et al. 2015). This includes careful methods comparisons conducted for grapevine, including data presented in the current chapter as well as in Jacobsen and Pratt (2012) and Tobin et al. (2013).

Measurements of hydraulic conductivity in grapevine are unaffected by the presence of open vessels when even a large number of vessels (40 %) are open. We measured the maximum vessel length on 1-year-old *V. vinifera* shoots using the air injection method (Greenidge 1952) ($n = 7$). We then collected 4 large 1-year-old branches, transported them to the laboratory, and excised stem segments greater than the longest measured maximum vessel length (Fig. 5.13a). Stem segments 1.05 m were vacuum flushed for 1 h using custom tubes filled with the same solution described above as being used for flushing samples and then both ends were trimmed to an initial measurement length of 1.00 m. These large segments were mounted into a tubing apparatus and both hydraulic resistance and conductivity determined. The segments were then shortened from the proximal end, with these same hydraulic parameters being measured at different lengths, until a final length of 0.05 m was obtained. This represented a length at which approximately 40 % of vessels were open through measured samples (Fig. 5.13a).

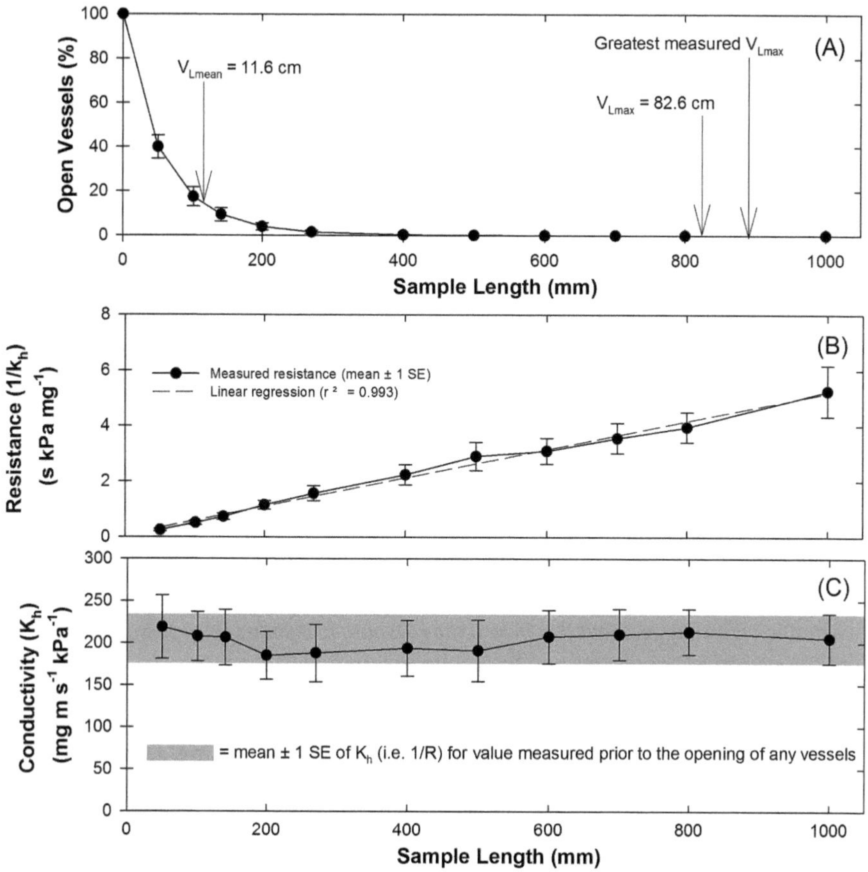

Fig. 5.13 One-year-old shoots of grapevine collected from the same individual and used to determine the proportion of open vessels using silicon injection ($n=6$) for the vessel length distribution (V_{Lmean}) and air injection ($n=7$) for the maximum vessel length (V_{Lmax}) (**a**) and to determine the hydraulic resistance (**b**) and conductivity (**c**) with varying sample length ($n=4$). The mean V_{Lmax} and the single greatest V_{Lmax} from among the measured samples and the mean V_{Lmean} are indicated. Resistance declined linearly as samples were shortened from 1 m to 0.05 m (**b**), while conductivity remained constant (**c**)

Hydraulic conductivity was not different between samples that had no open vessels and those that contained approximately 40 % of their vessel open (Fig. 5.13c). As samples were shortened, the conductivity remained constant (Fig. 5.13c). The resistance, which does not include path length as part of the calculation of this parameter, declined linearly with sample length (Fig. 5.13b) as expected and this also indicates that the opening of increasing numbers of vessels was not affecting resistance.

This result is similar to previously published studies, including a study that examined the effect of segment length on a different liana species and also found no effect of open vessels when segments were shortened from no vessels open to approximately 40 % of vessels open (Chiu and Ewers 1993). When samples are very short (less than 1–2 cm in length) and have many vessels open, conductivity has been shown to be impacted (Sperry et al. 2005), but samples this short are generally not utilized in hydraulic experiments and perhaps should be avoided because of potential impacts on conductivity estimates. Thus, for samples within the range of those typically measured (>5 cm), hydraulic conductivity estimates are reliable and are not impacted by the presence of open vessels.

6 Conclusion

Study of the xylem structure and function of grapevines presents unique challenges, but has the potential to increase our understanding of plant hydraulics, particularly through increased appreciation for the dynamic nature of processes occurring within xylem vessels. Vessels are not static and go through a lifespan that includes a period of prefunctional development, a limited period of hydraulic function, and a period of postfunctionality that may include the formation of either gels or tyloses within the vessel lumen. Large seasonal changes in the cohort of active vessels and plant hydraulic traits suggest that the xylem may be able to dynamically respond throughout the season as conditions and hydraulic requirements change.

Acknowledgments Two California State University, Bakersfield (CSUB) Research Council of the University (RCU) grants (awarded in 2010 and 2012) and NSF-IOS 1252232 awarded to ALJ are thanked for providing support to student researchers. Jacqueline M. Lopez, Estevan R. Ramirez, and Ryan S. Romero, participants in the summer 2014 CSUB Research Experience Vitalizing Science University Program (REVS-UP), are thanked for their assistance in measuring maximum vessel length in several grapevine varieties.

References

Akachuku AE (1987) A study of lumen diameter variation along the longitudinal axis of wood vessels in *Quercus rubra* using cinematography. IAWA Bulletin 8:41–45

Alder NN, Pockman WT, Sperry JS, Nuismer S (1997) Use of centrifugal force in the study of xylem cavitation. J Exp Bot 48:665–674

Almeida-Rodriguez AM, Hacke UG (2012) Cellular localization of aquaporin mRNA in hybrid poplar stems. Am J Bot 99:1249–1254

Alsina MM, de Herralde F, Aranda X, Savé R, Biel C (2007) Water relations and vulnerability to embolism are not related: experiments with eight grapevine cultivars. Vitis 46:1–6

Barnett JR (1982) Plasmodesmata and pit development in secondary xylem elements. Planta 155:251–260

Benayoun J, Catesson AM, Czaninski Y (1981) A cytochemical study of differentiation and breakdown of vessel end walls. Ann Bot 47:687–698

Bollhöner B, Prestele J, Tuominen H (2012) Xylem cell death: emerging understanding of regulation and function. J Exp Bot 63:1081–1094

Brodersen CR, McElrone AJ, Choat B, Matthews MA, Shackel KA (2010) The dynamics of embolism repair in xylem: in vivo visualizations using high-resolution computed tomography. Plant Physiol 154:1088–1095

Broderson CR, Lee EF, Choat B, Jansen S, Phillips RJ, Shackel KA, McElrone AJ, Matthews MA (2011) Automated analysis of three-dimensional xylem networks using high-resolution computed tomography. New Phytol 191:1168–1179

Brodersen CR, Choat B, Chatelet DS, Shackel KA, Matthews MA, McElrone AJ (2013) Xylem vessel relays contribute to radial connectivity in grapevine stems (*Vitis vinifera* and *V. arizonica*; Vitaceae). Am J Bot 100:314–321

Chaffey NJ, Barnett JR, Barlow PW (1997) Cortical microtubule involvement in bordered pit formation in secondary xylem vessel elements of *Aesculus hippocastanum* L. (Hippocastanaceae): a correlative study using electron microscopy and indirect immunofluorescence microscopy. Protoplasma 197:67–75

Chamberlain CJ (1901) Methods in plant histology. University of Chicago, Chicago

Chatelet DS, Matthews MA, Rost TL (2006) Xylem structure and connectivity in grapevine (*Vitis vinifera*) shoots provides a passive mechanism for the spread of bacteria in grape plants. Ann Bot 98:484–494

Chiu S-T, Ewers FW (1993) The effect of segment length on conductance measurements in *Lonicera fragrantissima*. J Exp Bot 44:175–181

Choat B, Drayton WM, Brodersen C, Matthews MA, Shackel KA, Wada H, McElrone AJ (2010) Measurement of vulnerability to water stress-induced cavitation in grapevine: a comparison of four techniques applied to long-vesseled species. Plant Cell Environ 33:1502–1512

Choné X, van Leeuwen C, Du Dobourdieu D, Gaudillére JP (2001) Stem water potential is a sensitive indicator of grapevine water status. Ann Bot 87:477–483

Cochard H, Delzon S, Badel E (2014) X-ray microtomography (micro-CT): a reference technology for high-resolution quantification of xylem embolism in trees. Plant Cell Environ DOI:10.1111/pce.12391

Cochard H, Herbette S, Barigah T, Badel E, Ennajeh M, Vilagrosa A (2010) Does sample length influence the shape of xylem embolism vulnerability curves? A test with the Cavitron spinning technique. Plant, Cell Environ 33:1543–1552

Domec J-C, Johnson DM (2012) Does homeostasis or disturbance of homeostasis in minimum leaf water potential explain the isohydric versus anisohydric behavior of *Vitis vinifera* L. cultivars? Tree Physiol 32:245–248

Drayton WM (2009) Embolism and stem hydraulic conductivity in cultivated grapevine. MS Thesis. University of California, Davis, p 51

Eames AJ, MacDaniels LH (1947) An introduction to plant anatomy, 2nd edn. McGraw-Hill, New York

Esau K (1936) Vessel development in celery. Hilgardia 10:479–488

Esau K (1953) Plant anatomy. Wiley, New York

Esau K, Charvat I (1978) On vessel member differentiation in the bean (*Phaseolus vulgaris* L.). Ann Bot 42:665–677

Esau K, Hewitt WB (1940) Structure of end walls in differentiating vessels. Hilgardia 13:229–244

Ewers FW, Fisher JB (1989) Variation in vessel length and diameter in stems of six tropical and subtropical lianas. Am J Bot 76:1452–1459

Ewers FW, Fisher JB, Chiu S-T (1990) A survey of vessel dimensions in stems of tropical lianas and other growth forms. Oecologia 84:544–552

Flexas J, Escalona JM, Medrano H (1999) Water stress induces different levels of photosynthesis and electron transport rate regulation in grapevines. Plant, Cell Environ 22:39–48

Fukuda H (1997) Tracheary element differentiation. Plant Cell 9:1147–1156

Greenidge KNH (1952) An approach to the study of vessel length in hardwood species. Am J Bot 39:570–574

Groover A, Jones AM (1999) Tracheary element differentiation uses a novel mechanism coordinating programmed cell death and secondary wall synthesis. Plant Physiol 119:375–384

Hacke UG, Venturas MD, MacKinnon ED, Jacobsen AL, Sperry JS, Pratt RB (2015) The standard centrifuge method accurately measures vulnerability curves of long-vesselled olive stems. New Phytol 205:116–127. doi:10.1111/nph.13017

Halis Y, Djehichi S, Senoussi MM (2012) Vessel development and the importance of lateral flow in water transport with developing bundles of current-year shoots of grapevine (Vitis vinifera L.). Trees 26:705–714

Handley WRC (1936) Some observations on the problem of vessel length determination in woody dicotyledons. New Phytol 35:456–471

Holbrook NM, Ahrens ET, Burns MJ, Zwieniecki MA (2001) *In vivo* observation of cavitation and embolism repaid using magnetic resonance imaging. Plant Physiol 126:27–31

Jacobsen AL, Pratt RB (2012) No evidence for an open vessel effect in centrifuge-based vulnerability curves of a long-vesselled liana (*Vitis vinifera*). New Phytol 194:982–990

Jacobsen AL, Pratt RB, Davis SD, Ewers FW (2007) Cavitation resistance and seasonal hydraulics differ among three arid Californian plant communities. Plant Cell Environ 30:1599–1609

Jacobsen AL, Pratt RB, Davis SD, Tobin MF (2014) Geographic and seasonal variation in chaparral vulnerability to cavitation. Madroño 61:317–327

Jacobsen AL, Pratt RB, Tobin MF, Hacke UG, Ewers FW (2012) A global analysis of xylem vessel length in woody plants. Am J Bot 99:1583–1591

Jaillon O, Aury J-M, Noel B, Policriti A, Clepet C, Casagrande A, Choisne N et al (2007) The grapevine genome sequence suggests ancestral hexaploidization in major angiosperm phyla. Nature 449:463–467

Jansen RK, Kaittanis C, Saski C, Lee S-B, Tomkins J, Alverson AJ, Daniell H (2006) Phylogenetic analyses of *Vitis* (Vitaceae) based on complete chloroplast genome sequences: effects of taxon sampling and phylogenetic methods on resolving relationships among rosids. BMC Evol Biol 6:32. doi:10.1186/1471-2148-6-32

Jaquish LL, Ewers FW (2001) Seasonal conductivity and embolism in the roots and stems of two clonal ring-porous trees, *Sassafras albidum* (Lauraceae) and *Rhus typhina* (Anacardiaceae). Am J Bot 88:206–212

Kolb KJ, Davis SD (1994) Drought tolerance and xylem embolism in co-occurring species of coastal sage and chaparral. Ecology 75:648–659

Kolb KJ, Sperry JS (1999) Transport constraints on water use by the Great Basin shrub, *Artemisia tridentata*. Plant Cell Environ 22:925–935

Loveys BR, Kriedemann PE (1973) Rapid changes in abscisic acid-like inhibitors following alterations in vine leaf water potential. Physiol Planta 28:476–479

Lovisolo C, Schubert A (1998) Effects of water stress on vessel size and xylem hydraulic conductivity in *Vitis vinifera* L. J Exp Bot 49:693–700

Lovisolo C, Perrone I, Hartung W, Schubert A (2008) An abscisic acid-related reduced transpiration promotes gradual embolism repair when grapevines are rehydrated after drought. New Phytol 180:642–651

Lovisolo C, Perrone I, Carra A, Ferrandino A, Flexas J, Medrano H, Schubert A (2010) Drought-induced changes in development and function of grapevine (*Vitis* spp.) organs and in their hydraulic and non-hydraulic interactions at the whole-plant level: a physiological and molecular update. Funct Plant Biol 37:98–116

Matthews MA, Anderson MM (1988) Fruit ripening in *Vitis vinifera* L.: responses to seasonal water deficits. Am J Enology Viticulture 39:313–320

McElrone AJ, Pockman WT, Martinez-Vilalta J, Jackson RB (2004) Variation in xylem structure and function in stems and roots of trees to a 20 m depth. New Phytol 163:507–517

McElrone AJ, Broderson CR, Alsina MM, Drayton WM, Matthews MA, Shackel KA, Wada H, Zufferey V, Choat B (2012) Centrifuge technique consistently overestimates vulnerability to water-stress induced cavitation in grapevines as confirmed with high resolution computed tomography. New Phytol 196:661–665

Meylan BA, Butterfield BG (1981) Perforation plant differentiation in the vessels of hardwoods. In: Barnett JR (ed) Xylem cell development. Castle House, Kent

Murmanis LL (1978) Breakdown of end walls in differentiating vessels of secondary xylem in *Quercus rubra* L. Ann Bot 42:679–682

Myles S, Chia J-M, Hurwitz B, Simon C, Zhong GY et al (2010) Rapid genomic characterization of the genus *Vitis*. PLoS ONE 5(1):e8219. doi:10.1371/journal.pone.0008219
Pérez-Donoso AG, Greve LC, Walton JH, Shackel KA, Labavitch JM (2007) *Xylella fastidiosa* infection and ethylene exposure result in xylem and water movement disruption in grapevine shoots. Plant Physiol 143:1024–1036
Ruzin SE (1999) Plant microtechnique and microscopy. Oxford University, New York
Salleo S, Lo Gullo MA, Siracusano L (1984) Distribution of vessel ends in stems of some diffuse- and ring-porous trees: the nodal regions as 'safety zones' of the water conducting system. Ann Bot 54:543–552
Schubert A, Lovisolo C, Peterlunger E (1999) Shoot orientation affects vessel size, shoot hydraulic conductivity and shoot growth rate in *Vitis vinifera* L. Plant Cell Environ 22:197–204
Schultz HR (2003) Differences in hydraulic architecture account for near-isohydric and anisohydric behavior of two field-grown *Vitis vinifera* L. cultivars during drought. Plant Cell Environ 26:1393–1405
Schultz HR, Matthews MA (1988) Resistance to water transport in shoots of *Vitis vinifera* L.: relation to growth at low water potential. Plant Physiol 88:718–724
Sperry JS, Holbrook NM, Zimmermann MH, Tyree MT (1987) Spring filling of xylem vessels in wild grapevine. Plant Physiol 83:414–417
Sperry JS, Hacke UG, Wheeler JK (2005) Comparative analysis of end wall resistivity in xylem conduits. Plant Cell Environ 28:456–465
Sperry JS, Christman MA, Torres-Ruiz JM, Taneda H, Smith DA (2012) Vulnerability curves by centrifugation: is there an open vessel artefact, and are 'r' shaped curves necessarily invalid? Plant Cell Environ 35:601–610
St. Aubin G, Canny MJ, McCully ME (1986) Living vessel elements in the late metaxylem of sheathed maize roots. Ann Bot 58:577–588
Sun Q, Rost TL, Matthews MA (2006) Pruning-induced tylose development in stems of current-year shoots of *Vitis vinifera* (Vitaceae). Am J Bot 93:1567–1576
Sun Q, Rost TL, Reid MS, Matthews MA (2007) Ethylene and not embolism is required for wound-induced tylose development in stems of grapevines. Plant Physiol 145:1629–1636
Sun Q, Rost TL, Matthews MA (2008) Wound-induced vascular occlusions in *Vitis vinifera* (Vitaceae): tyloses in summer and gels in winter. Am J Bot 95:1498–1505
Sun Q, Sun Y, Walker MA, Labavitch JM (2013) Vascular occlusions in grapevines with Pierce's disease make disease symptom development worse. Plant Physiol 161:1529–1541
Thorne ET, Young BM, Young GM, Stevenson JF, Labavitch JM, Matthews MA, Rost TL (2006) The structure of xylem vessels in grapevine (Vitaceae) and a possible passive mechanism for the systemic spread of bacterial disease. Am J Bot 93:497–504
Tibbetts TJ, Ewers FW (2000) Root pressure and specific conductivity in temperate lianas: exotic *Celastrus orbiculatus* (Celastraceae) vs. native *Vitis riparia* (Vitaceae). Am J Bot 87:1272–1278
Tobin MF, Pratt RB, Jacobsen AL, De Guzman ME (2013) Xylem vulnerability to cavitation can be accurately characterized in a species with long vessels using a centrifuge method. Plant Biol 15:496–504
Turner A, Gallois P, Brown D (2007) Tracheary element differentiation. Annu Rev Plant Biol 58:407–433
Velasco R, Zharkikh A, Troggio M, Cartwright DA, Cestaro A et al (2007) A high quality draft consensus sequence of the genome of a heterozygous grapevine variety. PLoS ONE 2:e1326. doi:10.1371/journal.pone.0001326
Wang R, Zhang L, Zhang S, Cai J, Tyree MT (2014) Water relations of *Robinia pseudoacacia* L.: do vessels cavitate and refill diurnally or are R-shaped curves invalid in *Robinia*? Plant Cell Environ 37:2667–2678. doi:10.1111/pce.12315
Wheeler JK, Sperry JS, Hacke UG, Hoang N (2005) Inter-vessel pitting and cavitation in woody Rosaceae and other vesselled plants: a basis for a safety versus efficiency trade-off in xylem transport. Plant Cell Environ 28:800–812
Yahya R, Koze K, Sugiyama J (2011) Fibre length in relation to the distance from vessels and contact with rays in *Acacia mangium*. IAWA J 32:341–350

Zanne AE, Westoby M, Falster DS, Ackerly DD, Loarie SR, Arnold SEJ, Coomes DA (2010) Angiosperm wood structure: global patterns in vessel anatomy and their relation to wood density and potential conductivity. Am J Bot 97:207–215

Zasada JC, Zahner R (1969) Vessel element development in the earlywood of red oak (*Quercus rubra*). Can J Bot 47:1965–1971

Zimmermann MH, Jeje AA (1981) Vessel-length distribution in stems of some American woody plants. Can J Bot 59:1882–1892

Zimmermann MH, Potter D (1982) Vessel-length distribution in branches, stem and roots of *Acer rubrum* L. IAWA Bulletin 3:103–109

Zufferey V, Cochard H, Ameglio T, Spring J-L, Viret O (2011) Diurnal cycles of embolism formation and repair in petioles of grapevine (*Vitis vinifera* cv. Chasselas). J Exp Bot 62:3885–3894

Zufferey V, Murisier F, Schultz HR (2000) A model analysis of the photosynthetic response of *Vitis vinifera* L. cvs Riesling and Chasselas leaves in the field: I. Interactions of age, light and temperature. Vitis 39:19–26

Chapter 6
Lianas as Structural Parasites

Frank W. Ewers, Julieta A. Rosell, and Mark E. Olson

1 Introduction

Lianas, or woody vines, have been referred to as structural or mechanical parasites (Putz 1984; Stevens 1987). Although they are rooted in the soil, they depend upon host plants or other external objects for their mechanical support. Studies suggest that lianas may be detrimental to the host trees. They receive "free" mechanical support and increase static mechanical loading on the hosts and also increase "sail area" exposed to wind forces. They compete with the host, both above ground for light, and below ground for water and mineral nutrients (Stevens 1987; Schnitzer 2005; Schnitzer et al. 2005). When compared to free-standing species, lianas may be able to provide water and minerals to their leaves without investing as much biomass in stem mechanical tissues; they use wide vessels to apparent hydraulic advantage (Fig. 6.1). They certainly have long stems; their tissue allocation strategy allows for greater stem extension per unit carbon invested in stems than can self-supporting plants (Gartner 1991a, b, c). Another feature of liana stems is they often have cambial variants such as successive cambia or multiple vascular cylinders (Fig. 6.2), which may allow for greater flexibility and effective wound healing following tree fall events (Fisher and Ewers 1989, 1992).

F.W. Ewers (✉)
California State Polytechnic University, Pomona, CA 91768, USA
e-mail: fwewers@cpp.edu

J.A. Rosell
Instituto de Ecología, Universidad Nacional Autónoma de México,
Tercer Circuito s/n de Ciudad Universitaria, México DF 04510, Mexico
e-mail: julieta.rosell@gmail.com

M.E. Olson
Instituto de Biología, Universidad Nacional Autónoma de México,
Tercer Circuito s/n de Ciudad Universitaria, México DF 04510, Mexico
e-mail: molson@ib.unam.mx

U. Hacke (ed.), *Functional and Ecological Xylem Anatomy*,
DOI 10.1007/978-3-319-15783-2_6

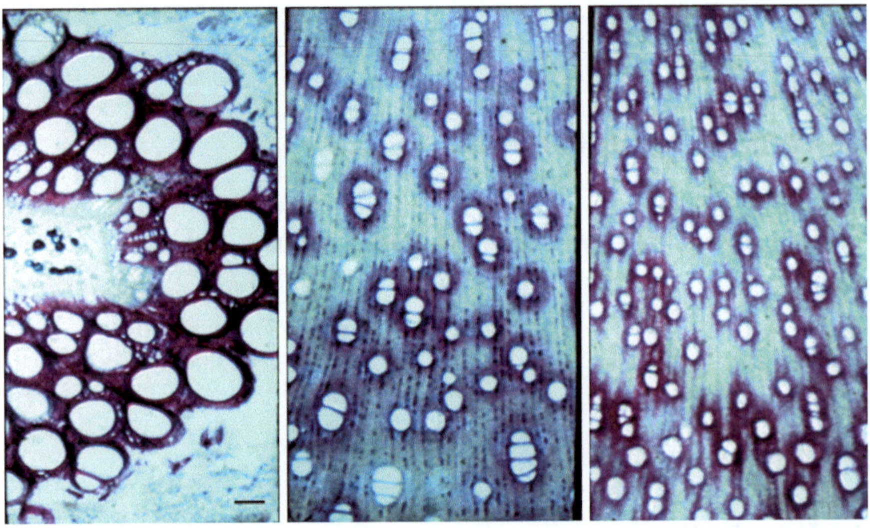

Fig. 6.1 Comparison of basal stem transverse sections of three growth forms within the genus *Bauhinia*. From *left* to *right*, the liana *B. fassoglensis*, the tree *B. blakeana*, and the shrub *B. galpinii*. The stems were each supplying about 3.5 m^2 of leaf area, but the liana was accomplishing this with much less xylem area and much greater maximum vessel diameter. All three at the same magnification, with the scale bar 200 μm. From Ewers et al. (1991)

Fig. 6.2 Lianas often have variant secondary growth, as shown in this transverse section of *Macfadyena unguis-cati* of the Bignoniaceae. The "cross vine" morphology with phloem wedges within the xylem is typical for lianas in this family. Very early in secondary growth (see area close to the pith) four phloem wedges were produced (P); the vascular cambium in those areas produced abundant phloem but very little xylem. Later four additional phloem wedges were produced. Normal portions of the vascular cambium (VC) produced much xylem to the inside (red staining here) and relatively little phloem to the outside of the stem. Such stems are flexible and effective at wound healing. Scale bar is 500 μm

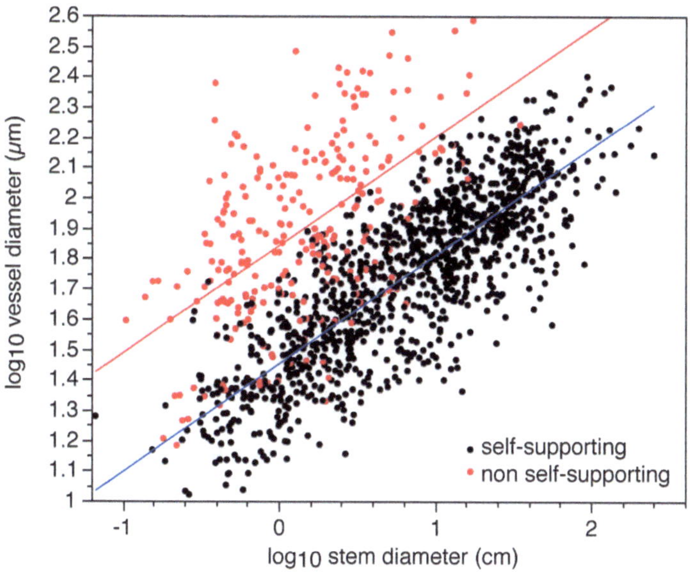

Fig. 6.3 Controlling for stem diameter, non-self-supporting species (mostly lianas), on average have wider vessels than in self-supporting species such as shrubs and trees. Data from 1,047 samples from 528 species from Olson and Rosell (2013), Olson et al. (2014), and Rosell and Olson (2014). $R^2=0.62$, slope 0.36

Most of the discussion in this chapter will be on the hydraulics of liana stems, which are better known than the roots and leaves of lianas. Lianas consistently have very narrow stems in relation to the leaf area they supply (Schenck 1893; Putz 1983; Ewers et al. 1991). Liana stems have exceptionally wide vessels for a given stem diameter as compared to self-supporting plants, often 500 μm or more in maximum diameter in stems less than 20 cm in diameter. We have found examples of individual liana vessels that were as wide as 700 μm (in large specimens of *Cissus hypoglauca* and *Entada phaseoloides*). Similarly, the monocotyledonous "woody" vine *Smilax panamensis* has vessels as wide as 700 μm (Tyree and Ewers 1996). In addition, for a given stem diameter liana vessels are exceptionally long, up to many meters in maximum length. They have much less xylem transverse area and many fewer fibers per unit stem transection than in trees or shrubs supplying similar leaf areas (Ewers et al. 1989, 1991; Ewers and Fisher 1991). Although liana stems have wider vessels than in similar diameter stems of trees and shrubs (Fig. 6.3), much of this can be explained by the relationship between stem length and vessel diameter. Hydraulic optimality models suggest that, for plants in general, longer stems should have wider vessels near the base (West et al. 1999; Petit and Anfodillo 2009; Savage et al. 2010; Sperry et al. 2012). Lianas may be most exceptional in having some of the longest stems in the plant kingdom, up to 200 m or more (Burkill 1966). Lianas are most common and abundant in the tropics and most of the examples in this chapter involve tropical lianas. Grapevines are perhaps the most important and well studied of the temperate lianas, and it serves for a case study in the chapter on "Grapevine xylem development, architecture, and function."

2 Root Systems of Lianas, Root Pressures

There are biomechanical and carbon investment reasons to expect lianas to have different root systems as compared to self-supporting plants. The root systems of self-supporting plants contribute in resisting mechanical loads (e.g., Niklas et al. 2002 and citations therein). Many large trees have deep bayonet, sinker, or tap roots which help anchor the plants and presumably resist the forces resulting from loads such as wind pushing the plants to one side (Fig. 6.4). Moreover, trees usually have multiple laterally spreading roots close to the surface that also likely act in transmitting lateral mechanical loads to the substrate. Many trees, especially tropical trees, also have root buttresses. Given that lianas are supported by the host plant along their lengths, and often trail on the ground before ascending into the canopy (Fig. 6.4), in most cases the mechanical loads transmitted to the root system are negligible. In lianas, freedom from mechanical selection pressures on stems and roots could make large amounts of carbon available for diversion to other structures. One possibility is that with greater carbon available for investment in roots, lianas could have deeper root systems on average than self-supporting plants of similar crown area, and thus proportionally greater access to water. It has also been observed that liana roots tend

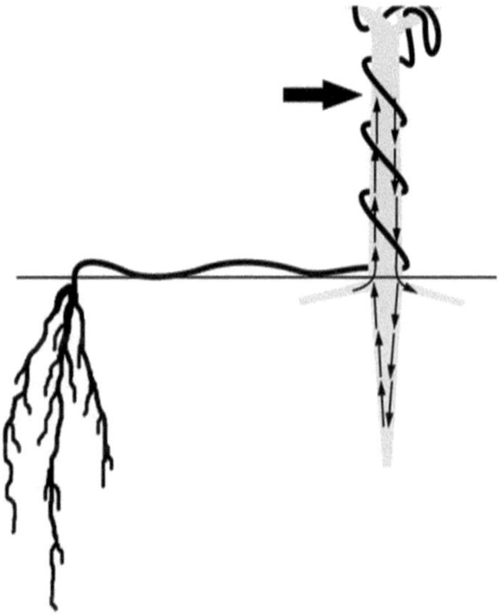

Fig. 6.4 Mechanical loading and potential rooting depths of trees versus lianas. The root system of a tree (gray) includes lateral support roots and vertical bayonet, sinker, or tap roots (tree drawing follows Niklas et al. 2002). These roots help resist mechanical loads, such as wind (*large horizonal arrow*). Tension and compression forces (*small black arrows*) are transmitted to the roots and through them to the ground (*black horizontal line* is ground level). In contrast, in lianas (*black*), essentially no mechanical loads are transmitted from the stem to the roots. Carbon that would otherwise be allocated to support roots can be deployed as resource acquiring roots. The drawing here uses the same number of pixels (carbon) to draw the belowground parts of the tree and the liana. Using a similar amount of "carbon" as the tree, the liana can potentially reach deeper water sources

Table 6.1 Root versus stem vessel diameters, with a comparison of 8 liana species and 11 free-standing species (trees and shrubs) of the Fabaceae

Vessel diameter	Roots ± SE	Stems ± SE
Mean, liana species	121 ± 9	164 ± 22[a]
Mean, free standing	120 ± 9	101 ± 11[a]
Maximum, liana species	240 ± 20	304 ± 42[a]
Maximum, free standing	197 ± 23	164 ± 16[a]

For liana species, vessels were significantly wider (in μm) in woody stems than woody roots. In contrast, for free-standing species, vessels were significantly wider in roots. Liana stems differed from stems of free-standing growth forms, but the root vessel diameters were not different between growth forms. From Ewers et al. (1997b)
[a]Significant difference between stems and roots based upon Wilcoxon's Singed-Rank test at $p < 0.05$

to have larger rays, more axial parenchyma, and more parenchyma in contact with vessels than what occurs in trees (Carlquist 1985a, 1991).

It has been suggested that lianas may have exceptionally deep roots compared to trees (Putz 1984; Schnitzer 2005; Chen et al. 2015), but this is not a consistent trait. Both shallow and deep rooted liana species can be found at a site (Tyree and Ewers 1996), and some lianas have very large basal tubers developed from the hypocotyl or from the root (Brenan 1967; Mooney and Gartner 1991; Hearn 2006). Based upon stable isotope composition of the xylem sap versus soil water at various depths, lianas may obtain their water from deeper soils than the trees they climb upon. However, this varies with the site and the season. In one study, liana species appeared to obtain water from similar depths as did co-occurring trees in both the wet and dry season (Andrade et al. 2005). In another study, at two sites with severe seasonal soil water deficit, lianas used a significantly higher proportion of "deep soil water" during the dry season (from 151 to 250 cm depth) than did co-occurring trees. However, the same study showed that at a third site, which had moist soils even in the dry season, lianas and trees appeared to obtain water from similar depths (Chen et al. 2015).

In terms of vessel diameter, a study of 51 liana species and 11 nonclimbing species (trees and shrubs) showed there was no significant difference between growth forms in the root vessel diameters (sampled woody roots were about 9 mm diameter for both growth forms). This same study showed consistently wider vessels in the stems of lianas than in similar diameter stems of trees and shrubs (Ewers et al. 1997b). A subset of those results, for eight liana species and 11 free-standing species all of the Fabaceae, are summarized in Table 6.1. Unfortunately, vessel lengths in roots of lianas have been little studied due to the general difficulty of sampling long roots of woody plants.

Root pressures are reported to be much more common in climbing plants than in trees (Fisher et al. 1997; Ewers et al. 1997a; Jiménez-Castillo and Lusk 2013). Some of the classic reports of root pressure involved lianas (Hales 1727; White 1938). Root pressures, caused by osmotic water uptake following solute taken into the root stele, are generally greatest during rainy seasons and when transpiration is minimal such as at night, during rainstorms, or just prior to leaf emergence in the case of deciduous lianas (Sperry et al. 1987; Cochard et al. 1994; Fisher et al. 1997; Tibbetts and Ewers 2000). Reported root pressures in lianas (e.g., maximum of 148 kPa near the stem base in *Cissus*, Vitaceae; Fisher et al. 1997) are usually modest, mostly well under 100 kPa, and probably not sufficient to reverse embolism in the upper stems in a forest canopy. Even without leakage or absorption, the gravity gradient would indicate a loss of 10 kPa per m of height. However, root pressures in lianas may be significant in reversing possible embolism in roots and in the lower stems. They may also reflect a high starch storage capacity in roots of lianas, given the high amounts of parenchyma in the roots and the common occurrence of storage tubers (Mooney and Gartner 1991). When the starch is broken down and mobilized, it could contribute to root pressures. However, the common occurrence of root pressures in lianas may also be due to lower capacitance in liana stems relative to trees. In trees, root pressures may be quickly dissipated with height as there is absorption of water under positive pressure into the various tissues of the wide stems. It may be that shallow rooted liana species are more likely to exhibit root pressures than deep rooted species. Root pressures may dissipate along an extended xylem pathway from deep soils.

3 Leaves of Lianas

The leaves of lianas, although highly variable, appear to differ from those of trees, especially in dry tropical forests. Compared to the host trees on which the stems climb, liana leaves can have lower leaf mass per area, they often have greater maximum photosynthetic capacity, they are shorter lived, and they have higher nitrogen and phosphorous content (Cai et al. 2009; Zhu and Cao 2009; Tang et al. 2012). One study showed that leaves of two liana species were more prone to embolism than the tree species that they were climbing upon (Johnson et al. 2013). However, in two studies liana leaves gave very similar results as tree leaves for many physical and physiological parameters (Santiago 2010; van der Sande et al. 2013). And when gas exchange measurements were made the day after a rainstorm in the canopy of a moist tropical forest, sun leaves of lianas underperformed compared to host trees, with lower rates of stomatal conductance and lower rates of carbon assimilation (Santiago and Wright 2007). Another study showed that liana leaves had greater photosynthetic capacity than the host trees at seasonally dry sites, but there was no apparent liana advantage at moist sites (Chen et al. 2015).

4 Stem Xylem Conductivity

Unlike free-standing growth forms, liana stems have both wide vessels and relatively high vessel frequency per mm^2 (Carlquist 1991; Baas et al. 2004). Since the mechanical requirements are much less for liana stems than for free-standing species, it follows that lianas tend to have lower wood density, lower percent fiber area,

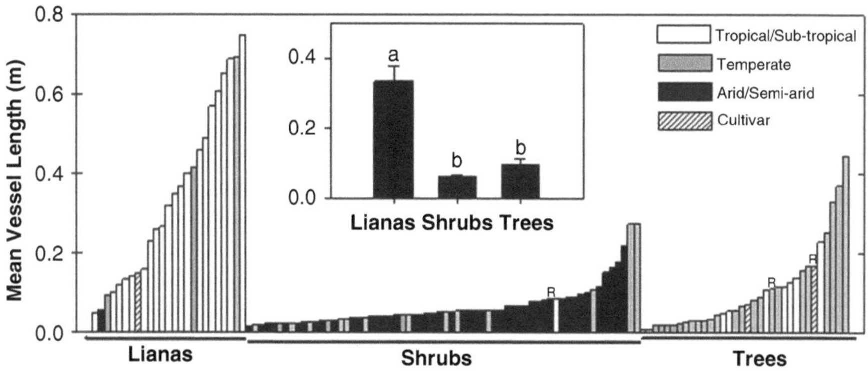

Fig. 6.5 Based upon a meta-analysis, lianas are shown to have long vessels when compared to self-supporting growth forms (shrubs and trees). Each bar represents a different species. From Jacobsen et al. (2012)

and vessel diameters and frequencies that are closer to the theoretical "packing limit" (Crivellaro et al. 2012).

Liana stems are well known to have wider vessels and higher specific conductivity (hydraulic conductivity per xylem transverse area, K_s) than in trees and shrubs of similar stem diameters (Ewers et al. 1989, 1991; Gartner et al. 1990; Chiu and Ewers 1992; Feild and Balun 2008; Zhu and Cao 2009; Jiménez-Castillo and Lusk 2013). Similarly, sap flow gauges have shown lianas to have much higher sap flux density (J_s in g m^{-2} s^{-1}) than trees, at least when soil water availability is high (Chen et al. 2015). Lianas also have greater average vessel lengths (Fig. 6.5) and greater maximum vessel lengths than in free-standing growth forms of similar stem diameters (Ewers and Fisher 1989a, b; Ewers et al. 1990). Given the wide vessel diameters, the high K_s and J_s values in liana stems can be explained by Poiseuille's Law for ideal capillaries. Conductivity should be proportional to vessel diameter to the fourth power. As an illustrative example, in a study of five liana species and nine tree species which served as hosts, mean vessel diameter was 2.2 times higher in the lianas, and the mean K_s was 18.3 times higher (Jiménez-Castillo and Lusk 2013). For perfect capillaries, a 2.2-fold increase in diameter would correspond to a 23-fold increase in conductivity, roughly similar to the 18.3-fold increase observed. As another example, in a common garden study, comparisons of lianas versus free-standing growth forms within the genus *Bauhinia* showed the liana species to have about six times greater K_s, but the free-standing species had about five times greater Huber values. The leaf-specific conductivity, LSC, was similar in the different growth forms (Table 6.2).

Although Poiseuille's Law provides a good first approximation, vessels do not behave as ideal capillaries, of course. For one thing, conductivity appears to be colimited by vessel length as well as vessel diameter; measured conductivity is typically about 50 % of that predicted by Poiseuille's Law. Vessel length is much less measured than vessel diameter, but lianas have among the longest vessels reported in the plant kingdom (Fig. 6.5).

Table 6.2 Xylem transport parameters: comparison of lianas versus free-standing growth forms (trees and shrubs) within the genus *Bauhinia* (Fabaceae), including Huber value (transverse xylem area per leaf area distal to the stem segment), sapwood-specific conductivity (K_s), leaf-specific conductivity (LSC), maximum transpiration rates, and maximum predicted pressure stem xylem pressure gradients (max dp/dx)

Growth form, species	Huber value in 10^{-5} m/m	Specific gravity of xylem	K_s in kg s^{-1} MPa^{-1} m^{-1}	LSC in 10^{-4} kg MPa^{-1} s^{-1} m^{-1}	E max in mmol m^{-2} s^{-1}	Max dp/dx in MPa m^{-1}
Liana						
B. fassoglensis	1.4±0.3	0.18±0.3	171.1±29.7	11.0±0.7	6.10±0.38	0.101
B. vahlii	2.4±0.6	0.40±0.1	35.3±3.7	6.8±0.7	4.45±0.38	0.119
Free standing						
B. aculeate	15.8±6.9	0.65±0.09	6.7±0.7	6.2±0.5	3.41±0.49	0.100
B. blakeana	11.8±3.1	0.33±0.08	14.1±1.9	11.2±1.4	3.24±0.33	0.053
B. galpinii	6.0±0.8	0.49±0.01	11.0±1.9	5.8±0.8	2.64±0.55	0.083
B. variegata	9.4±1.0	0.42±0.03	15.6±1.7	12.3±1.2	5.71±0.60	0.085

$N>12$ shoots per species. From Ewers et al. (1989, 1991)

It has been well established for lianas that within a stem are a wide range of vessel diameters and lengths, with more short and narrow conduits than long and wide ones (Carlquist 1985a; Ewers and Fisher 1989a, b; Ewers et al. 1990). The same may apply, but to a lesser extent, to plants in general (Zimmermann and Jeje 1981; Zimmennann and Potter 1982; Tyree and Zimmermann 2002).

Mean vessel diameter provides only very simplified prediction of the conductivity. The hydraulically weighted vessel diameter (Kolb and Sperry 1999) is a much better predictor, but that is also a simplification since stems also vary in vessel frequency per transverse area. Vessel frequency is generally inversely proportional to mean vessel diameter in plants (Zanne et al. 2010). However, in addition to greater vessel diameter, lianas are sometimes reported to have greater vessel frequency than in related trees and shrubs (Carlquist 1975, 1985a, 1991; Ewers and Fisher 1991; Gasson and Dobbins 1991). All of these sources of variation affect measures of hydraulic performance in lianas.

The K_s values correct for xylem transverse area whereas leaf-specific conductivity, LSC, controls for the leaf area supplied by a stem. Although K_s are consistently high in lianas, the LSC values have been reported as similar or lower in lianas than in trees and shrubs (Ewers et al. 1991; Gartner 1991a; Tyree and Ewers 1996; Feild and Balun 2008; Zhu and Cao 2009). In those studies the only consistent advantage that liana stems had is a low Huber value, that is, low xylem transverse area per leaf area supplied. However, K_s, LSC, and Huber value do not correct for the length of the pathway. If we consider the greater internode elongation for lianas, and the long and rambling pathways that liana stems often take along the ground and through the canopy, the hydraulic advantage for liana stems may disappear. Liana stems can be up to 200 m in length (Burkill 1966), much longer than the pathways that water travels in the host tree. Given the low Huber values there would appear to be little redundancy per leaf area supplied in the liana stem compared to the host trees on which they climb. When liana vessels become embolized the alternate routes for water transport may be severely diminished, unless narrow vessels and tracheids, mentioned below, can partially compensate. In addition, the greater lengths of liana stems can mask the total biomass allocation. In a study comparing lianas with free-standing shrub forms of the same species, the lianas had greater stem lengths but that same above ground biomass (Gartner 1991b, c).

It has been argued that the evolution of the vessel may have been essential to the widespread evolution of the liana habit in angiosperms (Feild et al 2012). Among gymnosperms the most prominent lianas are among the Gnetales, which have vessels that are functionally very similar to those of angiosperms (Fisher and Ewers 1995). Conifers do not include lianas in the usual sense, although there are some prostrate or scrambling conifers. The limitation of tracheids as conduits may be related to the fact that, in a tracheid-based transport system, the tracheid provides mechanical support as well as transport (see chapter on "The hydraulic architecture of conifers"). The mechanical function of tracheids may limit the hydraulic potential of tracheid systems (Pittermann et al. 2006). In contrast, vessel systems depend on fibers for mechanical reinforcement. Xylem fibers are important for stem support for plants in general, but perhaps also for the prevention of implosion of vessels when the water columns are under tension (Carlquist 1975; Hacke et al. 2001; Jacobsen et al. 2005). This is not mainstream thinking, but in lianas fiber presence and distribution may be mostly favored in the prevention of vessel implosion, since the stem support needs are largely provided by the host plant.

5 Stem Length and Vessel Diameter in Lianas

From Nehemiah Grew (1682) to the present day (e.g., Schenck 1893; Ewers et al. 1990; Fisher and Ewers 1995; Fisher et al. 2002; Baas et al. 2004; Gartner et al. 1990; Carlquist 1991; Gartner 1991b; Cai et al. 2009; Isnard and Silk 2009; Zhu and Cao 2009; Angyalossy et al. 2012; Jiménez-Castillo and Lusk 2013; van der Sande et al. 2013; Wyka et al. 2013), virtually every published work on liana anatomy or hydraulics remarks that lianas have exceptionally wide vessels. These studies usually take stem diameter as their basis for comparison, and it is clear that lianas have much wider mean vessel diameters for a given stem diameter than self-supporting plants do. What is not clear is whether stem diameter is always the appropriate metric for making these comparisons.

There are reasons to expect mean vessel diameter to be predicted by stem length across plant habits. The biological significance of Poiseuille's law, published in the mid-1800s, was not lost on plant biologists, who soon invoked notions of hydraulic resistance to explain the wide vessels of lianas (e.g., Westermaier and Ambronn 1881; Haberlandt 1914). Lengthening of a vessel of constant diameter adds fluid-wall friction and therefore hydraulic resistance. As a consequence, with no vessel diameter increase, flow rate would drop as the stem and therefore the conductive path becomes longer. Vessels tend to be narrow closer to stem apices, and even narrower in terminal leaf veins. The average diameter of vessels widens with distance from the stem tip. The phenomenon is readily seen in transverse view as a radial increase in vessel diameter in going from the pith out to the vascular cambium. Narrow vessels near the stem center correspond to when the stem was short (see Mencuccini et al. 2007). Although often attributed to Sanio (1872), the pattern of conduit increase with growth was first described by Grew (1682). Although the idea that conductive path length should be related to vessel diameter has been present in the literature for a long time (Westermaier and Ambronn 1881; Haberlandt 1914; Ewart 1904–1905; Jost 1907; Cowan 1965), it was not modeled until fairly recently.

The vessel diameter–stem length relationship has been modeled thoroughly by West et al. (1999), Becker and Gribben (2001), McCulloh et al. (2003), Mencuccini (2002), Enquist (2003), Mencuccini et al. (2007), Coomes et al. (2007), Petit and Anfodillo (2009), Petit and Anfodillo (2011), Savage et al. (2010), Sperry et al. (2012), and von Allmen et al. (2012). Recent data seem congruent with the expectation that conductive path length, as reflected by stem length, predicts average basal vessel diameter in both self-supporting and non-self-supporting plants. In their comparative study of mean vessel diameter–stem length relations, Anfodillo et al. (2006) included one non-self-supporting plant, *Hedera helix*. They plotted the average hydraulically weighted conduit diameter against stem length across 31 species.

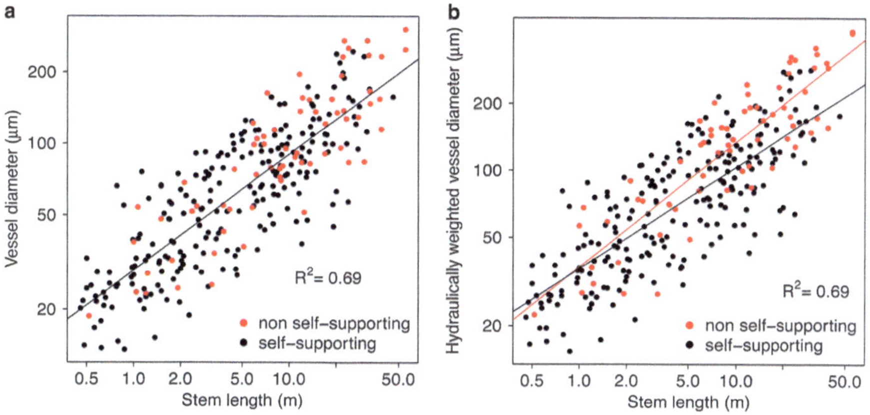

Fig. 6.6 The relationship between mean vessel diameter and stem length. Lianas are traditionally regarded as having wide vessels, but they have the same mean natural (nonhydraulically weighted) vessel diameter that self-supporting plants do for a given stem length (**a**). Because lianas tend to have greater maximum and minimum vessel diameters (greater extremes within a stem), their hydraulic vessel diameter tends to be very slightly greater than in self-supporting plants when stem length is controlled for (**b**). After Rosell and Olson (2014)

They found that mean vessel diameter was predictable given stem length, with taller plants having predictably wider vessels basally. *Hedera* fell where it would be expected to given its stem length. Rosell and Olson found that the slope of the vessel diameter–stem length relationship was the same in both lianas and self-supporting plants (Fig. 6.6a). Moreover, for a given stem length, lianas had slightly greater average hydraulically weighted diameters as compared to self-supporting plants. This appears to be achieved by a widening of the vessel diameter variance in lianas. That is, lianas have more extremes in vessel diameter within a stem. The occurrence of just a few wider vessels is sufficient to make the mean hydraulically weighted vessel diameter of lianas slightly wider than those of similar length self-supporting plants. In contrast to the common perception, when standardized by stem length, the difference in mean vessel diameter between lianas and self-supporting plants is very subtle (Fig. 6.6b). This comparison underscores the need to standardize by stem length in morphological and physiological studies of the xylem (see Sect. 6.11). In addition, this result highlights an often overlooked aspect of lianas, mentioned earlier, their abundance of narrow vessels.

6 The Importance of Narrow Vessels and Tracheids in Lianas

One of the interesting/complicating things about liana stems is that they tend to have an exceptionally high number of narrow vessels in addition to their wide vessels (Carlquist 1985a; Ewers and Fisher 1989a, b; Ewers et al. 1990), as illustrated by an example in Fig. 6.7. In addition, conductive imperforate cells (various types of tracheids) are common in liana stems.

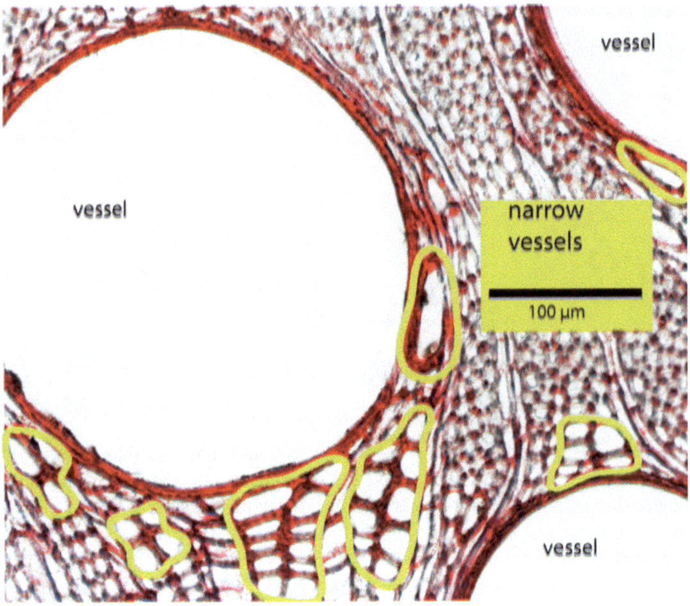

Fig. 6.7 In addition to wide vessels (portions of three are shown here), lianas tend to have a large number of narrow vessels, as in this stem of *Gouania lupuloides* (Rhamnaceae). The narrow vessels (*circled* in *yellow*) may be important for redundancy of the hydraulic pathway. The narrow vessels and fibers may also function in mechanical support of the transport system, preventing implosion under water stress

The narrowest vessels in a stem can be difficult to detect without corroborative observations with longitudinal sections or macerations. Narrow vessels may be confused with tracheids or even fibers in some cases. Figure 6.8a, b shows different tissue allocations in a liana and tree stem, and the impact of narrow vessels on measurements of mean, median, and hydraulically weighed mean vessel diameters in a tree and a liana. Narrow vessels of lianas can lead to a misleading value for mean vessel diameter; hydraulic means should be used if the interest is in conductive potential. However, as noted by Carlquist (1985a, b) the narrow vessels may be important for hydraulic redundancy and alternate pathways in the event of vessel dysfunction. Narrow vessels may also have a mechanical role, along with the fibers, in helping to prevent vessel implosion by the wide vessels.

Conductive imperforate tracheary elements have been classified as true tracheids, vasicentric tracheids, and vascular tracheids (Carlquist 1985b; Rosell et al. 2007). Vasicentric tracheids (see also the chapter on "Integrative xylem analysis of chaparral shrubs") are particularly common in lianas and are found in vessel groupings, often large ones, apparently serving as maximally safe conductive cells that both connect vessels laterally as well as maintain a minimal conductive stream in case of embolization of the wide vessels. In contrast to the large groups of vessels and vasicentric tracheids in lianas, when true tracheids are present, vessels are solitary as seen in transverse view.

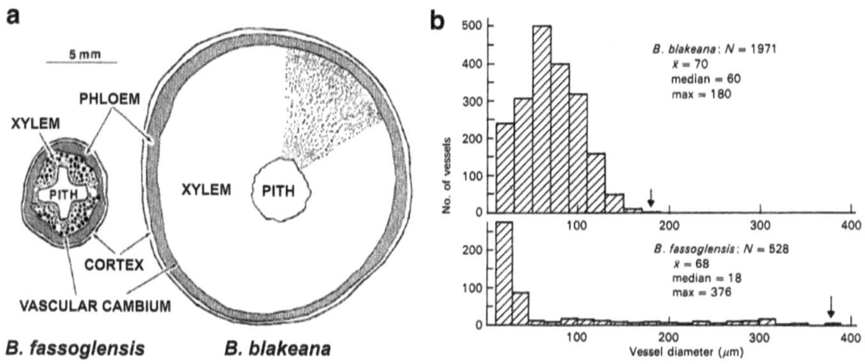

Fig. 6.8 Stem cross sections of a stem of a liana (*Bauhinias. fassoglensis*) and a tree (*B. blakeana*) each supplying the same leaf area. (**a**) Comparison of tissue distributions of stems, arrow indicates maximum vessel diameter. (**b**) Comparison of vessel diameter frequency distributions from same stems. Note that in this case the mean vessel diameter was essentially the same, but the maximum vessel diameter was much greater in the liana. The hydraulic mean vessel diameter was 112 μm for the tree versus 286 μm for the liana. (**a**) is From Ewers and Fisher (1991), (**b**) is from Ewers et al. (1991)

7 Hormonal Mechanisms of Vascular Differentiation and Hydraulic Adaptation

The global pattern of tip-to-base vessel widening that occurs across the flowering plants is consistent with what is known about the hormonal control of vascular development. Vessels appear to develop in the context of an auxin gradient along the length of stems. Auxin is apparently produced in abundance in developing leaves (Uggla et al. 1998; Scarpella and Meijer 2004), and directional movement of auxin along the sites of differentiating secondary xylem cells establishes a high–low concentration gradient leaf to root (Aloni and Zimmermann 1983; Aloni 1987; Lovisolo et al. 2002). In the case of basipetal vessel widening, lower concentrations of auxin correlate with the developmental outcome of wider vessels. Wider conduits appear to develop because of a longer time of development rather than an accelerated widening rate (Anfodillo et al. 2012, 2013).

Models of vascular differentiation highlight missing developmental information vital to testing hypotheses of the adaptive causes of tip-to-base vessel widening. It is clear that there must be *some* developmental mechanism controlling vessel diameter, but the particulars are hardly important when trying to elucidate *why* the patterns are observed. Even if every species of plant on the planet had a completely different developmental mechanism, the adaptive prediction of West et al. (1999) would remain valid, as would efforts to test it with anatomical data. In fact, the majority of studies examining vessel-stem size scaling, such as West et al. (1999), Becker and Gribben (2001), McCulloh et al. (2003), Mencuccini (2002), Mencuccini et al. (2007), Coomes et al. (2007), Savage et al. (2010), Gleason et al. (2012), Sperry et al. (2012), and von Allmen et al. (2012) cite no developmental information at all and require none. This is because developmental mechanisms do not predict *why* a global vessel diameter–stem diameter relationship should be observed across the angiosperms, much less that a specific exponent should be observed. Natural selection acts on heritable variation between individuals. Establishing the adaptive basis of vessel widening therefore involves understanding the degree to which heritable variation in vessel widening exponents is possible, and to study directly how these different exponents vary in their hydraulic performance (cf. Olson 2012).

8 Lability Between Self-Supporting and Climbing Habits

Depending upon their climbing mode, lianas can be characterized as shoot twiners, tendril climbers, root climbers, and scramblers, the latter of which tend to fall upon their host plants (Reddy and Parthasarathy 2006; Campanello et al. 2007). Lianas may have thorns, spines, or hooks to latch onto the host. The scrambler mode would seem to require the fewest evolutionary steps.

The tendency for mechanically supported stems to become liana like in xylem properties may be inherent to all woody plants, as can be observed by well-known "thigmotrophic" responses. Within a species, or even within a clone, when plants are provided mechanical support by staking the stems, the stems become liana like compared to stems that are either self-supporting or stems subjected to mechanical perturbation. In a wide range of gymnosperms and angiosperms, it has been shown that staked stems are longer and narrower than free-standing stems, with larger leaves (Neel and Harris 1971; Pruyn et al. 2000; Kern et al. 2005). The xylem properties of unperturbed staked stems also become more liana like, with greater K_s greater mean, maximum and hydraulic vessel diameters, and greater vessel lumen area per xylem area. It is of particular interest that in staked plants, although the vessel diameters increase, the vessel frequency per transverse area remains the same (Gartner 1991a; Kern et al. 2005). Selection for a mechanically weaker stem would be concomitant with a longer, narrower axis with wider vessels near the base.

An interesting case study involved a species, *Toxicodendron diversilobum*, which grows as a liana when provided external mechanical support but grows as a shrub when free standing (Gartner 1991a, b, c). Supported stems had lower Huber values, greater K_s, greater vessel diameters, and greater vessel lumen areas than free-standing stems (Gartner 1991a, b). However, the leaf biomass, leaf number, and stem biomass were not different between growth forms. Supported stems were longer and had a more gradual taper. They had greater internode elongation than free-standing stems (Gartner 1991b, c). Apparently the liana habit mostly involved "stretching out" the stem. As discussed in Sect. 6.5, with greater stem lengths, the wider vessels in supported stems would be as expected.

Plastic variation between self- and non-self-supporting forms within species is paralleled by extreme evolutionary liability between self- and non-self-supporting plants. The liana habit has arisen many times in the flowering plants (Fig. 6.9). For example, *Trimenia moorei* is closely related to self-supporting *Trimenia* species, and *Bursera instabilis*, an unusual tree with lianescent branches, is found within an

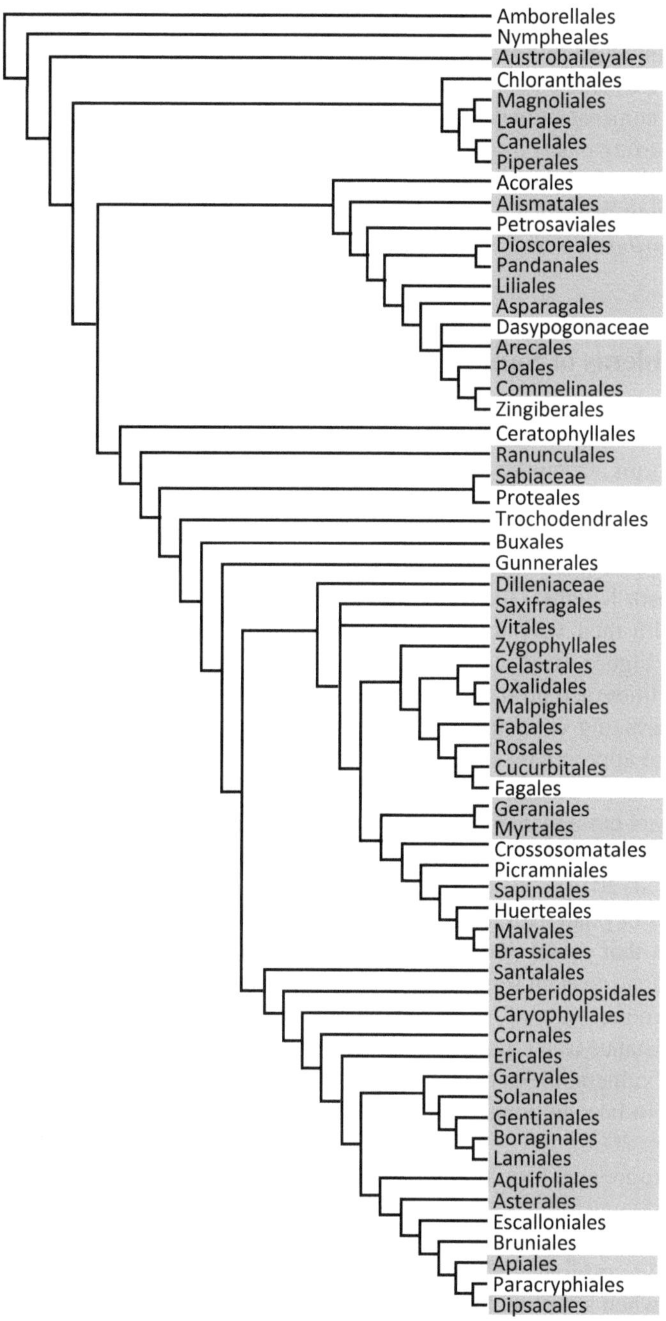

Fig. 6.9 Ordinal distribution of lianas, following Stevens (2001). Orders that include climbing plants are boxed in *gray* to show that they are present in almost all orders of angiosperms. Within orders, it seems that there have been very frequent independent derivations of the liana habit from self-supporting ancestors. Some orders, such as Cucurbitales, Malpighiales, Solanales, and Vitales include self-supporting species that have apparently evolved from liana ancestors

otherwise self-supporting genus (Rosell et al. 2012). Every time lianescence arises, it is accompanied by the attendant shifts in vessel diameter–stem diameter proportions. By the same token, the shift from liana to self-supporting plant seems to have occurred numerous times (e.g., Lahaye et al. 2005). One route to the evolution of self-supporting plants from lianas likely is the cooption of the very abundant storage parenchyma (Carlquist 1985a) to form self-supporting water storing trees, often with twining or tendril-bearing branches (e.g., *Dendrosicyos*, some tree *Ipomoea, Adenia karibaensis*, etc., see Olson 2003; Hearn 2009).

9 Problems of Sampling Vessels in Long Stems

In most cases, stem diameter has been used as a basis for comparison between growth forms. Comparisons based upon supplied leaf area have also been done. However, as noted above, lianas have much longer stems for a given diameter than do self-supporting plants. If mean vessel diameter is a reflection of stem length, then comparisons that are standardized by stem diameter will reflect very different conductive path lengths. Given a constant rate of basipetal vessel widening, then the lianas, with their very long stems, will inevitably have much wider vessels and therefore higher mean vessel diameters, higher conductivities, greater embolism vulnerabilities, and probably longer vessels as well.

Standardizing comparisons by stem length and by leaf area supplied would seem more appropriate, especially for main trunk comparisons. Comparing a 60 m long liana and a 60 m long self-supporting tree shows that these both have similar mean vessel basal diameters (Rosell and Olson 2014). Use of the residuals of mean vessel diameter regressed on stem length is one way of "factoring out" stem length (Olson et al. 2014).

Studies of plant hydraulics need to take conductive path length into account. To the extent that vessel diameter is related to vulnerability to cavitation and embolism, comparing stem segments drawn from different distances to the stem tip certainly introduces a confounding factor. The wider vessels of the stem taken from a greater distance to the tip will have many wider vessels and therefore higher flow rates and vulnerabilities to embolism (Petit and Anfodillo 2009, 2011). Vessels are expected to become longer with increasing distance from the stem tip (Comstock and Sperry 2000).

The exponential form of tip-to-base vessel widening must be taken into account. Vessel widening follows a power law in which the slope of the mean vessel diameter–stem length curve becomes less pronounced with distance from the tip. As a result, variation in vessel diameter per unit stem length increment will be much more pronounced when sampling terminal twigs as compared to main trunks. The best strategy for standardizing in these cases is not entirely clear, mostly due to the potential role of leaf size and its influence on terminal twig vessel diameter. Larger leaves have longer conductive paths within the lamina and so we would predict them to have wider vessels in their petiole bases.

Following that logic, plants with larger leaves should have wider terminal twig vessels. Some work remains to be done to identify ways of generating truly comparable measurements of hydraulic performance in twigs (Petit and Anfodillo 2011). Moreover, recent results suggest that terminal twig vessel diameters scale positively with tree height and liana stem length (Olson et al. 2014). This would mean that stem length could potentially be a confounding factor not only in comparisons between liana and self-supporting plant twigs but leaves as well. This is because wide terminal twig vessels might indicate wider vessels entering into petioles. If plants of many different lengths were sampled in these comparisons, then the results might reflect the variation introduced by stem length rather than inherent liana/self-supporting differences. As a result, studies that standardize by stem length are needed.

Studies of hydraulic performance on lianas also face the challenge of taking into account the changing stem proportionalities that lianas undergo tip to base. The terminal parts of liana stems are often mostly self-supporting. These "searcher" stem segments help lianas make their way across open spaces between host branches. The stem length–diameter relations in self-supporting plants and lianas less than 1 m long are quite similar (Fig. 6.10). Therefore, studies of liana hydraulics based only on slender terminal stems may not accurately reveal the differences between lianas and self-supporting plants.

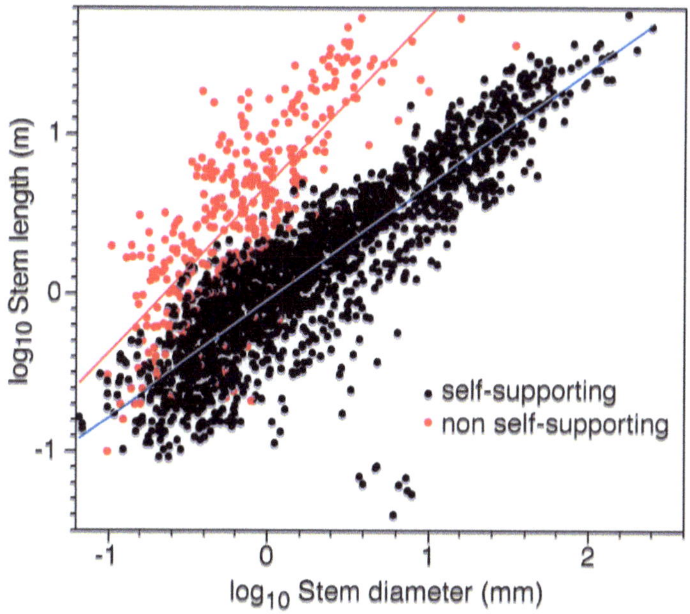

Fig. 6.10 Self- versus non-self-supporting plant stem length–diameter allometry, based on data from 2,388 samples from 472 species. From Olson et al. 2014, Rosell and Olson 2014 and unpublished data. $R^2=0.76$, self-supporting slope 0.90, non-self-supporting slope 1.065. There is extensive overlap at short stem lengths, but as stem length becomes longer, there are fewer and fewer individuals occupying the space between the main liana and self-supporting scaling lines

10 Xylem Dysfunction in Lianas

Since LSCs are not exceptional in lianas, the predicted stem pressure gradients in lianas are similar to those for trees and shrubs with similar transpiration rates (Table 6.2). Thus, lianas with very long stems could actually be at a hydraulic disadvantage in a mature forest.

Considering the great lengths that liana stems typically have, the few long and wide vessels in the basal parts of liana stems may represent low-cost high-risk investments. Such conduits should be more vulnerable to embolism caused by freeze–thaw, water stress, bending stress, pathogens, and herbivores. An additional level of risk is the low level of redundancy in liana conduits (Carlquist 1975; Ewers et al. 2007). Low Huber values in liana stems indicate there are fewer alternate pathways when some vessels become air blocked—this adds to the risk in the liana transport system, which may or may not be sufficiently ameliorated by abundant narrow vessels and tracheids in the xylem.

As would be predicted from their wide vessels and low vessel redundancy, liana stems tend to have higher P_{50} values than the trees they climb upon, that is, they are more vulnerable to water stress-induced embolism (Zhu and Cao 2009; van der Sande et al. 2013). In addition, lianas are clearly more prone to freeze–thaw embolism than are trees in temperate areas (Zimmermann 1983; Sperry et al. 1987; Tibbetts and Ewers 2000; Jiménez-Castillo and Lusk 2013). In fact, vulnerability to freeze–thaw embolism may be the major reason that lianas are much more common in tropical than in temperate forests.

11 What Lianas Tell Us About Self-Supporting Plants

Here, we briefly explain how lianas might help to identify features that are likely adaptive, both in lianas as well as in self-supporting plants, as opposed to those due to "constraint," that is, a lack of developmental alternatives (Olson 2012). The theoretical packing limit is a prime example. The packing limit refers to the limited range of values that the relationship between conduit number and conduit diameter must necessarily occupy. The proportion of a given trans-sectional area of stem that is occupied by vessels is maximally 1 and minimally 0. As a result, because of the inescapable space trade-off involved in packing many small or few large conduits into a given area, the conduit number and diameter relationship can be thought of as "constrained" (Zanne et al. 2010). This space trade-off is associated with the negative relationship between vessel diameter and vessel density across the flowering plants, as well as the decay in vessel density with distance from the stem tip (Zanne et al. 2010; Olson et al. 2014). Lianas have higher vessel densities for a given average vessel diameter than self-supporting plants do. This can be due both to the lack of need to pack in as many fibers in the liana stem, as well as the tendency for lianas to have many narrow vessels in addition to wide ones. While variation is bounded by the packing limit, the exact position of the vessel density–diameter intercept in self-supporting plants should be determined by natural selection (Crivellaro et al. 2012; Rosell and Olson 2014).

For stems longer than 2 m, the slope of the liana length–diameter relationship is statistically indistinguishable from that in self-supporting plants, with a significantly higher intercept in lianas. At shorter stem lengths, especially 2 m or shorter, stem length–diameter relations between lianas and self-supporting plants are very similar. For stems of up to 4 m long or so, many plants can be found with stems that lean or rest on other plants, without twining or climbing. Such intermediate morphologies are never observed at long stem lengths, beyond 10 m long. Figure 6.10 illustrates this tendency. This pattern suggests that leaning or semi-self-supporting plants of very long stem lengths are not favored by selection. This pattern of selection then leads to the empty space shown on the right of Fig 6.10.

12 Global Distribution of Lianas and Liana Strategies

Lianas are most diverse and abundant in tropical forests, especially dry tropical forests (Schnitzer and Bongers 2002; Schnitzer 2005). In such areas freeze–thaw embolism is generally not a risk. It has also been suggested that the thriving of lianas in dry tropical forests may be due to an exceptionally deep rooting system that taps into water resources that are not available to trees (Chen et al. 2015). Lianas have high levels of native embolism in their stems—they may operate with a narrow hydraulic safety margin (Tibbetts and Ewers 2000; Jiménez-Castillo and Lusk 2013; Johnson et al. 2013). If stems are "expendable" and roots are conserved, then a high-risk strategy for the stems and leaves would be adaptive.

Although lianas can survive at the top of mature forest canopies, they are most competitive and abundant in secondary forests and, in mature forests, in tree fall gaps (Putz 1984; Schnitzer 2005; Schnitzer et al. 2005; Schnitzer and Bongers 2011; Jiménez-Castillo and Lusk 2013). Based upon liana exclusion studies, at tree fall gaps lianas dramatically decreased the biomass production of trees (Schnitzer et al. 2014). In an extensive study of Costa Rican forests, the number of liana stems declined significantly with age of the forest, while the number of self-supporting stems did not change. With advancing forest age, species richness and biomass increased significantly for self-supporting plants, while lianas held steady or sometimes declined in those parameters (Letcher and Chazdon 2009). The predicted xylem pressure gradients are similar in stems of lianas as in free-standing growth forms (Table 6.2), which could mean that extremely long stems of lianas would eventually be at a hydraulic disadvantage. However, the relationship between vessel diameter and stem length should adjust for that in large plants. Such predictions also do not account for possible foliar absorption of water during periods of mist or rain (Rundel 1982; Burgess and Dawson 2004; Breshears et al. 2008; Limm et al. 2009).

We are not aware of studies of foliar absorption of water by liana leaves, but that phenomenon would dampen the pressure gradients and otherwise mitigate water stress.

In addition to the well-known "mechanical parasite" strategy, lianas might be categorized as ruderal plants (*sensu* Grime 1977) considering that they thrive most in early successional situations. This may be surprising given their great stem lengths. The inexpensive stems (on a per unit stem length basis) with throw-away leaves may be central to the ruderal liana strategy. This is especially evident in dry tropical forests, tree fall gaps, and in secondary tropical forests, where lianas most thrive (Schnitzer and Bongers 2002; Schnitzer 2005). Although lianas appear to be exceptional for their long and wide vessels, the great vessel widths may be only commensurate with their remarkable stem lengths. The anomalous anatomical features of many liana stems, such as successive cambia surrounded by unlignified conjunctive tissue, wide unlignified vascular rays, multiple steles and sectored xylem, all tend to increase stem flexibility and regeneration capacity in the event of tree falls and other disturbance events (Dobbins and Fisher 1986; Fisher and Ewers 1989; Ewers and Fisher 1991; Fisher and Ewers 1992; Carlquist 2007). Many lianas are very effective at sprouting from the basal parts of the shoot. For example, Putz (1984) reported that 90 % of lianas at a site in Panama sprouted within 12 months after their host trees fell. Some lianas also are capable of splitting along sectored xylem to form new individuals (Caballe 1994). When mature liana stems touch ground, they are very capable or rooting and in some cases forming new individuals by vegetative spread (Fisher and Ewers 1991). The root systems of lianas may be long lasting, and as noted above, the root xylem has abundant storage parenchyma. In contrast, the shoots may be expendable.

13 Future Research Directions

Many questions remain to be addressed, making liana biology a fertile field for research. The distinctness of roots systems of lianas versus self-supporting plants is one. Foliar absorption could ameliorate the water relations of liana shoots, but this has not been studied. There is a need to establish methodologies for making conductive path length-standardized comparisons of vessel dimensions and hydraulic performance between lianas and self-supporting plants, including the need to take leaf size into account. For analysis of risk, structure and function studies need to take into account the total length of the xylem pathways. The relative level of hydraulic redundancy in lianas is at question given the low Huber values but abundant narrow vessels and frequent presence of vasicentric tracheids. A clear priority is also determining whether variation in the taper exponent is heritable and how this variation affects hydraulic performance. Lastly, if lianas are regarded as ruderal, then they are the world's longest weeds. Although much remains to be studied, it is clear that when the proverbial "tree falls in the forest," whether someone hears it or not, lianas have the hydraulic capacity to rapidly exploit the newly opened space.

References

Aloni R, Zimmermann MH (1983) The control of vessel size and density along the plant axis: a new hypothesis. Differentiation 24:203–208

Aloni R (1987) Differentiation of vascular tissues. Ann Rev Plant Physiol 38:179–204

Andrade JL, Meinzer FC, Goldstein G, Schnitzer SA (2005) Water uptake and transport in lianas and co-occurring trees of a seasonally dry tropical forest. Trees 19(3):282–289

Anfodillo T, Carraro V, Carrer M, Fior C, Rossi S (2006) Convergent tapering of xylem conduits in different woody species. New Phytol 169:279–290

Anfodillo T, Deslauriers A, Menardi R, Tedoldi L, Petit G, Rossi S (2012) Widening of xylem conduits in a conifer tree depends on the longer time of cell expansion downwards along the stem. J Exp Bot 63:837–845

Anfodillo T, Petit G, Crivellaro A (2013) Axial conduit widening in woody species: a still neglected anatomical pattern. IAWA J 34(4):352–364

Angyalossy V, Angeles G, Pace MR, Lima AC, Dias-Leme CL, Lohmann LG, Madero-Vega C (2012) An overview of the anatomy, development and evolution of the vascular system of lianas. Plant Ecol Diversity 5:167–182

Baas P, Ewers FW, Davis SD, Wheeler EA (2004) Evolution of xylem physiology. In: Hemsley AR, Poole I (eds) The evolution of plant physiology: from whole plants to ecosystems. Elsevier, London, pp 273–295

Becker P, Gribben RJ (2001) Estimation of conduit taper for the hydraulic resistance model of West et al. Tree Physiol 21:697–700

Brenan JPN (1967) Flora of Tropical East Africa. Part 2, Caesalpinioideae. Crown Agents for Oversea Governments and Administrations, London, p 215

Breshears DD, McDowell NG, Goddard KL, Dayem KE, Martens SN, Meyer CW, Brown KM (2008) Foliar absorption of intercepted rainfall improves woody plant water status most during drought. Ecology 89(1):41–47

Burgess SSO, Dawson TE (2004) The contribution of fog to the water relations of Sequoia sempervirens (D. Don): foliar uptake and prevention of dehydration. Plant Cell Environ 27(8): 1023–1034

Burkill IH (1966) A dictionary of the economic products of the Malay Peninsula. Ministry of Agriculture and Cooperative, Kuala Lumpur

Caballe G (1994) Ramet proliferation by longitudinal splitting in the Gabonese rain forest liana Dalhousiea africana S. Moore (Papilionaceae). Biotropica 26:266–275

Cai ZQ, Schnitzer SA, Bongers F (2009) Seasonal differences in leaf-level physiology give lianas a competitive advantage over trees in a tropical seasonal forest. Oecologia 161(1):25–33

Campanello PI, Garibaldi JF, Gatti MG, Goldstein G (2007) Lianas in a subtropical Atlantic Forest: host preference and tree growth. Forest Ecol Manage 242(2):250–259

Carlquist SJ (1975) Ecological strategies of xylem evolution. University of California, Berkeley

Carlquist S (1985a) Observations on functional wood histology of vines and lianas: vessel dimorphism, tracheids, vasicentric tracheids, narrow vessels, and parenchyma. Aliso 11:139–157

Carlquist S (1985b) Vasicentric tracheids as a drought survival mechanism in the woody flora of southern California and similar regions: review of vasicentric tracheids. Aliso 11:37–68

Carlquist S (1991) Anatomy of vine and liana stems: a review and synthesis. In: Putz FE, Mooney HA (eds) The biology of vines. Cambridge University Press, Cambridge, pp 53–71

Carlquist S (2007) Successive cambia revisited: ontogeny, histology, diversity, and functional significance. J Torrey Botanical Soc 134(2):301–332

Castorena M, Rosell JA, Olson ME (unpublished data) An empirical morphospace for woody plant habit diversity. Ann Bot

Chen YJ, Cao KF, Schnitzer SA, Fan ZX, Zhang JL, Bongers F (2015) Water-use advantage for lianas over trees in tropical seasonal forests. New Phytol 205:128–136

Chiu ST, Ewers FW (1992) Xylem structure and water transport in a twiner, a scrambler, and a shrub of Lonicera (Caprifoliaceae). Trees 6(4):216–224

Cochard H, Ewers FW, Tyree MT (1994) Water relations of a tropical vine-like bamboo (Rhipidocladum racemiflorum): root pressures, vulnerability to cavitation and seasonal changes in embolism. J Exp Bot 45(8):1085–1089
Comstock JP, Sperry JS (2000) Theoretical considerations of optimal conduit length for water transport in vascular plants. New Phytol 148:195–218
Coomes DA, Jenkins KL, Cole LES (2007) Scaling of tree vascular transport systems along gradients of nutrient supply and altitude. Biol Lett 3:87–90
Cowan IR (1965) Transport of water in the soil-plant-atmosphere system. J Appl Ecol 2:221–239
Crivellaro A, McCulloh K, Jones FA, Lachenbruch B (2012) Anatomy and mechanical and hydraulic needs of woody climbers contrasted with subshrubs on the island of Cyrpus. IAWA J 33:355–373
Dobbins DR, Fisher JB (1986) Wound responses in girdled stems of lianas. Bot Gaz 147:278–289
Enquist BJ (2003) Cope's Rule and the evolution of long-distance transport in vascular plants: allometric scaling, biomass partitioning and optimization. Plant Cell Environ 26:151–161
Ewart AJ (1904–1905) The ascent of water in trees. Proc R Soc London 74:554–556
Ewers FW, Carlton MR, Fisher JB, Kolb KJ, Tyree MT (1997a) Vessel diameters in roots versus stems of tropical lianas and other growth forms. IAWA J 18:261–279
Ewers FW, Cochard H, Tyree MT (1997b) A survey of root pressures in vines of a tropical lowland forest. Oecologia 110(2):191–196
Ewers FW, Ewers JM, Jacobsen AL, López-Portillo J (2007) Vessel redundancy: modeling safety in numbers. IAWA J 28(4):373
Ewers FW, Fisher JB (1989a) Techniques for measuring vessel lengths and diameters in stems of woody plants. Am J Bot 76:645–656
Ewers FW, Fisher JB (1989b) Variation in vessel length and diameter in stems of six tropical and subtropical lianas. Am J Bot 76:1452–1459
Ewers FW, Fisher JB (1991) Why vines have narrow stems: histological trends in Bauhinia (Fabaceae). Oecologia 88(2):233–237
Ewers FW, Fisher JB, Chiu ST (1990) A survey of vessel dimensions in stems of tropical lianas and other growth forms. Oecologia 84(4):544–552
Ewers FW, Fisher JB, Chiu ST (1989) Water transport in the liana Bauhinia fassoglensis (Fabaceae). Plant Physiol 91(4):1625–1631
Ewers FW, Fisher JB, Fichtner K (1991) Water flux and xylem structure in vines. In: Putz FE, Mooney HA (eds) The biology of vines. Cambridge University Press, Cambridge, pp 127–160
Feild TS, Balun L (2008) Xylem hydraulic and photosynthetic function of Gnetum (Gnetales) species from Papua New Guinea. New Phytol 177(3):665–675
Feild TS, Chatelet DS, Balun L, Schilling EE, Evans R (2012) The evolution of angiosperm lianescence without vessels—climbing mode and wood structure—function in Tasmannia cordata (Winteraceae). New Phytol 193(1):229–240
Fisher JB, Guillermo Angeles A, Ewers FW, López-Portillo J (1997) Survey of root pressure in tropical vines and woody species. Int J Plant Sci 158:44–50
Fisher JB, Ewers FW (1989) Wound healing in stems of lianas after twisting and girdling injuries. Bot Gaz 150:251–265
Fisher JB, Ewers FW (1991) Structural responses to stem injury in vines. In: Putz FE, Mooney HA (eds) The biology of vines. Cambridge University Press, Cambridge, pp 99–124
Fisher JB, Ewers FW (1995) Vessel dimensions in liana and tree species of Gnetum (Gnetales). Am J Bot 82:1350–1357
Fisher JB, Ewers FW (1992) Xylem pathways in liana stems with variant secondary growth. Botanical J Linnean Soc 108(2):181–202
Fisher JB, Tan HT, Toh LP (2002) Xylem of rattans: vessel dimensions in climbing palms. Am J Bot 89(2):196–202
Gartner BL (1991a) Structural stability and architecture of vines vs. shrubs of poison oak, Toxicodendron diversilobum. Ecology 72:2005–2015
Gartner BL (1991b) Stem hydraulic properties of vines vs. shrubs of western poison oak, *Toxicodendron diversilobum*. Oecologia 87(2):180–189

Gartner BL (1991c) Relative growth rates of vines and shrubs of western poison oak, *Toxicodendron diversilobum* (Anacardiaceae). Am J Bot 78:1345–1353
Gartner BL, Bullock SH, Mooney HA, Brown VB, Whitbeck JL (1990) Water transport properties of vine and tree stems in a tropical deciduous forest. Am J Bot 77:742–749
Gasson P, Dobbins DR (1991) Wood anatomy of the Bignoniaceae, with a comparison of trees and lianas. IAWA Bull (NS) 12:389–417
Gleason SM, Butler DW, Zieminska K, Waryszak P, Westoby M (2012) Stem xylem conductivity is key to plant water balance across Australian angiosperm species. Funct Ecol 26:343–352
Grew N (1682) The anatomy of plants. W. Rawlins, London
Grime JP (1977) Evidence for the existence of three primary strategies in plants and its relevance to ecological and evolutionary theory. Am Nat 111:1169–1194
Haberlandt G (1914) Physiological plant anatomy. Macmillan, London
Hacke UG, Sperry JS, Pockman WT, Davis SD, McCulloh KA (2001) Trends in wood density and structure are linked to prevention of xylem implosion by negative pressure. Oecologia 126:457–461
Hales S (1727) Vegetable staticks; or, an account of some statical experiments on the sap in vegetables. W. and J. Innys, London
Hearn DJ (2006) *Adenia* (Passifloraceae) and its adaptive radiation: phylogeny and growth form diversification. Syst Bot 31:805–821
Hearn DJ (2009) Developmental patterns in anatomy are shared among separate evolutionary origins of stem succulent and storage root-bearing growth habits in *Adenia* (Passifloraceae). Am J Bot 96(11):1941–1956
Isnard S, Silk W (2009) Moving with climbing plants from Charles Darwin's time into the 21st century. Am J Bot 96:1205–1221
Jacobsen AL, Ewers FW, Pratt RB, Paddock WA, Davis SD (2005) Do xylem fibers affect vessel cavitation resistance? Plant Physiol 139(1):546–556
Jacobsen AL, Pratt RB, Tobin MF, Hacke UG, Ewers FW (2012) A global analysis of xylem vessel length in woody plants. Am J Bot 99(10):1583–1591
Jiménez-Castillo M, Lusk CH (2013) Vascular performance of woody plants in a temperate rain forest: lianas suffer higher levels of freeze–thaw embolism than associated trees. Funct Ecol 27(2):403–412
Johnson DM, Domec JC, Woodruff DR, McCulloh KA, Meinzer FC (2013) Contrasting hydraulic strategies in two tropical lianas and their host trees. Am J Bot 100(2):374–383
Jost L (1907) Lectures on plant physiology. Clarendon, Oxford
Kern KA, Ewers FW, Telewski FW, Koehler L (2005) Mechanical perturbation affects conductivity, mechanical properties and aboveground biomass of hybrid poplars. Tree Physiol 25(10):1243–1251
Kolb K, Sperry JS (1999) Differences in drought adaptation between subspecies of sagebrush (*Artemisia tridentata*). Ecology 80:2373–2384
Lahaye R, Civeyrel L, Speck T, Rowe NP (2005) Evolution of shrub-like growth forms in the lianoid subfamily Secamonoideae (Apocynaceae s.l.) of Madagascar: phylogeny, biomechanics, and development. Am J Bot 92:1381–1396
Letcher SG, Chazdon RL (2009) Lianas and self-supporting plants during tropical forest succession. Forest Ecol Manage 257(10):2150–2156
Limm EB, Simonin KA, Bothman AG, Dawson TE (2009) Foliar water uptake: a common water acquisition strategy for plants of the redwood forest. Oecologia 161(3):449–459
Lovisolo C, Hartung W, Schubert A (2002) Whole-plant hydraulic conductance and root-to-shoot flow of abscisic acid are independently affected by water stress in grapevines. Funct Plant Biol 29:1349–1356
McCulloh KA, Sperry JS, Adler FR (2003) Water transport in plants obeys Murray's law. Nature 421:939–942
Mencuccini M (2002) Hydraulic constraints in the functional scaling of trees. Tree Physiol 22(8):553–565

Mencuccini M, Hölttä T, Petit G, Magnani F (2007) Sanio's laws revisited. Size-dependent changes in the xylem architecture of trees. Ecol Lett 10:1084–1093

Mooney HA, Gartner BL (1991) Reserve economy of vines. In: Putz FE, Mooney HA (eds) The biology of vines. Cambridge University Press, Cambridge, pp 161–179

Neel PL, Harris RW (1971) Motion-induced inhibition of elongation and induction of dormancy in Liquidambar. Science 173(3991):58–59

Niklas KJ, Molina-Freaner F, Tinoco-Ojanguren C, Paolillo DJ (2002) The biomechanics of *Pachycereus pringlei* root systems. Am J Bot 89(1):12–21

Olson ME (2003) Stem and leaf anatomy of the arborescent Cucurbitaceae Dendrosicyos socotrana, with comments on the evolution of pachycauls from lianas. Plant Syst Evol 239(3 and 4):199–214

Olson ME (2012) The developmental renaissance in adaptationism. Trends Ecol Evol 27:278–287

Olson ME, Rosell JA (2013) Vessel diameter–stem diameter scaling across woody angiosperms and the ecological causes of xylem vessel diameter variation. New Phytol 197:1204–1213

Olson ME, Anfodillo T, Rosell JA, Petit G, Crivellaro A, Isnard S, León-Gómez C, Alvarado-Cárdenas LO, Castorena M (2014) Universal hydraulics of the flowering plants: vessel diameter scales with stem length across angiosperm lineages, habits and climates. Ecol Lett 17:988–997

Petit G, Anfodillo T (2009) Plant physiology in theory and practice: an analysis of the WBE model for vascular plants. J Theoret Biol 259:1–4

Petit G, Anfodillo T (2011) Comment on "The blind men and the elephant: the impact of context and scale in evaluating conflicts between plant hydraulic safety and efficiency" by Meinzer et al. (2010). Oecologia 165:271–274

Pittermann J, Sperry JS, Wheeler JK, Hacke UG, Sikkema EH (2006) Mechanical reinforcement of tracheids compromises the hydraulic efficiency of conifer xylem. Plant Cell Environ 29:1618–1628

Pruyn ML, Ewers BJ, Telewski FW (2000) Thigmomorphogenesis: changes in the morphology and mechanical properties of two Populus hybrids in response to mechanical perturbation. Tree Physiol 20(8):535–540

Putz FE (1983) Liana biomass and leaf area of a" tierra firme" forest in the Rio Negro Basin, Venezuela. Biotropica 15:185–189

Putz FE (1984) The natural history of lianas on Barro Colorado Island, Panama. Ecology 65(6):1713–1724

Reddy MS, Parthasarathy N (2006) Liana diversity and distribution on host trees in four inland tropical dry evergreen forests of peninsular India. Tropical Ecol 47(1):109–124

Rosell JA, Olson ME, Aguirre R, Carlquist S (2007) Logistic regression in comparative wood anatomy: tracheid types, wood anatomical terminology, and new inferences from the Carquist & Hoekman southern California dataset. Botanical J Linnean Soc 154(3):331–351

Rosell JA, Olson ME, Aguirre-Hernández R, Sánchez-Sesma FJ (2012) Ontogenetic modulation of branch size, shape, and biomechanics produces diversity across habitats in the *Bursera simaruba* clade of tropical trees. Evol Develop 14:437–449

Rosell JA, Olson ME (2014) Do lianas really have wide vessels? Vessel diameter-stem length scaling in non self-supporting plants. Perspect Plant Ecol Evol Syst 16:288–295

Rundel PW (1982) Water uptake by organs other than roots. In: Lange OL, Nobel PS, Osmond CB, Ziegler H (eds) Physiological plant ecology II. Springer, Berlin, pp 111–134

Sanio K (1872) Ueber die Grösse der Holzzellen bei der gemeinen der Kiefer (*Pinus silvestris*). Jahrbuecher fuer wissenscafliche Botanik 8:401–420

Santiago LS (2010) Can growth form classification predict litter nutrient dynamics and decomposition rates in lowland wet forest? Biotropica 42(1):72–79

Santiago LS, Wright SJ (2007) Leaf functional traits of tropical forest plants in relation to growth form. Funct Ecol 21(1):19–27

Savage VM, Bentley LP, Enquist BJ, Sperry JS, Smith DD, Reich PB, von Allmen EI (2010) Hydraulic trade-offs and space filling enable better predictions of vascular structure and function in plants. Proc Natl Acad Sci U S A 107:22722–22727

Scarpella E, Meijer AH (2004) Pattern formation in the vascular system of monocot and dicot plant species. New Phytol 164:209–242

Schenck H (1893) Beiträge zur biologie und anatomie der lianen, in beson- deren der in Brasilien einheimishe arten. 2. Beiträge zur anatomie der lianen. In: Schimpers AFW (ed) Botanische Mittheilungen aus der Tropen, vol 5. Fischer, Jena, pp 1–271

Schnitzer SA (2005) A mechanistic explanation for global patterns of liana abundance and distribution. Am Nat 166(2):262–276

Schnitzer SA, Bongers F (2002) The ecology of lianas and their role in forests. Trends Ecol Evol 17(5):223–230

Schnitzer SA, Bongers F (2011) Increasing liana abundance and biomass in tropical forests: emerging patterns and putative mechanisms. Ecol Lett 14:397–406

Schnitzer SA, van der Heijden GMF, Mascaro J, Carson WP (2014) Lianas in gaps reduce carbon accumulation in a tropical forest. Ecology 95:3008–3017

Schnitzer SA, Kuzee ME, Bongers F (2005) Disentangling above-and below-ground competition between lianas and trees in a tropical forest. J Ecol 93(6):1115–1125

Sperry JS, Holbrook NM, Zimmermann MH, Tyree MT (1987) Spring filling of xylem vessels in wild grapevine. Plant Physiol 83(2):414–417

Sperry JS, Smith DD, Savage VM, Enquist BJ, McCulloh KA, Reich PB, Bentley LP, von Allmen EI (2012) A species-level model for metabolic scaling in trees I. Exploring boundaries to scaling space within and across species. Funct Ecol 26:1054–1065

Stevens GC (1987) Lianas as structural parasites: the Bursera simaruba example. Ecology 68:77–81

Stevens PF (2001 onwards). Angiosperm phylogeny website. Version 12. http://www.mobot.org/MOBOT/research/APweb/

Tang Y, Kitching RL, Cao M (2012) Lianas as structural parasites: a re-evaluation. Chin Sci Bull 57(4):307–312

Tibbetts TJ, Ewers FW (2000) Root pressure and specific conductivity in temperate lianas: exotic Celastrus orbiculatus (Celastraceae) vs. native Vitis riparia (Vitaceae). Am J Bot 87(9):1272–1278

Tyree MT, Ewers FW (1996) Hydraulic architecture of woody tropical plants. In: Mulkey SS, Chazdon RL, Smith AP (eds) Tropical forest plant ecophysiology. Chapman & Hall, New York, pp 217–243

Tyree MT, Zimmermann M (2002) Xylem structure and the ascent of sap, 2nd edn. Springer, Berlin

Uggla C, Mellerowicz EJ, Sundberg B (1998) Indole-3-acetic acid controls cambial growth in Scots pine by positional signaling. Plant Physiol 117:113–121

van der Sande MT, Poorter L, Schnitzer SA, Markestein L (2013) Are lianas more drought-tolerant than trees? A test for the role of hydraulic architecture and other stem and leaf traits. Oecologia 172(4):961–972

von Allmen EI, Sperry JS, Smith DD, Savage VM, Enquist BJ, Reich PB, Bentley LP (2012) A species level model for metabolic scaling of trees II. Testing in a ring- and diffuse-porous species. Funct Ecol 26:1066–1076

West G, Brown JH, Enquist BJ (1999) A general model for the structure and allometry of plant vascular systems. Nature 400:664–667

Westermaier M, Ambronn H (1881) Beziehungen zwischen Lebensweise und Struktur der Schling- und Kletterpflanzen. Flora 69:417–436

White PR (1938) "Root-pressure"—an unappreciated force in sap movement. Am J Bot 25:223–227

Wyka TP, Oleksyn J, Karolewski P, Schnitzer SA (2013) Phenotypic correlates of the lianescent growth form: a review. Ann Bot 112:1667–1681

Zanne AE, Westoby M, Falster DS, Ackerly DD, Loarie SR, Arnold SEJ, Coomes DA (2010) Angiosperm wood structure: global patterns in vessel anatomy and their relations to wood density and potential conductivity. Am J Bot 97:207–215

Zimmermann MH, Jeje AA (1981) Vessel-length distribution in stems of some American woody plants. Can J Bot 59:1882–1892

Zimmennann MH, Potter D (1982) Vessel-length distribution in branches, stem and roots of *Acer rubrum* L. IAWA Bull 3:103–109

Zimmermann MH (1983) Xylem structure and the ascent of Sap. Springer, Berlin

Zhu SD, Cao KF (2009) Hydraulic properties and photosynthetic rates in co-occurring lianas and trees in a seasonal tropical rainforest in southwestern China. Plant Ecol 204(2):295–304

Chapter 7
Integrative Xylem Analysis of Chaparral Shrubs

R. Brandon Pratt, Marta I. Percolla, and Anna L. Jacobsen

1 Chaparral Shrublands

1.1 The Ecological and Evolutionary Context of a Model Community for Physiological Studies

The chaparral shrub community is dominated by evergreen sclerophyllous shrubs and primarily occurs in southern California (Fig. 7.1a) (Parker et al. in press). The chaparral has probably been more studied in an ecophysiological context than any other natural vegetation type in the world. The species that dominate this vegetation type have proven to be excellent model organisms for a range of physiological studies including studies of photosynthesis (Field and Mooney 1986), water relations (Morrow and Mooney 1974; Poole and Miller 1975), drought tolerance (Harrison et al. 1971; Kolb and Davis 1994), and xylem structure and function (Carlquist and Hoekman 1985; Hargrave et al. 1994; Wagner et al. 1998). This chapter will focus on xylem traits and for a recent broader review see Parker et al. (in press).

There are important reasons why chaparral vegetation has drawn the attention of physiologists (reviewed in Field and Davis 1989). One chief reason is that the environment where chaparral occurs is predictably stressful, thus providing a clear ecological context for understanding stress resistance traits. The source of the stress is the Mediterranean-type climate that characterizes the region where chaparral occurs in California. The precipitation falls as rain in the winter months when the temperatures are cool and the summers are hot with a predictable 4–6 months rainless period (Cowling et al. 2005). During the summer and fall, many species develop considerable water deficits and consistently have some of the lowest water potentials ever measured (Oechel 1988; Davis et al. 2002; Jacobsen et al. 2007a, b).

R.B. Pratt (✉) • M.I. Percolla • A.L. Jacobsen
Department of Biology, California State University, Bakersfield, Bakersfield, CA 93311, USA
e-mail: rpratt@csub.edu; mpercolla@gmail.com; ajacobsen@csub.edu

U. Hacke (ed.), *Functional and Ecological Xylem Anatomy*,
DOI 10.1007/978-3-319-15783-2_7

Fig. 7.1 A typical chaparral shrubland in the winter months in the western Santa Monica Mountains with bright green redshank (*Adenostoma sparsifolium*) and white flowering big-pod ceanothus (*Ceanothus megarcarpus*) in the foreground and background (panel **a**). (**b**) A photograph taken March 2014 of a chaparral shrubland in the Cold Creek of the Santa Monica Mountains suffering under extreme drought stress during the current multiyear drought gripping California. The plants in the photo are mostly evergreens so the lack of green and nongreen colors indicates shrubs that have dieback or that are dead. The green plant in the foreground is on the side of the road and receives extra runoff water and the oaks in the middleground are trees with access to stream water. Photos taken by R. Brandon Pratt

Beyond the summer rainless period, there is wide natural variability in annual rainfall from year to year and decade to decade. These extreme events place considerable stress on shrubs beyond an average year (Jacobsen et al. 2007a, b) and it is not uncommon to observe dieback and mortality of adult shrubs during these droughts (Fig. 7.1b) (Horton and Kraebel 1955). This readily observable stress leads to clear and frequent selection for traits related to drought (Davis et al. 2002; Pratt et al. 2014). Moreover, when droughts happen, they create opportunities for natural experiments to study mechanisms of drought response and survival (Parsons et al. 1981; Davis et al. 2002; Paddock et al. 2013; Pratt et al. 2014), which has also been the case in other Mediterranean-type ecosystems (Peñuelas et al. 2001; Vilagrosa et al. 2003; Lloret et al. 2004).

Field and Davis (1989) describe the "chaparral era" of physiological ecology studies as one typified by integration. One important theme of these studies has been cross-ecosystem comparisons of convergent evolution between chaparral and the four other regions of the world that have a Mediterranean-type climate: the cape region of South Africa, central Chile, Southwest Australia, and the Mediterranean basin (Mooney and Dunn 1970; Jacobsen et al. 2009). A related approach has been to compare chaparral communities to other non-Mediterranean-type shrub communities to examine which traits are unique to Mediterranean-type species (Bhaskar et al. 2007; Jacobsen et al. 2007a, b). Other studies have sought to integrate functional traits over different spatial and temporal scales (Miller and Poole 1979; Miller 1981; Davis and Mooney 1985; Cornwell and Ackerly 2009). These studies have included analyses over evolutionary timescales that have examined the evolutionary impact of the onset of the Mediterranean-type climate as well as the role of ecological sorting processes on functional traits of extant species (Ackerly 2004a, 2009).

1.2 Implications of Mediterranean-Type Climate for Vascular Function

In the context of vascular function, being evergreen means that chaparral shrubs have to maintain hydraulic supply to leaves during the dry season and during drought. This has been a key selective force for xylem resistance to cavitation and some chaparral shrubs are among the most cavitation resistant woody species known (Jacobsen et al. 2007a, b). Co-occurring chaparral shrub species can have rooting depths ranging from shallow to deep leading to a range of functional types arrayed along a drought avoidance to tolerance continuum (Hellmers et al. 1955; Thomas and Davis 1989; Ackerly 2004a, b; Parker et al. in press). Minimum seasonal water potential experienced by a species is strongly correlated to resistance of xylem to cavitation (Fig. 7.2). This suggests that xylem cavitation is a key trait that is linked to the level of water deficit experienced by adult shrubs. It has also been shown that seedling mortality during drought is correlated to cavitation resistance, which is another indication that cavitation resistance is an important drought tolerance trait (Fig. 7.3). Xylem refilling is unlikely to be important for chaparral shrubs

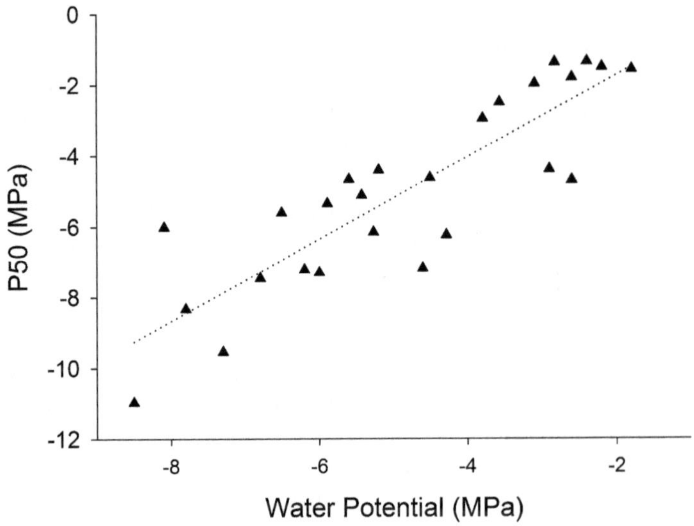

Fig. 7.2 The relationship between the minimum water potential experienced during the dry season and the resistance of branches to xylem cavitation (P50). Cavitation resistance was measured as the point where 50 % of hydraulic conductivity has been lost (P50) relative to the maximum conductivity with emboli removed by a flushing treatment. Data are means of both cavitation resistance and water potential measured at predawn (Jacobsen et al. 2007a, b; Pratt et al. 2007a, b)

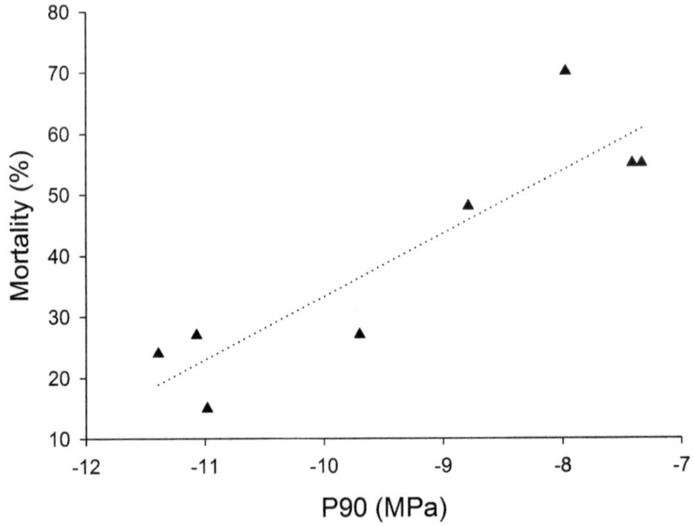

Fig. 7.3 Mortality of anisohydric-type seedlings during drought plotted against cavitation resistance of stems (P90). Figure modified from Pratt et al. (2008)

during the dry season because most shrubs will be beyond the water potential where refilling can occur (Hacke and Sperry 2003); moreover, the protracted rainless season means that plants will stay predictably dry until the fall and winter months when rains come, thus refilling would be futile.

Freeze/thaw stress has also been found to be an important force affecting chaparral shrub physiology and distributions (Ewers et al. 2003; Davis et al. 2005, 2007; Pratt et al. 2005). When the sap in the xylem freezes, air bubbles come out of solution and can expand upon thawing leading to gas-filled or embolized vessels. Species that have larger vessel diameters are more vulnerable to freeze/thaw-induced cavitation (Davis et al. 1999a, b). Freeze/thaw stress can also interact with drought. If the fall/winter rains do not occur before December (often the month with the lowest annual temperature) then the vascular system will be exposed to freeze/thaw stress when the xylem is under significant tension, which has been shown to lead to even greater levels of embolism than freezing in a hydrated state (Langan et al. 1997; Davis et al. 2005, 2007; Pittermann and Sperry 2006). Freeze/thaw stress at colder sites has been a factor limiting the maximum vessel diameter of a species (Ewers et al. 2003), and may have affected xylem structure in other ways such as patterns of vessel grouping (Carlquist 1984), and the presence and abundance of tracheids.

1.3 Integrative Studies of Xylem Traits

There are many scales over which xylem has been studied in chaparral and an exhaustive review of all of them is beyond the scope of this chapter. Some of these studies have been aimed at furthering our understanding of plant vascular function, while others have used xylem traits to test hypotheses about ecology and evolution of chaparral shrubs. Of course these two study aims are not mutually exclusive and many studies shed light on both.

One important theme has been the examination of xylem traits across multiple life stages. The crown fire regime in chaparral means that plants are top-killed by fire, which has a strong influence on community dynamics. Most species either resprout or recruit seedlings from a dormant seed bank after fire, and some species even do both (Keeley et al. 2012). The seedling stage, being small and shallowly rooted, is often the one exposed to the most extreme levels of water deficits and is one of the most vulnerable stages to mortality (Frazer and Davis 1988; Thomas and Davis 1989; Williams et al. 1997; Pratt et al. 2005). Thus, the xylem traits of seedlings are under keen selection to survive the summer dry season and stiff competition for resources in their first year of life. Selection at the seedling stage may determine many important xylem adaptations of a species and, without ontogenetic plasticity or allometric shifts, the xylem traits we measure on adults may be more a reflection of what happens at the seedling stage than conditions experienced at the adult stage (Pratt et al. 2007a, b, 2008).

In addition to seedlings, recent studies have suggested that xylem resistance to cavitation is important for drought survival of postfire resprouting chaparral species (Utsumi et al. 2010; Pratt et al. 2014). Interestingly, resprouts generally have reduced resistance to cavitation compared to unburned adult plants (Ramirez et al. 2012), and the anatomical/mechanistic reasons for this are currently under study (Anna Jacobsen, unpublished manuscript). The different ways that species cope with fire (resprouting vs. seedling recruitment) shift their allocation of resources and lead to trade-offs with implications for their physiological adaptations, including vascular traits (Pratt et al. 2007a, b, 2012a, b). Studies across different life stages will continue to be an important area of research and will help shed light on xylem plasticity across these stages, as well as how selection operates across the whole life cycle (Poorter 2007).

Another area of integration are studies examining xylem traits across multiple organs. This has been done many times comparing branches and roots (Davis et al. 2002; Pratt et al. 2007a, b, 2008); some have compared leaves and stems (Méndez-Alonzo, unpublished manuscript; Pratt, unpublished manuscript); and others have examined leaves, shoots, branches, fine roots, and root systems (Pratt et al. 2010). An important result from these studies is that roots are typically more vulnerable to cavitation than stems; however, because roots have much greater maximum conductivity than stems even when they are highly embolized they can still achieve similar conductivity to stems that are far less embolized (Pratt et al. 2008). Selection operates on intact functioning individuals so studies of the vascular system as an integrated unit are ultimately important to understand vascular adaptations.

Few studies have sought to examine xylem traits across space and time (Anderegg and Meinzer 2015). It has been shown that xylem cavitation resistance can shift over the course of a season (Jacobsen et al. 2007a, b). This shift is not one-directional and some species shift to being more vulnerable, but more commonly species shift to being more resistant during the dry season (Jacobsen et al. 2007a, 2014). Examination of cavitation resistance across moisture gradients has found that chaparral at drier sites has greater resistance to cavitation than those at moister sites (Davis et al. 1999a), whereas for some taxa, e.g., *Arctostaphylos* spp., there seems to be very little change (Vasey et al. 2012; Jacobsen and Pratt 2013).

Studies of trade-offs in xylem function has been an active area of chaparral research. One of the earliest of these studies to examine a trade-off other than between safety and efficiency was the relationship between transport efficiency and mechanical strength (Wagner et al. 1998). Since then, many important studies have expanded the study of mechanical strength and transport traits (Hacke et al. 2001; Jacobsen et al. 2005). Greater mechanical strength of chaparral xylem is generally associated with greater cavitation resistance (Jacobsen et al. 2007a, b). Water storage (capacitance) and carbohydrate storage are also negatively correlated with cavitation resistance (Pratt et al. 2007a, b; Anderegg and Meinzer 2015). The trade-off between xylem safety from cavitation and efficiency in transport is an influential model and forms the basis for new analyses in the following sections of this chapter.

A final area of study has been to compare the xylem of chaparral to the xylem of other woody communities in California and to the fynbos in South Africa. The goals of these studies have been to identify traits unique to the chaparral, which would indicate that they are important in the context of the stressful Mediterranean-type climate (Carlquist and Hoekman 1985; Carlquist 1989; Hacke et al. 2009). Other studies have addressed the question of convergent evolution of vascular traits to assess if chaparral are more similar to species inhabiting Mediterranean-type climate regions on other continents than species inhabiting regions with non-Mediterranean-type climates (Jacobsen et al. 2009; Pratt et al. 2007a, b). If a trait is found to be convergent in similar environments when compared across distantly related taxa that are separated by large geographical expanses, then this is among the strongest demonstrations that a trait is adaptive in the context of that environment. Key xylem traits such as cavitation resistance, xylem efficiency, and xylem density are convergent among South African and Californian Mediterranean-type shrublands when compared to other semiarid and arid shrublands (Jacobsen et al. 2009).

The remainder of this chapter will integrate long-standing ideas gleaned from comparative xylem studies regarding the importance of various xylem traits for contributing to xylem safety of chaparral shrubs with current models of xylem safety and efficiency.

2 Ecological Xylem Anatomy of Chaparral

Chaparral species were a key part of a seminal ecological xylem anatomy study by Sherwin Carlquist and David Hoekman (Carlquist and Hoekman 1985). This study, and other related ones (Carlquist 1989; Rosell et al. 2007), demonstrated many important trends regarding xylem anatomical traits linked to xylem safety. Many of these discoveries are only recently being fully appreciated among physiologists for their insight and importance for xylem function generally (Lens et al. 2011). At the time Carlquist conducted many of his studies of chaparral xylem there was not a database of xylem safety and efficiency measures that could be used to test some of his ideas. This is no longer the case and there is an extensive database of these traits for chaparral shrubs (Jacobsen et al. 2007a, b; Pratt et al. 2007a, b; Hacke et al. 2009). In the following sections, and in the spirit of the "integrative era" of chaparral studies, we review some of Carlquist's findings and, where possible, integrate these data with published and unpublished data on xylem safety and efficiency to independently and directly test some of Carlquist's hypotheses about chaparral xylem safety.

There are many xylem anatomical features of chaparral shrub xylem that reflect the water stress regime that these angiosperms experience. One feature is the presence of tracheids and their role in safe transport of water. Carlquist and Hoekman (1985) distinguish between three types of tracheids: true tracheids, vascular tracheids, and vasicentric tracheids. True tracheids are ancestral and along with parenchyma make up the background tissue of the xylem. Vascular and vasicentric tracheids are found among species in lineages that lost tracheids through the evolution of fiber-tracheids and libriform fibers. Thus, in these species tracheids were reevolved and represent a derived character. Vascular tracheids are located at the end of a growth ring and are hypothesized to be a safe transport pathway to keep the cambium hydrated, especially in drought-deciduous species. Being evergreen, these

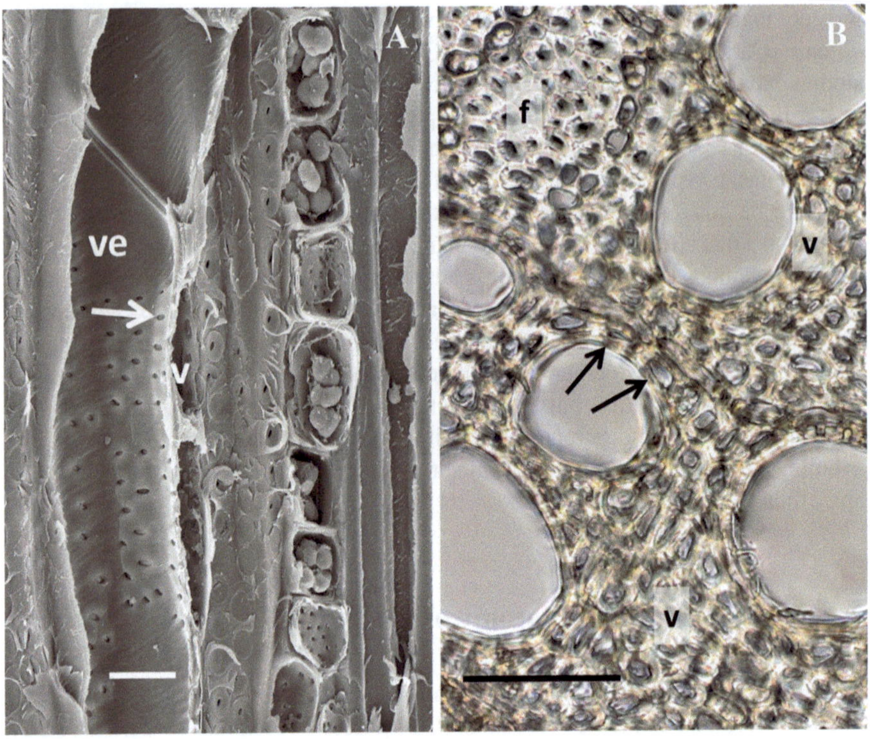

Fig. 7.4 (**a**) Scanning electron micrograph of xylem of big-pod ceanothus (*Ceanothus megacarpus*). Of note is the imperforate vasicentric tracheid (v) that shares a wall with the vessel element (ve) left of center, and this wall has numerous large circular bordered pits indicating the role of transporting water for the tracheids (*arrow*). Scale bar is 20 μm. Micrograph taken by Anna L. Jacobsen. (**b**) A light micrograph (×400 magnification) showing the abundant vasicentric (v) tracheids that surround the large vessels in scrub oak (*Quercus berberidifolia*). Also seen are libriform fibers (f) and large pits between vasicentric tracheids and vessels (*arrows*). The scale bar is about 42 μm. Micrograph was taken by Marta Percolla

tracheids are not common among chaparral shrubs and will not be emphasized in this review (Carlquist 1989). Vasicentric tracheids are located around vessels and these are hypothesized to aid in transport safety for many chaparral shrubs (Fig. 7.4a, b). These tracheids are hypothesized to form a water filled sheath around vessels that minimize air-filled compartments from which air might seed. The different tracheid types are generally mutually exclusive and the presence of true tracheids precludes the presence of vasicentric or vascular tracheids (Carlquist 1989).

To determine the importance of these tracheids, Carlquist compared the prevalence of the different types of imperforate tracheary elements among chaparral species to the prevalence in other California plant communities (Carlquist 1989). Among chaparral species, 43.9 % have vasicentric tracheids and among species that lacked true tracheids and vascular tracheids (i.e., those species that could potentially have vasicentric tracheids), 75 % had vasicentric tracheids. By contrast, compared to a sample of species taken across a range of environments from within California,

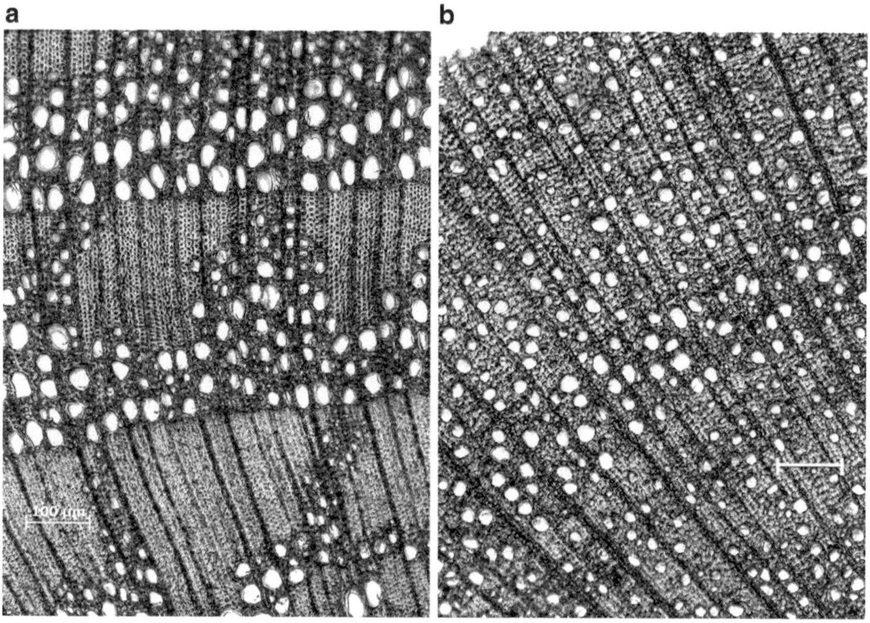

Fig. 7.5 (**a**) Light micrograph showing grouped vessels surrounded by vasicentric tracheids in cup leaf ceanothus (*Ceanothus greggi* var. *vestitus*) and (**b**) the generally solitary and nongrouped vessels generally surrounded by true tracheids of toyon (*Heteromeles arbutifolia*). Both images are ×100 and the scale bars are 100 μm. Micrographs taken by Brandon Pratt

these same numbers are lower at 33.3 % and 52.3 %, respectively. Among riparian species, none have vasicentric or vascular tracheids (Carlquist 1989).

The importance of vasicentric tracheids is that they maintain connections between vessels that are separated by an embolized vessel. This maintains the vessel network under times of drought. Tracheids have narrow diameters and are not able to transport water as efficiently as vessels, but during the dry season water use is greatly restricted, so the inefficient tracheids may be sufficient to supply transpiration. Similar to vasicentric tracheids, the presence of true tracheids is hypothesized to maintain the vessel network, which precludes a need for vasicentric and vascular tracheids.

Tracheid presence has been hypothesized to affect vessel grouping among chaparral species and species more generally (Carlquist 1989, 2009; Rosell et al. 2007). The presence of nonconductive imperforate tracheary elements (fiber-tracheids and libriform fibers) in the background xylem tissue is associated with greater groupings of vessels (Fig. 7.5). On the other hand, the presence of abundant true tracheids in the background tissue may minimize vessel grouping (Fig. 7.5) (Rosell et al. 2007). The reasons for these patterns were studied by Carlquist (1984) who hypothesized that the benefit of vessel groups is to increase connections among conduits because if one vessel embolizes then it is connected to other vessels in the network. In other words, in species lacking the network connections afforded by true tracheids, vessel groups lead to a safer vascular system in the context of cavitation. Angiosperm tracheids of any type may be safer than vessels because they are more abundant and they may be more resistant to cavitation (Hargrave et al. 1994). Species with vasicentric tracheids that only have tracheids around their vessels have to form vessel groups to take advantage of the connections offered by the tracheids (Fig. 7.5), but they show wide variation in this trait (Rosell et al. 2007), which may be related to the abundance of vasicentric tracheids they have.

2.1 *Imperforate Tracheary Element Type and Safety: A Test of Carlquist's Hypothesis*

When examined using published data, and consistent with Carlquist's hypotheses, chaparral species that had libriform fibers were least resistant to cavitation (Fig. 7.6a). The next least resistant group were those species with vascular tracheids, which were all drought-deciduous species. Deciduous chaparral shrubs are generally more vulnerable to cavitation than evergreens (Parker et al. in press), thus tracheid type and leaf habit are confounded for this group; moreover, this analysis is

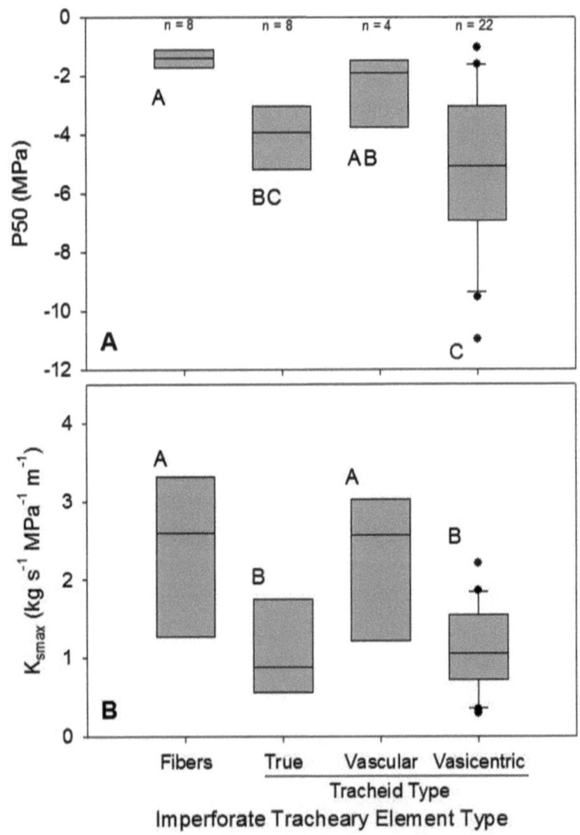

Fig. 7.6 (**a**) Box plots plotted for cavitation resistance (P50) and (**b**) xylem specific conductivity, i.e., efficiency (K_{smax}) plotted for species that differ in the background imperforate tracheary element type. For the box plots the middle line represents the median and lower and upper lines of the box represent the 25th and 75th percentiles, respectively. The error bars (whiskers) are the 5th and 95th percentiles and data points shown are outliers that are beyond that. Error bars are only calculated for a sample size of 9 or greater. Boxes with unique letters are significantly different (1-way ANOVA on log transformed data)

limited by sample size (data are available for only four species containing vascular tracheids) (Fig. 7.6a). Species with true tracheids had relatively safe xylem compared to the other groups (Fig. 7.6a). Species with vasicentric tracheids had the most resistant xylem; although, they were also the most variable group indicating that the presence of vasicentric tracheids does not necessarily lead to high levels of safety (Fig. 7.6a). A trade-off between safety and efficiency may explain why species with vasicentric tracheids do not uniformly have highly safe xylem.

These data are broadly supportive of Carlquist's hypotheses about the importance of tracheids in conferring xylem safety to drought. A key challenge remains to understand the mechanism of how tracheids contribute to xylem safety during drought (Hargrave et al. 1994). The tracheids may lead to greater safety because there are so many of them that they create redundancy (Ewers et al. 2007) or the pit and pit membrane characteristics of these tracheids may be modified for safety (Jansen et al. 2009; Plavcová et al. 2013). The small diameter of tracheids precludes rapid uptake of dye tracers and they are not generally resolvable with *en planta* imaging techniques, thus significant challenges remain in the study of these tracheids.

2.2 *Trade-off Between Safety and Efficiency: The Role of Tracheids*

Xylem safety generally comes at a cost of xylem efficiency and this may also relate to imperforate tracheary element type. The species that had true and vasicentric tracheids had the least efficient xylem compared to species with libriform fibers and vascular tracheids. These patterns point to a trade-off between safety and efficiency that may be linked to tracheid type (Figs. 7.6b and 7.7). It is possible that the trade-off between safety and efficiency is related to having tracheids with narrow lumens that are not efficient (Tyree and Ewers 1991) and this limits the maximum efficiency for species with tracheids. Alternatively, species with tracheids may be those that inhabit the driest sites and thus have reduced efficiency for evolutionary/ecological reasons, not simply because they have tracheids.

The rare pit hypothesis (a.k.a the pit area hypothesis) is currently an important framework for understanding the trade-off between xylem safety and efficiency (Hargrave et al. 1994; Wheeler et al. 2005; Christman et al. 2009). According to this hypothesis, species that have greater pit area or pit numbers for a vessel have a greater chance of having a large pore in a pit membrane that is vulnerable to cavitation by air-seeding (Lens et al. 2011). The pit area hypothesis has been examined in southern California shrubs (Hacke et al. 2009), but chaparral shrubs have not been separately analyzed. A plot of pit area/vessel area against cavitation resistance appears to show a strong relationship between these two traits that is consistent with the rare pit hypothesis; however, there is one extreme outlier that defies the overall trend (Fig. 7.8).

The outlier in this analysis, *Quercus beberidifolia*, has low pit area and low resistance to cavitation. This combination of traits should lead to high safety and low

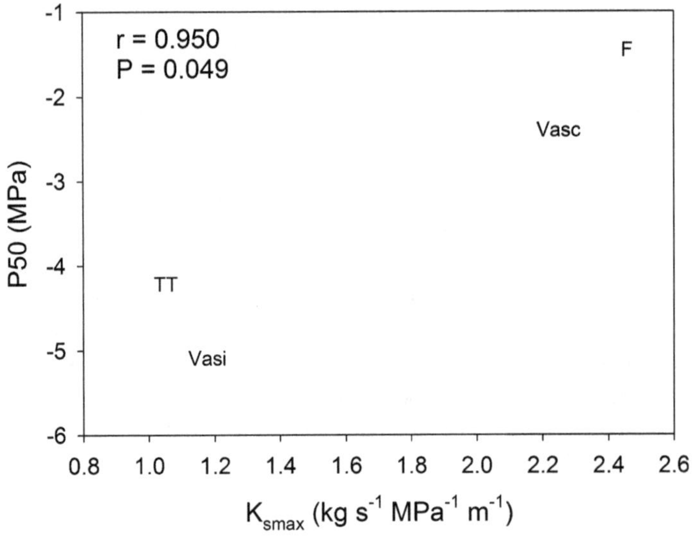

Fig. 7.7 Cavitation resistance (P50) plotted against xylem efficiency (K_{smax}) illustrating a trade-off between safety and efficiency for species with different imperforate tracheary element types: true tracheids (TT), vasicentric tracheids (Vasi), vascular tracheids (Vasc), and libriform fibers (F)

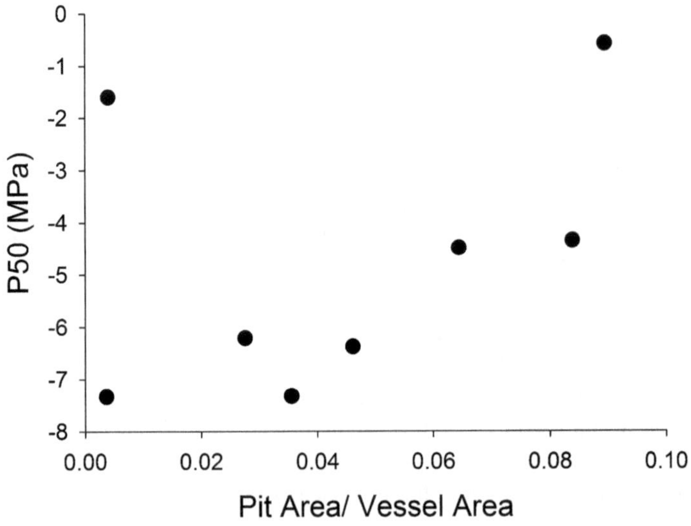

Fig. 7.8 The relationship between cavitation resistance (P50) and intervessel pit area/vessel area. The outlier in the upper left part of the figure is *Q. berberidifolia*

efficiency, but just the opposite is observed and this species has relatively high efficiency and low safety. The main xylem structural feature that leads to this species having low pit area is that it has many solitary vessels and thus a low proportion of vessel wall area in contact with adjacent vessels (Fig. 7.4b). Solitary vessels are common to *Quercus* spp. (Wheeler and Thomas 1981; Carlquist 1984). This anatomy is made possible because of abundant vasicentric tracheids (Fig. 7.4b). Interestingly, *Q. berberidifolia* has relatively high K_s compared to other chaparral species (only 7 species have higher values out of 47 species in the database). Thus, these tracheids can apparently achieve high levels of transport efficiency, and one reason they can do so is because they are highly abundant (Carlquist and Hoekman 1985; Rosell et al. 2007). It is also possible that they have modified pit and pit membrane traits such as membrane thickness, pit pore aperture, or pit membrane area that contribute to their efficiency (Hacke and Jansen 2009; Jansen et al. 2009; Lens et al. 2011). *Quercus crispula* was documented to have large pores in the pit membranes of its vasicentric tracheids, which could increase efficiency and decrease safety (Sano and Jansen 2006). Whatever modifications have been made they have compromised safety, as these oaks are vulnerable to water stress-induced cavitation when compared to other chaparral species (Jacobsen et al. 2007a, b; Tobin et al. 2013). These considerations highlight an unappreciated role that anatomical constraints may play in the trade-off between safety and efficiency and offer an explanation of why oak species may be generally vulnerable to cavitation (Tobin et al. 2013; Martin-StPaul et al. 2014). It would be valuable in future studies to examine pit area between vessels and tracheids on *Quercus* spp. in the context of the rare pit hypothesis, and such an analysis should lead to *Q. berberidifolia* falling farther to the right on the *x*-axis in Fig. 7.8.

The narrow diameters of the tracheids should render them resistant to freeze/thaw stress (Davis et al. 1999a, b; Pittermann and Sperry 2003). This suite of xylem anatomical traits may be linked to the evolutionary history of oak species. These species have an arcto-tertiary biogeographical origin whereby they colonized the warmer and milder regions of California from more northern regions (Axelrod 1983; Manos and Stanford 2001). Tracheids are an asset for safety in areas of freeze/thaw stress (Chap. 3), which would have been common in the regions occupied by ancestral *Quercus* spp., and also characterizes many of the habitats that oaks presently occupy.

2.3 *Integrating Vessel Grouping with Current Models of Vessel Connectivity*

The connections between vessels as part of the xylem network have been the topic of recent studies (Schenk et al. 2008; Martínez-Vilalta et al. 2012). Vessel connectivity (defined here as the number of unique vessels a vessel shares a wall with) has been hypothesized to be associated with safety and efficiency, with greater connectivity being associated with greater efficiency (xylem specific conductivity) and lower safety (Loepfe et al. 2007). This is hypothesized to be the case because greater connectivity increases the pathways for air to spread among vessels once one becomes embolized. It also creates more pathways for water to move radially, thus increasing overall efficiency. Carlquist's ideas of vessel grouping are partially at odds with the connectivity hypothesis of Loepfe et al. (2007). He suggested that greater vessel grouping is associated with greater safety because if one vessel embolized then connections to other vessels can maintain continuity of the pathway (Carlquist 1984). However, the type of imperforate tracheary elements present are important in Carlquist's conception of the hypothesis, and vessel grouping only becomes important for safety in species with libriform fibers or vasicentric tracheids in the background tissue. It is important to note that species with vascicentric tracheids surrounding vessels may tend to form vessel groups, but they could still have low connectivity with respect to vessel-to-vessel connections (e.g., *Ceanothus greggi* var. *vestitus* Fig. 7.5). If tracheids are safer, i.e., more cavitation resistant (this will certainly be the case for freeze/thaw stress), then vessels that are grouped could be safe if they have abundant vessel-to-tracheid connectivity, but not vessel-to-vessel connectivity (Carlquist 2001).

Vessel connectivity was examined in the context of imperforate tracheary element type. As expected, connectivity was greatest in species with nonconductive fibers and vascular tracheids in the background tissue (Fig. 7.9). Species with true

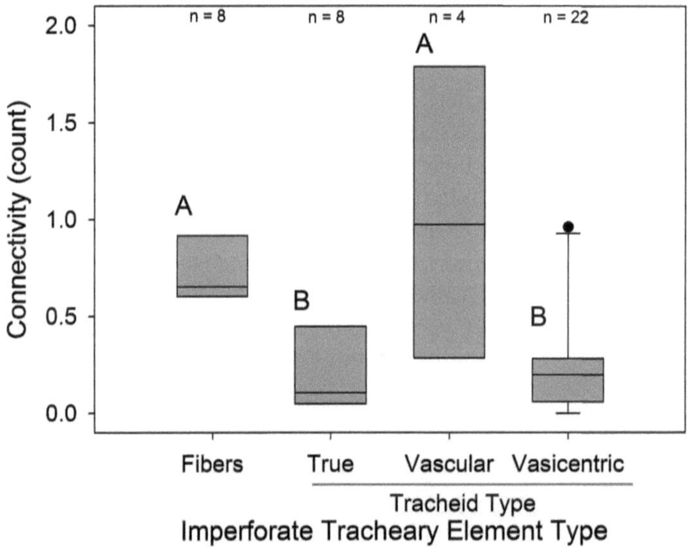

Fig. 7.9 Box plots plotted for connectivity (the number of unique vessels a vessel shares a wall with) plotted for species that differ in the background imperforate tracheary element type (Percolla and Pratt, unpublished data). For the box plots the middle line represents the median and lower and upper lines of the box represent the 25th and 75th percentiles, respectively. The error bars (whiskers) are the 5th and 95th percentiles and data points shown are outliers that are beyond that. Error bars are only calculated for a sample size of 9 or greater. Boxes with unique letters are significantly different (1-way ANOVA on log transformed data)

tracheids and vasicentric tracheids were not significantly different. These results suggest that estimates of connectivity are linked to the type of imperforate tracheary element in the background tissue of xylem and the presence of true or vasicentric tracheids limits connectivity. We did not have enough data on vessel grouping to do a robust analysis of the relationship between vessel grouping and connectivity, but preliminarily the association is not significant. This is not surprising because species with vasicentric tracheids can have high numbers of vessels per group and have low levels of connectivity. Mean connectivity for the different imperforate tracheary element groups was strongly and positively correlated to xylem efficiency (compare Fig. 7.6b and 7.9; $r=0.917$, $P=0.083$), which supports the hypothesis of Loepfe et al. (2007).

A recent study examined the relationship between connectivity and cavitation resistance in *Acer* spp. (Lens et al. 2011). *Acer* wood lacks conductive tracheids, thus greater vessel-to-vessel connectivity should confer safety according to Carlquist's hypothesis and should be more vulnerable according to Loepfe et al. (2007). Lens et al. (2011) found support for Carlquist's hypothesis and more connectivity was associated with greater cavitation resistance.

3 Conclusions

The xylem of chaparral species will continue to be a model system for ecological xylem anatomy. A fertile area of research will be the continued effort to integrate concepts developed from ecological xylem anatomy studies with current models of xylem function (Lens et al. 2011). For example, it is clear from consideration of *Q. berberidifolia* that tracheids, especially vasicentric ones, are going to have to be integrated into analyses examining pit area and pit numbers in the context of a trade-off between safety and efficiency (Hacke et al. 2009). Moreover, to fully understand the role of vessel connectivity, much more will need to be learned about tracheids and their role in connectivity and sectoriality (Schenk et al. 2008; Martínez-Vilalta et al. 2012). Other areas that are interesting fronts for further work are the helical thickenings on the inside of vessel elements that are associated with xylem safety (Carlquist 1989; Lens et al. 2011). The mechanistic importance of this trait remains unknown. The pit characteristics of chaparral shrubs have not been extensively studied and will likely yield valuable information regarding drought tolerance traits (Jarbeau et al. 1995). Trade-offs in xylem function, be they safety vs. efficiency, safety and biomechanics, or trade-offs with storage continue to be a fruitful area of research (Jacobsen et al. 2005; Pratt et al. 2007a, b; Anderegg and Meinzer 2015).

Acknowledgements Stephen Davis, Michael Tobin, and Uwe Hacke are gratefully acknowledged for the data they collected that were used in this chapter. The author gratefully acknowledges NSF funding (IOS-0845125) and funding from The Andrew Mellon Foundation.

References

Ackerly DD (2004a) Adaptation, niche conservatism, and convergence: comparative studies of leaf evolution in the California chaparral. Am Nat 163:654–671

Ackerly DD (2004b) Functional strategies of chaparral shrubs in relation to seasonal water deficit and disturbance. Ecol Monogr 74:25–44

Ackerly DD (2009) Evolution, origin and age of lineages in the Californian and Mediterranean floras. J Biogeogr 36:1221–1233

Anderegg WRL, Meinzer FC (2015) Wood anatomy and plant hydraulics in a changing climate. In: Hacke UG (ed) Functional and ecological xylem anatomy. Springer, Berlin

Axelrod DI (1983) Biogeography of oaks in the Arcto-Tertiary province. Ann Mo Bot Gard 70:629–657

Bhaskar R, Valiente-Banuet A, Ackerly DD (2007) Evolution of hydraulic traits in closely related species pairs from mediterranean and nonmediterranean environments of North America. New Phytol 176:718–726

Carlquist S (1984) Vessel grouping in dicotyledon wood: significance and relationship to imperforate tracheary elements. Aliso 10:505–525

Carlquist S (1989) Adaptive wood anatomy of chaparral shrubs. In: Keeley SC (ed) The California chaparral: paradigms reexamined. Science series no. 34. Natural History Museum of Los Angeles County, Los Angeles, pp 25–35

Carlquist S (2001) Comparative wood anatomy: systematic, ecological, and evolutionary aspects of dicotyledon wood. Springer, Berlin

Carlquist S (2009) Non-random vessel distribution in woods: patterns, modes, diversity, correlations. Aliso 27:39–58

Carlquist S, Hoekman DA (1985) Ecological wood anatomy of the woody southern California flora. Int Assoc Wood Anat Bull 6:319–347

Christman MA, Sperry JS, Adler FR (2009) Testing the 'rare pit' hypothesis for xylem cavitation resistance in three species of Acer. New Phytol 182:664–674

Cornwell WK, Ackerly DD (2009) Community assembly and shifts in plant trait distributions across an environmental gradient in coastal California. Ecol Monogr 79:109–126

Cowling RM, Ojeda F, Lamont B, Rundel PW, Lechmere-Oertel R (2005) Rainfall reliability, a neglected factor in explaining convergence and divergence of plant traits in fire-prone mediterranean-climate ecosystems. Glob Ecol Biogeogr 14:509–519

Davis SD, Mooney HA (1985) Comparative water relations of adjacent California shrub and grassland communities. Oecologia 66:522–529

Davis SD, Ewers FW, Wood J, Reeves JJ, Kolb KJ (1999a) Differential susceptibility to xylem cavitation among three pairs of *Ceanothus* species in the Transverse Mountain Ranges of southern California. Ecoscience 6:180–186

Davis SD, Sperry JS, Hacke UG (1999b) The relationship between xylem conduit diameter and cavitation caused by freezing. Am J Bot 86:1367–1372

Davis SD, Ewers FW, Sperry JS, Portwood KA, Crocker MC, Adams GC (2002) Shoot dieback during prolonged drought in *Ceanothus* (Rhamnaceae) chaparral of California: a possible case of hydraulic failure. Am J Bot 89:820–828

Davis SD, Ewers FW, Pratt RB, Brown PL, Bowen TJ (2005) Interactive effects of freezing and drought on long distance transport: a case study of chaparral shrubs of California. Vascular transport in plants. Elsevier, Amsterdam, pp 425–435

Davis SD, Helms AM, Heffner MS, Shaver AR, Deroulet AC, Stasiak NL, Vaughn SM, Leake CB, Lee HD, Sayegh ET (2007) Chaparral zonation in the Santa Monica Mountains: the influence of freezing temperatures. Fremontia 35:12–15

Ewers FW, Lawson MC, Bowen TJ, Davis SD (2003) Freeze/thaw stress in *Ceanothus* of southern California chaparral. Oecologia 136:213–219

Ewers FW, Ewers JM, Jacobsen AL, López-Portillo J (2007) Vessel redundancy: modeling safety in numbers. IAWA J 28:373–388

Field CB, Davis SD (1989) Physiological ecology. In: Keeley SC (ed) The California chaparral: paradigms reexamined. Science series no. 34. Natural History Museum of Los Angeles County, Los Angeles, pp 154–164

Field C, Mooney HA (1986) The photosynthesis—nitrogen relationship in wild plants. In: Givnish TJ (ed) On the economy of plant form and function. Cambridge University Press, Cambridge, pp 25–55

Frazer JM, Davis SD (1988) Differential survival of chaparral seedlings during the first summer drought after wildfire. Oecologia 76:215–221

Hacke UG, Jansen S (2009) Embolism resistance of three boreal conifer species varies with pit structure. New Phytol 182:675–686

Hacke U, Sperry J (2003) Limits to xylem refilling under negative pressure in *Laurus nobilis* and *Acer negundo*. Plant Cell Environ 26:303–311

Hacke UG, Sperry JS, Pockman WT, Davis SD, McCulloh KA (2001) Trends in wood density and structure are linked to prevention of xylem implosion by negative pressure. Oecologia 126:457–461

Hacke U, Jacobsen A, Pratt R (2009) Xylem function of arid-land shrubs from California, USA: an ecological and evolutionary analysis. Plant Cell Environ 32:1324–1333

Hargrave KR, Kolb KJ, Ewers FW, Davis SD (1994) Conduit diameter and drought-induced embolism in *Salvia mellifera* Greene (Labiatae). New Phytol 126:695–705

Harrison AT, Small E, Mooney HA (1971) Drought relationships and distribution of two Mediterranean-climate California plant communities. Ecology 52:869–875

Hellmers H, Horton JS, Juhren G, O'Keefe J (1955) Root systems of some chaparral plants in southern California. Ecology 36:667–678

Horton JS, Kraebel CJ (1955) Development of vegetation after fire in the chamise chaparral of southern California. Ecology 36:244–262

Jacobsen AL, Pratt RB (2013) Vulnerability to cavitation of central California Arctostaphylos (Ericaceae): a new analysis. Oecologia 171:329–334

Jacobsen AL, Pratt RB, Davis SD, Tobin MF (2014) Geographic and seasonal variation in chaparral vulnerability to cavitation. Madroño 61:317–327

Jacobsen AL, Ewers FW, Pratt RB, Paddock WA, Davis SD (2005) Do xylem fibers affect vessel cavitation resistance? Plant Physiol 139:546–556

Jacobsen AL, Pratt RB, Davis SD, Ewers FW (2007a) Cavitation resistance and seasonal hydraulics differ among three arid California plant communities. Plant Cell Environ 30: 1599–1609

Jacobsen AL, Pratt RB, Ewers FW, Davis SD (2007b) Cavitation resistance among twenty-six chaparral species of southern California. Ecol Monogr 77:99–115

Jacobsen AL, Esler KJ, Pratt RB, Ewers FW (2009) Water stress tolerance of shrubs in Mediterranean-type climate regions: Convergence of fynbos and succulent karoo communities with California shrub communities. Am J Bot 96:1445–1453

Jansen S, Choat B, Pletsers A (2009) Morphological variation of intervessel pit membranes and implications to xylem function in angiosperms. Am J Bot 96:409–419

Jarbeau JA, Ewers FW, Davis SD (1995) The mechanism of water-stress-induced embolism in two species of chaparral shrubs. Plant Cell Environ 18:189–196

Keeley JE, Bond WJ, Bradstock RA, Pausas JG, Rundel PW (2012) Fire in Mediterranean ecosystems: ecology, evolution and management. Cambridge University Press, New York

Kolb KJ, Davis SD (1994) Drought tolerance and xylem embolism in co-occurring species of coastal sage and chaparral. Ecology 75:648–659

Langan SJ, Ewers FW, Davis SD (1997) Xylem dysfunction caused by water stress and freezing in two species of co-occurring chaparral shrubs. Plant Cell Environ 20:425–437

Lens F, Sperry JS, Christman MA, Choat B, Rabaey D, Jansen S (2011) Testing hypotheses that link wood anatomy to cavitation resistance and hydraulic conductivity in the genus Acer. New Phytol 190:709–723

Lloret F, Siscart D, Dalmases C (2004) Canopy recovery after drought dieback in holm-oak Mediterranean forests of Catalonia (NE Spain). Glob Chang Biol 10:2092–2099

Loepfe L, Martinez-Vilalta J, Pinol J, Mencuccini M (2007) The relevance of xylem network structure for plant hydraulic efficiency and safety. J Theor Biol 247:788–803
Manos PS, Stanford AM (2001) The historical biogeography of Fagaceae: tracking the tertiary history of temperate and subtropical forests of the Northern Hemisphere. Int J Plant Sci 162:S77–S93
Martínez-Vilalta J, Mencuccini M, Álvarez X, Camacho J, Loepfe L, Piñol J (2012) Spatial distribution and packing of xylem conduits. Am J Bot 99:1189–1196
Martin-StPaul N, Longepierre D, Huc R, Delzon S, Burlett R, Joffre R, Rambal S, Cochard H (2014) How reliable are methods to assess xylem vulnerability to cavitation? The issue of 'open vessel' artifact in oaks. Tree Physiol 34(8):787–791. doi:10.1093/treephys/tpu1059
Miller PC (ed) (1981) Resource use by chaparral and matorral. Springer, New York
Miller PC, Poole DK (1979) Patterns of water use by shrubs in southern California. For Sci 25:84–98
Mooney HA, Dunn EL (1970) Convergent evolution of Mediterranean climate evergreen sclerophyll shrubs. Evolution 24:292–303
Morrow PA, Mooney HA (1974) Drought adaptations in two Californian evergreen sclerophylls. Oecologia 15:205–222
Oechel WC (1988) Minimum non-lethal water potentials in Mediterranean shrub seedlings. Time scales and water stress. In: di Castri F, Floret C, Rambal S, Roy J (eds) Proceedings of the 5th international conference on Mediterranean ecosystems (MEDECOS V). International Union of Biological Sciences, Paris, pp 125–131
Paddock WAS, Davis SD, Pratt RB, Jacobsen AL, Tobin MF, Lopez-Portillo J, Ewers FW (2013) Factors determining mortality of adult chaparral shrubs in an extreme drought year in California. Aliso 31:49–57
Parker, VT, Pratt RB, Keeley JE (in press) Chaparral. In: Mooney H, Zavaleta E (eds) Terrestrial ecosystems of California. University of California Press, Oakland
Parsons DJ, Rundel PW, Hedlund R, Baker GA (1981) Survival of severe drought by a non-sprouting chaparral shrub. Am J Bot 68:973–979
Peñuelas J, Lloret F, Montoya R (2001) Severe drought effects on Mediterranean woody flora in Spain. For Sci 47:214–218
Pittermann J, Sperry JS (2003) Tracheid diameter is the key trait determining the extent of freezing-induced embolism in conifers. Tree Physiol 23:907–914
Pittermann J, Sperry JS (2006) Analysis of freeze-thaw embolism in conifers. The interaction between cavitation pressure and tracheid size. Plant Physiol 140:374–382
Plavcová L, Jansen S, Klepsch M, Hacke UG (2013) Nobody's perfect: can irregularities in pit structure influence vulnerability to cavitation? Front Plant Sci 4:453
Poole DK, Miller PC (1975) Water relations of selected species of chaparral and coastal sage communities. Ecology 56:1118–1128
Poorter L (2007) Are species adapted to their regeneration niche, adult niche, or both? Am Nat 169:433–432
Pratt RB, Ewers FW, Lawson MC, Jacobsen AL, Brediger MM, Davis SD (2005) Mechanisms for tolerating freeze-thaw stress of two evergreen chaparral species: *Rhus ovata* and *Malosma laurina* (Anacardiaceae). Am J Bot 92:1102–1113
Pratt RB, Jacobsen AL, Ewers FW, Davis SD (2007a) Relationships among xylem transport, biomechanics and storage in stem and roots of nine Rhamnaceae species of the California chaparral. New Phytol 174:787–798
Pratt RB, Jacobsen AL, Golgotiu KA, Sperry JS, Ewers FW, Davis SD (2007b) Life history type and water stress tolerance in nine California chaparral species (Rhamnaceae). Ecol Monogr 77:239–253
Pratt RB, Jacobsen AL, Mohla R, Ewers FW, Davis SD (2008) Linkage between water stress tolerance and life history type in seedlings of nine chaparral species (Rhamnaceae). J Ecol 96:1252–1265
Pratt RB, North GB, Jacobsen AL, Ewers FW, Davis SD (2010) Xylem root and shoot hydraulics is linked to life history type in chaparral seedlings. Funct Ecol 24:70–81

Pratt RB, Jacobsen AL, Hernandez J, Ewers FW, North GB, Davis SD (2012a) Allocation tradeoffs among chaparral shrub seedlings with different life history types (Rhamnaceae). Am J Bot 99:1464–1476
Pratt RB, Jacobsen AL, Jacobs SM, Esler KJ (2012b) Xylem transport safety and efficiency differ among fynbos shrub life history types and between two sites differing in mean rainfall. Int J Plant Sci 173:474–483
Pratt RB, Jacobsen AL, Ramirez AR, Helms AM, Traugh CA, Tobin MF, Heffner MS, Davis SD (2014) Mortality of resprouting chaparral shrubs after a fire and during a record drought: physiological mechanisms and demographic consequences. Glob Chang Biol 20:893–907
Ramirez A, Pratt R, Jacobsen A, Davis S (2012) Exotic deer diminish post-fire resilience of native shrub communities on Santa Catalina Island, southern California. Plant Ecol 213:1037–1047
Rosell JA, Olson ME, Aguirre-Hernández R, Carlquist S (2007) Logistic regression in comparative wood anatomy: tracheid types, wood anatomical terminology, and new inferences from the Carlquist and Hoekman southern Californian data set. Bot J Linn Soc 154:331–351
Sano Y, Jansen S (2006) Perforated pit membranes in imperforate tracheary elements of some angiosperms. Ann Bot 97:1045–1053
Schenk HJ, Espino S, Goedhart CM, Nordenstahl M, Cabrera HIM, Jones CS (2008) Hydraulic integration and shrub growth form linked across continental aridity gradients. Proc Natl Acad Sci 105:11248–11253
Thomas C, Davis S (1989) Recovery patterns of three chaparral shrub species after wildfire. Oecologia 80:309–320
Tobin MF, Pratt RB, Jacobsen AL, De Guzman ME (2013) Xylem vulnerability to cavitation can be accurately characterised in species with long vessels using a centrifuge method. Plant Biol 15:496–504
Tyree MT, Ewers FW (1991) The hydraulic architecture of trees and other woody plants. New Phytol 119:345–360
Utsumi Y, Bobich EG, Ewers FW (2010) Photosynthetic, hydraulic and biomechanical responses of *Juglans californica* shoots to wildfire. Oecologia 164(2):331–338
Vasey MC, Loik ME, Parker VT (2012) Influence of summer marine fog and low cloud stratus on water relations of evergreen woody shrubs (*Arctostaphylos*: Ericaceae) in the chaparral of central California. Oecologia 170(2):325–337
Vilagrosa A, Bellot J, Vallejo VR, Gil-Pelegrín E (2003) Cavitation, stomatal conductance, and leaf dieback in seedlings of two co-occurring Mediterranean shrubs during an intense drought. J Exp Bot 54:2015–2024
Wagner KR, Ewers FW, Davis SD (1998) Tradeoffs between hydraulic efficiency and mechanical strength in the stems of four co-occurring species of chaparral shrubs. Oecologia 117:53–62
Wheeler E, Thomas R (1981) Ultrastructural characteristics of mature wood of southern red oak (*Quercus falcata* Michx.) and white oak (*Quercus alba* L.). Wood Fiber Sci 13:169–181
Wheeler JK, Sperry JS, Hacke UG, Hoang N (2005) Inter-vessel pitting and cavitation in woody Rosaceae and other vesselled plants: a basis for a safety versus efficiency trade-off in xylem transport. Plant Cell Environ 28:800–812
Williams JE, Davis SD, Portwood KA (1997) Xylem embolism in seedlings and resprouts of *Adenostoma fasciculatum* after fire. Aust J Bot 45:291–300

Chapter 8
The Role of Xylem Parenchyma in the Storage and Utilization of Nonstructural Carbohydrates

Lenka Plavcová and Steven Jansen

1 The Structure, Abundance, and Function of Ray and Axial Parenchyma in Wood

1.1 The Structure of Ray and Axial Parenchyma in Wood

The fact that most of the cells present in mature wood are dead is often highlighted. Indeed, all xylem conduits (vessels and tracheids) found in the functional sapwood undergo cell autolysis, forming hollow tubes made of lignified secondary cell wall. However, cells with a protoplasm are also present in secondary xylem. These cells are referred to as wood parenchyma because their cell wall is often much thinner than that of fibers. Depending on their arrangement and orientation with respect to the main stem axis, parenchyma cells can be divided into two distinct types—radial (ray) and axial parenchyma.

Ray parenchyma consists of ribbon-like aggregates of cells that are produced by ray initials, extending radially from the cambial zone in the xylem and phloem. Xylem rays can be classified depending on their width as uniseriate (Fig. 8.1a–d), biseriate (Fig. 8.1e), and multiseriate, referring to rays that are one-, two-, or more cells wide. Rays can also be subdivided according to the dimensions of individual cells viewed in a radial section. While most of the ray parenchyma cells have their longest axis oriented radially (procumbent cells), vertically elongated (upright) or isodiametric (square) cells also occur. Rays comprised exclusively of procumbent cells are called homocellular, while rays made of more than one parenchyma cell type are termed heterocellular (Carlquist 2001; Evert 2006). The entire ray system can consist of a single ray type, but a combination of different ray types commonly

L. Plavcová (✉) • S. Jansen
Institute for Systematic Botany and Ecology, Ulm University,
Albert-Einstein-Allee 11, Ulm 89081, Germany
e-mail: lenka.plavcova@gmail.com

U. Hacke (ed.), *Functional and Ecological Xylem Anatomy*,
DOI 10.1007/978-3-319-15783-2_8

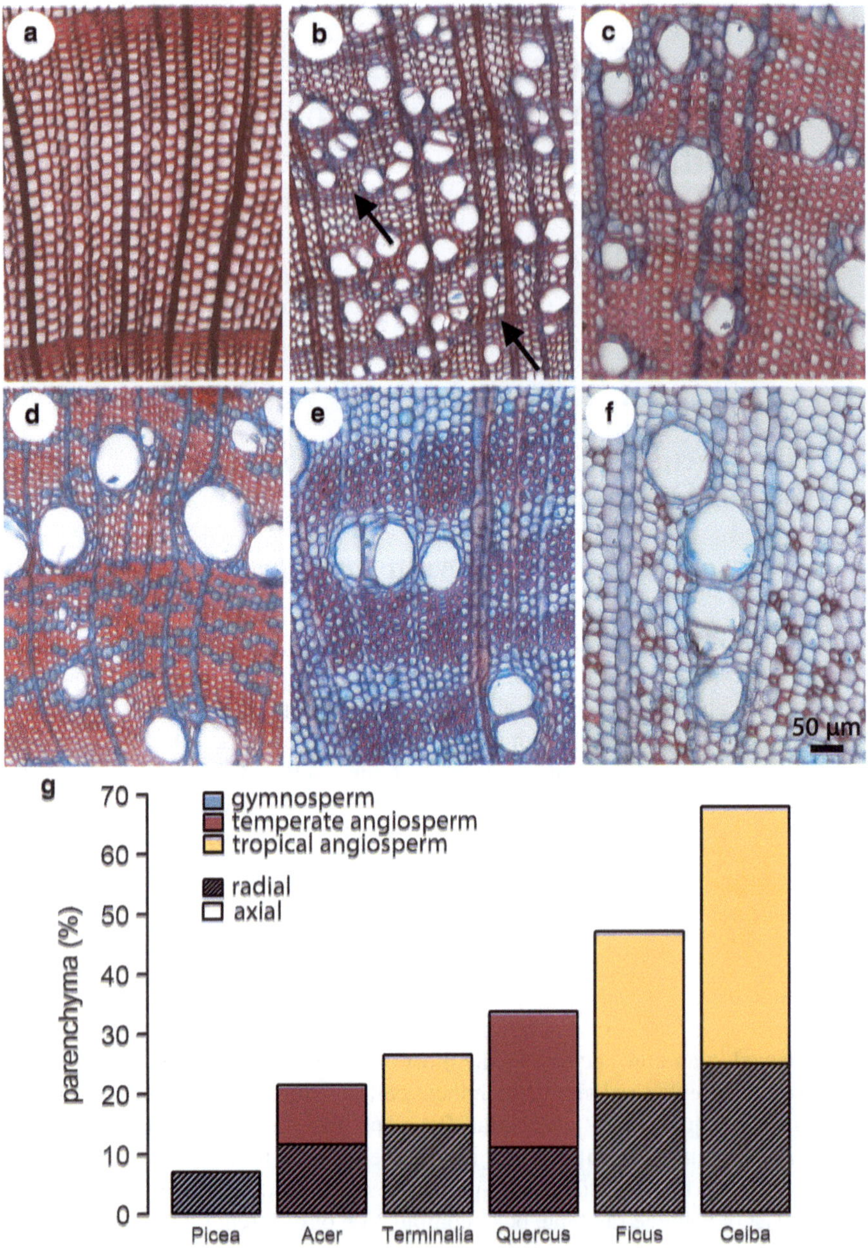

Fig. 8.1 Diversity in ray and axial parenchyma patterns as viewed in transverse wood sections stained with *safranin* and *alcian blue*. Thick lignified secondary cell walls of fibers and vessels stain *pink*, while protoplasts and less extensively lignified cell walls of parenchyma cells appear *blue*. All images were taken at the same magnification; a scale bar is shown in (**f**). (**a**) A conifer species (*Picea abies*) showing the lowest proportion of parenchyma; rays are uniseriate, axial parenchyma is absent, (**b**) a temperate, diffuse-porous tree (*Acer pseudoplatanus*)

occurs. In addition to the ray types described above, other specialized ray systems such as aggregated and interconnected rays have been identified. Interestingly, a temporal or permanent absence of rays occurs in the wood of several species (Barghoorn 1941; Carlquist 1970). Nevertheless, the complete absence of rays is a rarity restricted to small plants, in which woodiness is not pronounced and has evolved secondarily (Carlquist 1970).

Axial parenchyma is produced by fusiform cambial initials that undergo transverse divisions, resulting in a parenchyma strand or axial series of two or more parenchyma cells. In some species, axial parenchyma is absent or sparse. This condition is characteristic of conifers (Fig. 8.1a); however, it can be found in many angiosperms as well. For instance, sparse axial parenchyma occurs in *Populus*, *Aesculus*, *Berberis*, *Magnolia*, and *Eucalyptus*, to name a few examples. If present, axial parenchyma can be arranged in different patterns as distinguished in transverse sections (Carlquist 2001; Kribs 1937). Traditionally, axial parenchyma is classified as apotracheal if it appears distributed without a direct connection to vessels (Fig. 8.1d), and paratracheal if it is strongly associated with xylem vessels (Fig. 8.1b, c). These two basic categories can be further subdivided. For instance, parenchyma cells can be randomly scattered within the vessels and fibers (diffuse apotracheal parenchyma), be in contact with vessels but not ensheathing them completely (scanty paratracheal parenchyma, Fig. 8.1b), form a complete sheath surrounding vessels (vasicentric paratracheal parenchyma, Fig. 8.1c), or be arranged in distinct tangential bands (banded parenchyma, Fig. 8.1d, e). In addition, an increased occurrence of parenchyma cells at the tree ring boundary is often found in temperate species and referred to as marginal parenchyma (Fig. 8.1b). Last but not least, wood of some trees such as *Adansonia* (Chapotin et al. 2006) or *Ceiba* (Fig. 8.1f) is extremely parenchymatous, with axial parenchyma comprising most of the matrix between vessels and rays. The categorization outline above is useful, but to a certain extent arbitrary. Thus, intermediary patterns and co-occurrence of more than one type of axial parenchyma are frequently observed. Besides thin-walled parenchyma cells, some thick-walled axially oriented cells, which could be hence termed fibers, also retain living protoplasts (Fahn and Leshem 1963). Axial cells that can be classified as living fibers rather than parenchyma occur, for instance, in *Acer* (Fig. 8.1b),

Fig. 8.1 (continued) and biseriate rays, axial cells that can be, due to their thick secondary cell wall, classified as living fibers are in a scanty paratracheal and marginal arrangement (*arrows*), (**c**) a tropical dry-deciduous tree (*Terminalia catappa*) with uniseriate rays and paratracheal vasicentric parenchyma, (**d**) a ring-porous temperate species (*Quercus robur*) with apotracheal axial parenchyma arranged in narrow bands and scanty paratracheal parenchyma contacting the vessels, rays are uniseriate (but multiseriate, up to 30-cells wide rays are common in older stems of this species), (**e**) a tropical evergreen tree (*Ficus rubiginosa*) with uniseriate and biseriate rays and axial parenchyma arranged in wide bands, (**f**) a tropical dry-deciduous tree (*Ceiba aesculifolia*) showing highly parenchymatous wood composed of multiseriate rays and thin-walled axial parenchyma cells. (**g**) The relative proportion of ray and axial parenchyma cells measured using transverse sections for all six species shown above

Robinia pseudoacacia (Yamada et al. 2011), or *Cactaceae* (Mauseth and Plemons-Rodriguez 1997), although the distinction between these two cell types has often been neglected. When these cell types are distinguished, it has been hypothesized that living fibers substitute for, or complement, the function of axial parenchyma cells (Carlquist 2001; Yamada et al. 2011; Wheeler et al. 2007).

1.2 How Much Ray and Axial Parenchyma Occurs in Wood?

Ray and axial parenchyma cells make up a substantial proportion of all wood cells (Fig. 8.1g). The volumetric content is hard to measure directly; however, the proportion of parenchyma cells can be estimated from the area measurements taken on a transversal or tangential section. In gymnosperms, the total parenchyma proportions are commonly between 5 and 10 % and compose mainly of radial parenchyma (Fig. 8.1a). In angiosperms, the total amount or parenchyma ranges typically between 20 and 40 % (Fig. 8.1b–d); however, values between 40 and 60 % are not uncommon among tropical angiosperms (Fig. 8.1e). The proportion of ray parenchyma is typically around 10–20 %, while the axial parenchyma proportions between 1 and 30 % are common in angiosperms. The aforementioned numbers represent values typically encountered in wood (Fig. 8.1g) (Von Frey-Wyssling and Aeberli 1942; Wagenführ 2007; Ruelle et al. 2006; Zieminska et al. 2013); however, more extreme values also occur. For example, very low ray proportions of around 7 % were reported for two *Acacia* species (Zieminska et al. 2013), while very high axial parenchyma proportions of 67 % were measured in *Ceiba aesculifolia* (Fig. 8.1f) and several species of the genus *Adenia* (Hearn 2009).

1.3 The Function of Ray and Axial Parenchyma in the Storage of Nonstructural Carbohydrates

The function of wood parenchyma in storage is often highlighted and put in contrast with the main role of vessels in facilitating water conduction and the role of fibers in providing the mechanical support. Nonstructural carbohydrates (NSC) represent the most abundant reserves stored in wood parenchyma. The importance of NSC storage for tree growth and functioning has been known for many decades (Kozlowski 1992; Kramer and Kozlowski 1979). Recently, this topic has received renewed attention because the size and the dynamics of the NSC pool might represent factors potentially limiting tree growth (Palacio et al. 2014) and affecting tree survival under drought stress (McDowell et al. 2008).

The total volume of ray and axial parenchyma can be viewed as a finite compartment potentially available for storage. Given the large volume of woody trunks, the size of this storage pool is substantial from the whole plant perspective. For instance,

Würth et al. (2005) calculated that the carbon stored in the above ground woody biomass accounts for 80 % of the total carbon pool present in a seasonally dry tropical forest and that this amount would be sufficient to completely regrow the entire canopy. However, the storage capacity provided by wood parenchyma is not always completely filled up. Instead, the NSC levels fluctuate, reflecting the dynamic balance between carbohydrate production and utilization.

Besides the total amount of NSCs, their partitioning into starch and soluble sugar fraction is of importance. Out of these two components, starch can be considered as the primary long-term storage form of NSCs. Its molecules are large and cannot move freely between cells; however, they can be readily hydrolyzed to produce soluble sugars. Soluble sugars constitute a plethora of mono- and oligosaccharides that are mobile and fulfill more active physiological roles.

As ray and axial parenchyma are the main sites of NSC accumulation in wood, the NSC status of these cells should be directly mirrored in the NSC content and composition measured in the bulk sapwood. The sapwood NSCs have been analyzed in a number of ecological studies (Hoch et al. 2003; Sauter and Wellenkamp 1998; Ashworth et al. 1993; Palacio et al. 2007; Carbone et al. 2013), providing insights into the size and dynamics of the wood parenchyma carbohydrate pool. In the following sections, we will review the NSC accumulation patterns observed in different tree species and different woody organs and discuss changes in NSC concentration and composition that occur throughout the growing season. These aspects of sapwood NSC dynamics will be linked to the anatomy and physiology of ray and axial parenchyma cells.

2 Patterns in NSC Accumulation Across Different Woody Species, Organs, and Time

2.1 Variation in NSC Across Different Tree Species

As already mentioned, a positive relationship between the NSC content and the proportion of ray and axial parenchyma in wood can be expected. To the best of our knowledge, this assumption has not been confirmed empirically, except for the notion that conifer wood tends to have lower NSC concentrations than the wood of angiosperms, which is in agreement with the lower proportion of parenchyma found in conifer wood (Johnson et al. 2012). Considering the large differences in wood parenchyma content across angiosperms, as illustrated by the more than threefold variation shown by the five angiosperms in Fig.8.1a–g, it would be interesting to see if the tendency for a higher NSC content with increasing volume of parenchyma holds true also within this group. While the meta-analysis of published data can provide useful insights (Johnson et al. 2012), the finer-scale patterns are likely to be confounded by the different sampling schemes employed by different authors and the high variation in NSC concentration found across different woody organs and

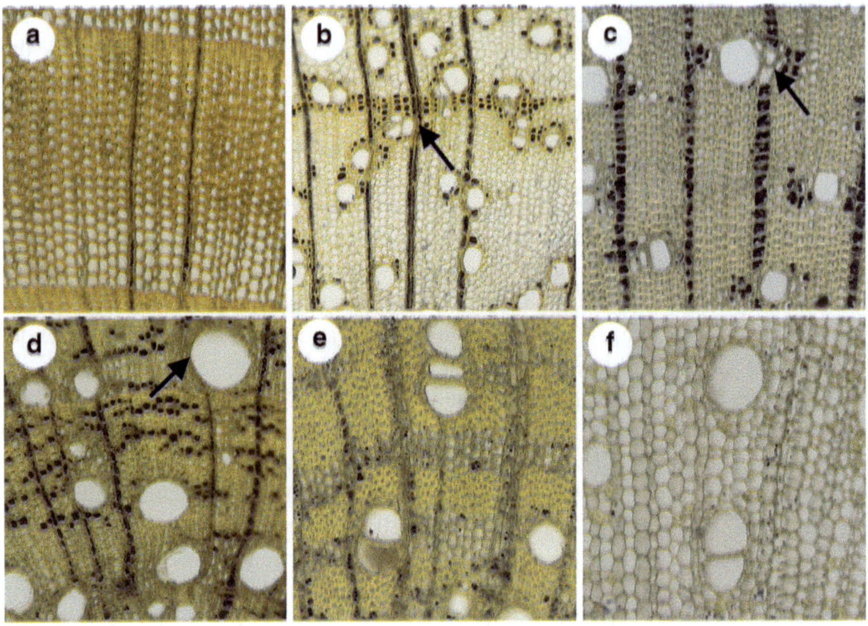

Fig. 8.2 Distribution of starch in wood parenchyma cells in the same species as shown in Fig. 8.1. The wood samples of the temperate species *Picea abies* (**a**), *Acer pseudoplatanus* (**b**), and *Quercus robur* (**d**) were collected at the Ulm University campus in October. In these species, most of the wood parenchyma cells appear densely packed with starch grains. The samples of the tropical plants *Terminalia catappa* (**c**), *Ficus rubiginosa* (**e**), and *Ceiba aesculifolia* (**f**) were obtained from the living collections of the botanical garden of Ulm University in early June. Despite growing under the same conditions of a tropical greenhouse, these species show marked differences in starch deposition. While *Terminalia* accumulates large amounts of starch in xylem parenchyma cells (**c**), the wood parenchyma in *Ficus* (**e**) and in particular in *Ceiba* (**f**) show much lower starch content. Note the absence of starch in contact cells in *Acer*, *Terminalia*, and *Quercus* (*arrows*)

seasons. In addition, the sapwood NSC content is usually expressed as a mass-based concentration, and therefore depends on wood density, which varies substantially between different species (Chave et al. 2009) and to some extent also within a single tree (Domec and Gartner 2002; McCulloh et al. 2012). This problem could be solved by expressing the NSC concentration on a wood volume rather than a wood mass basis. Unfortunately, the studies looking at the sapwood NSC content do not usually report wood density values to allow this conversion. Thus, additional research is required to demonstrate conclusively whether the proportion of ray and axial parenchyma is an important driver of the amount of NSC stored in wood. Here, we will provide some initial hints on answering this question.

Using the same species as shown in Fig. 8.1a–g, we visualized the starch deposition by staining with Lugol's solution (Fig. 8.2a–f). The iodine test revealed obvious differences in starch accumulation in these species that differ greatly in the amount of wood parenchyma. The parenchyma was packed with starch in all three temperate species sampled at the end of the growing season (Fig. 8.2a, b, d). In contrast, the tropical plants showed varying patterns in starch accumulation, despite growing under the same conditions in a tropical greenhouse (Fig. 8.2c, e, f). The starch abundance was highest in *Terminalia* in which almost all parenchyma cells were filled with starch grains (Fig. 8.2c). The amount of starch was much lower in *Ficus* (Fig. 8.2e). In this species, the starch grains were sparse and almost absent in the regions closer to the cambium. The lowest starch accumulation was observed in the highly parenchymatous wood of *Ceiba aesculifolia* (Fig. 8.2f). Based on this simple iodine test, we cannot say whether most of the NSC were present in the form of soluble sugars in the wood of *Ficus* and *Ceiba*, or whether the total levels of NSC were low in spite of the high parenchyma content found in these two species. More than threefold differences in wood NSC concentration have recently been reported in a study encompassing 17 tropical trees from 10 different families (Würth et al. 2005). It would be interesting to see if this variability is at least partially explained by the amount of wood parenchyma. Marked differences in NSC concentration were observed not only between species but also between different tree parts. Thus, we will continue this review by comparing the patterns in NSC content found between woody branches, trunks, and roots.

2.2 *Variation in NSC Across Different Woody Organs*

Most studies monitoring the NSC levels in more than one woody organ report values for the main trunk and small terminal branches. Three times higher NSC concentrations in branch wood than in trunk wood were observed in eleven temperate trees (Hoch et al. 2003; Sala and Hoch 2009); however, such trend was much less pronounced in a tropical environment (Würth et al. 2005; Newell et al. 2002). The proportion of parenchyma does not differ greatly between branches and trunks (Bhat et al. 1985; Koch 1985); therefore, any differences in NSC concentrations between these two tissues are likely caused by physiological rather than anatomical drivers. Alternatively, it is possible that the branch wood NSC values are overestimated because of the inclusion of pith in the samples used for branch wood NSC measurements. Besides affecting the total branch biomass, medullary (i.e., pith) tissue is known to accumulate starch (Essiamah and Eschrich 1985). Both of these phenomena could bias the NSC concentrations toward higher values.

Nevertheless, the branch wood NSC pool consistently appears to be more dynamic than the trunk wood pool. This was manifested by a more dramatic seasonal change (Würth et al. 2005; Hoch et al. 2003; Newell et al. 2002), a higher proportion of the soluble sugar fraction (Sala and Hoch 2009), and a steeper increase in NSC levels with increasing tree height (Sala and Hoch 2009; Woodruff and Meinzer 2011) in branches as compared to trunks. The aforementioned differences between branches and trunks are consistent with the more proximal position of branches to the source of photoassimilates and developing buds and fruits that act as strong carbon sinks. Therefore, it can be suggested that branch wood parenchyma helps to buffer against a short-term imbalance between carbon supply and demand.

In contrast, the trunk wood parenchyma might be more specialized for long-term storage. This idea is supported by a recent study showing surprisingly long residence times of NSC extracted from the outermost 2 cm of sapwood in red maple and eastern hemlock. Using radiocarbon dating, the mean age of sapwood NSC has been estimated to be about a decade, although a substantial fraction of carbon seemed to have much faster turnover (Richardson et al. 2013). This faster fraction likely represents carbohydrates used to support parenchyma metabolism and cambial growth (Carbone et al. 2013; Hill et al. 1995).

In older trunks, the NSC concentration is known to decline radially with increasing distance from the cambium (Würth et al. 2005; Hoch et al. 2003). This decline is likely linked with parenchyma cell death and transition to heartwood. While low values of NSC concentrations are typically reached at the depth of 15–20 cm in trunks that showed a stem diameter of 30–100 cm, the radial patterns show interesting interspecific differences and variation with tree age (Würth et al. 2005; Hoch et al. 2003; Barbaroux and Bréda 2002). More specifically, the decrease in NSC concentration with increasing sapwood depth was particularly sharp in ring-porous oak, while diffuse porous trees exhibited a more gradual decline (Hoch et al. 2003; Barbaroux and Bréda 2002). The tropical tree, *Luehea seemanii*, exhibited remarkably constant NSC content throughout the entire 12-cm-thick sapwood (Würth et al. 2005).

Besides above-ground xylem biomass, woody roots also accumulate large amounts of NSC (Würth et al. 2005; Palacio et al. 2007; Loescher et al. 1990; Pate et al. 1990). High NSC storage capacity of belowground woody tissue is in accordance with a higher proportion of ray and axial parenchyma typically found in small woody roots compared to branches (Lens et al. 2000; Pratt et al. 2007). However, in coarser roots, the relative proportion of wood parenchyma might not be significantly different from stems due to an increase in inter-ray distance with increasing root diameter (Koch 1985; Wargo 1976).

The below-ground storage pool is particularly important when coping with disturbances that destroy a substantial portion of the above-ground plant biomass. A greater dependence on below-ground storage is characteristic for plants resprouting after disturbance, as opposed to plants that regenerate from seeds. In agreement, a higher proportion of wood parenchyma, paralleled by a greater amount of starch reserves, have been observed in resprouters than seeders growing in fire-prone habitats of Western Australia and South Africa (Pate et al. 1990; Verdaguer and Ojeda 2002). Interestingly, however, the overall starch tissue content was more strongly driven by the starch packing density than the amount of parenchyma tissue (Pate et al. 1990). Furthermore, not all the plants under study, including some of the resprouters, accumulated starch in ray and axial parenchyma cells. While some species accumulated starch only in their root cortex, other plants deposited starch in both or only one of the wood parenchyma subsystems. Yet other species had starch grains distributed in all three tissues. Unfortunately, it is not known if the same patterns occurred consistently throughout the season and what the levels of soluble sugars in the roots of these plants were.

Taken together, the comparison between species and woody organs suggests that the amount of parenchyma is important for the overall capacity to store carbohydrates;

however, the differences in starch accumulation patterns, possibly tied to the concentration of soluble sugars, provide another layer of complexity.

2.3 Seasonal Variation in NSC

The NSC content and its partitioning between starch and soluble sugars is also known to vary seasonally. The seasonal dynamics of wood NSC is most widely studied in temperate winter-deciduous trees (Kozlowski 1992; Ashworth et al. 1993; Sauter and van Cleve 1994), providing the following general picture. The total NSC concentration usually peaks at the end of the growing season and declines throughout winter, reaching its minimum during or shortly after bud break. Importantly, starch is often almost completely hydrolyzed during winter months in response to low temperatures, resulting in an increased concentration of soluble sugars (Fig. 8.3a) (Sauter and Wellenkamp 1998; Schoonmaker 2013). At the end of winter, the starch is transiently resynthesized, only to be hydrolyzed again shortly before bud break (Essiamah and Eschrich 1985; Sauter and van Cleve 1994). Starch and the total NSC levels then recover over the growing season. In contrast to deciduous trees, the peak in NSC concentration commonly occurs before bud break and remains low throughout the growing season in both temperate and boreal conifers (Hoch et al. 2003; Schoonmaker 2013). The aforementioned patterns make intuitive sense in terms of the typical progression of photosynthetic activity and growth and likely hold true on a large scale. However, recent studies indicate that this view might be an oversimplification and that various modifications of this general pattern can be found across different woody tissues, species, sites, and seasons (Hoch et al. 2003; Richardson et al. 2013). For instance, Hoch et al. (2003) did not observe a considerable reduction of NSC concentration during bud break in most of the angiosperm species studied. In some years, Richardson et al. (2013) even found higher NSC concentrations in March than in November in maple and oak, suggesting that a redistribution of NSC took place during the dormant season.

The patterns in carbohydrate concentration appear even more variable in a seasonally dry tropical forest. In a study on four trees differing in their leaf phenology (Newell et al. 2002), the branch wood of a truly drought-deciduous species (*Cecropia longipes*) exhibited a fourfold higher NSC concentration during the dry season, which was driven by a large increase in starch concentration (Fig. 8.3b). In contrast, the brevi-deciduous trees such as *Anacardium excelsum* showed much smaller seasonal variation in their NSC levels and a slight increase in soluble sugar fraction during the dry season (Fig. 8.3c). The tendency for higher levels of NSC in branch wood during the dry season was confirmed by a follow-up study encompassing 17 species (Würth et al. 2005). However, the seasonal effect was relatively weak in comparison with a striking interspecific variability.

The seasonal changes in sapwood NSC reflect the balance between carbon supply by photosynthesis and carbon utilization for various physiological needs such as growth, reproduction, or stress mitigation. The structure of ray and axial parenchyma

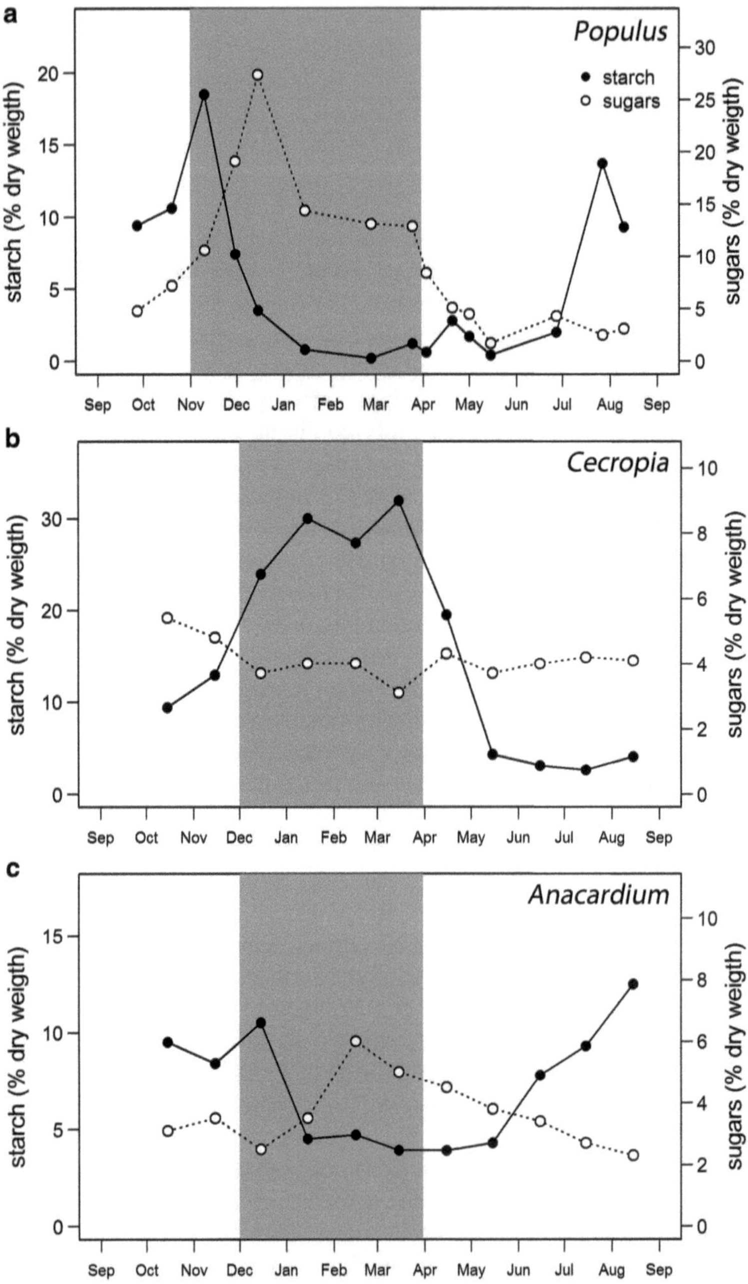

Fig. 8.3 Seasonal course of starch and soluble sugar concentrations in branch wood of winter-deciduous *Populus × canadensis* (**a**), drought-deciduous *Cecropia longipes* (**b**), and brevi-deciduous *Anacardium excelsum* (**c**). The shaded area in each graph indicates the period with unfavorable growing conditions (i.e., the winter in case of *Populus* and the dry season in the case of *Anacardium* and *Cecropia*). Data redrawn from Sauter and van Cleve (1994) (**a**) and Newell et al. (2002) (**b, c**)

and their biochemical machinery likely have a great influence on the NSC dynamics described above. The total parenchyma volume should be the parameter most closely related to the seasonal maximum in NSC content, assuming that the storage capacity of parenchyma cells is fully used during this period. In contrast, the dynamics of NSC is driven by physiological activity of ray and axial parenchyma cells and surrounding source and sink tissues (e.g., leaves, flushing buds, developing fruits, the cambium). Furthermore, the spatial proximity and connectivity between wood parenchyma and these sinks and sources is important for facilitating the NSC translocation within the plant body. While mechanisms underlying the buildup of NSC stores remain poorly understood, several studies have focused on processes involved in the mobilization and utilization of starch stored in wood parenchyma cells. The underlying cellular processes will be discussed in the following section.

3 Metabolic Activity of Wood Parenchyma Underlying the Dynamics of Sapwood NSC

3.1 Starch Mobilization and Metabolism of Soluble Sugars

The mobilization of starch reserves is initiated by the depolymerization of its molecules (Fig. 8.4). In plants, starch breakdown can be catalyzed by various enzymes such as amylases, glucosidases, and glucanohydrolases (Zeeman et al. 2010). To the best of our knowledge, only one study has looked at the starch hydrolyzing machinery acting in wood parenchyma (Witt et al. 1995). In this study, numerous enzymes with a potential amylolytic activity have been investigated in the ray parenchyma of *Populus × canadensis* and the main effect on starch degradation has been attributed to α-amylase and it has been hypothesized that the high temperature sensitivity of this enzyme underlies the mid-winter starch degradation and its resynthesis in early spring.

Simple sugars originating from starch hydrolysis are metabolized in a myriad of ways. Typically, the sugar molecules need to be phosphorylated before they can participate in further biochemical reactions. The phosphorylation of sugars is catalyzed by phosphotransferases. One of these enzymes is a plant hexokinase catalyzing the phosphorylation of hexoses, most importantly glucose. In addition, hexokinase also plays a prominent role in sugar-mediated signaling (Jang et al. 1997). While we are not aware of any studies on hexokinase activity in wood parenchyma, a gene encoding for this enzyme has recently been shown to exhibit a xylem parenchyma-specific expression in leaf petioles of tobacco (Giese et al. 2005).

Phosphorylation of glucose is the first step of glycolysis that can be followed by aerobic respiration. Respiration represents the key process through which the chemical energy contained in nutrients is released and made available for fueling the cellular metabolism. Respiration of wood parenchyma has been studied in a series of interesting experiments (Spicer and Holbrook 2005, 2007a, b). The respiration rates expressed per unit sapwood volume were between 0.4 and 1 μmol $O_2 \, cm^{-3} \, h^{-1}$

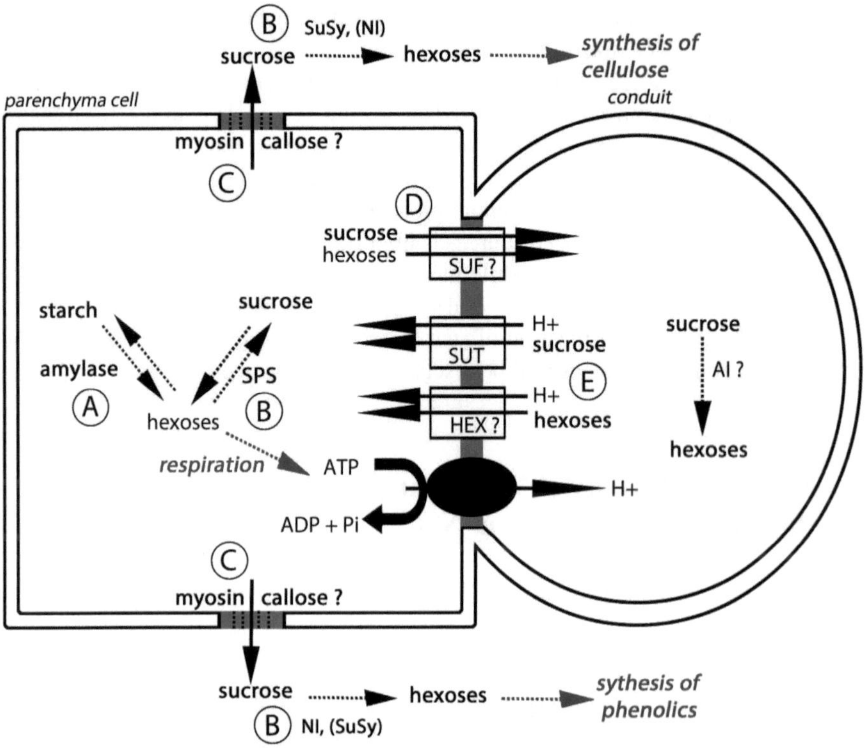

Fig. 8.4 NSC dynamics at the cellular level. Enzymes and transporters involved in the biochemical transformation and intercellular trafficking of NSC are shown. Starch hydrolysis is catalyzed by α-amylase (*A*). The synthesis of sucrose at the sites of starch mobilization is catalyzed by sucrose-6-phosphate-synthase (SPS). Sucrose breakdown at the sink sites is catalyzed by sucrose synthase (SuSy) and neutral invertase (NI) (*B*). Sucrose might also be hydrolyzed in the apoplast by acid invertase (AI). Soluble sugars are transported within the parenchyma network symplastically through pit plasmodesmata. The permeability of plasmodesmata might be affected by myosin contraction and callose deposition (*C*). Soluble sugars can move across a parenchyma-vessel pit membrane to the apoplast. The efflux of sugars to a conduit lumen is passive and driven by the concentration gradient. Putative membrane channels (SUF) likely facilitates the efflux (*D*). Soluble sugars can also be retrieved from the conduit apoplast to the parenchyma symplast by active transport via proton-sugar symporters (SUT, HEX). The electrochemical proton gradient required for the sugar uptake is generated by ATP-dependent proton pumps (*E*)

and did not differ greatly between three angiosperm and two conifer species studied. However, when expressed per living parenchyma volume, the two conifers showed one order of magnitude higher respiration than the angiosperms, with the respiration rates being around 12 and 3 μmol $O_2 cm^{-3} h^{-1}$, respectively. Such respiration rates are much lower than those typically found in meristems, but considerably higher than those measured in tissues purely devoted to storage, such as tubers. This suggests that wood parenchyma cells have a more active role than just being a simple storage compartment. Based on cytochemical staining, respiratory activity appears particularly high in parenchyma cells that are in a direct contact with xylem conduits. These so-called contact or vessel-associated cells are characterized by high mitochondrial counts, high activity of respiratory enzymes (Sauter et al. 1973; Alves et al. 2001), and reduced starch accumulation (Fig. 8.2b, c, d, Braun 1984). Interestingly, wood parenchyma cells of some species contain chloroplasts and are photosynthetically active (Langenfeld-Heyser 1989; Cocoletzi et al. 2013; Larcher et al. 1988). The carbon assimilation rates exhibited by woody stems are low and usually not sufficient to result in a net carbon uptake; nevertheless, they may be involved in the refixation of CO_2 released during parenchyma respiration, thereby reducing respiratory carbon loss (Pfanz et al. 2002).

Sucrose is usually the most abundant component of soluble sugar fraction found in wood. There are three key enzymes governing the metabolism of sucrose in plants—sucrose-6-phosphate-synthase (SPS), sucrose-synthase (SuSy), and invertase (Fig. 8.4). While SPS catalyzes the synthesis of sucrose molecules from phosphorylated monomers, the other two enzymes, SuSy and invertase, are responsible for sucrose catabolism. The difference between these two enzymes is that SuSy catalyzes the conversion of sucrose into fructose and UDP-glucose, while invertase catalyzes the hydrolysis of sucrose into nonphosphorylated monomers. In plants, several types of invertases can be distinguished based on their subcellular localization and pH optimum. While netural invertase (NI) is localized in the cytoplasm, acid invertases (AI) are found in vacuoles and cell walls (Sturm 1999).

Seasonal changes in the activity of all three enzymes, SPS, SuSy, and invertase, which play an important role in sucrose metabolism, were studied along a radial profile in the sapwood of *Robinia pseudoacacia*, providing interesting insights into the coordination of NSC mobilization and utilization (Hauch and Magel 1998). An increased activity of sucrose synthesizing SPS was indicative of starch mobilization. Thus, a high SPS activity was observed throughout the entire width of sapwood during cold winter months when starch is being hydrolyzed into soluble sugars. The high SPS activity persisted in the middle and outermost sapwood during bud break. In contrast, the activities of SuSy and neutral invertase peaked in sink tissues to which sucrose was transported and subsequently catabolized. In spring, SuSy was highly active in sapwood regions close to the cambium, producing UDP-glucose for the synthesis of cell walls of newly developing xylem cells. By contrast, NI was mostly active in the sapwood-to-heartwood transition zone during autumn, likely providing precursors for the synthesis of heartwood phenolic. Similar patterns in SPS and SuSy activity as observed in *Robinia pseudoaccacia* have also been detected in the wood of *Populus × canadensis* (Schrader and Sauter 2002).

3.2 *Translocation of Soluble Sugars*

In order to supply carbon and energy to the cambium and the sapwood-to-heartwood transition zone, sucrose and other soluble sugars arising from starch mobilization have to move radially within the stem (Fig. 8.4). Rays provide an ideal path for such a translocation (Van Bel 1990). Ray parenchyma cells are interconnected via pits

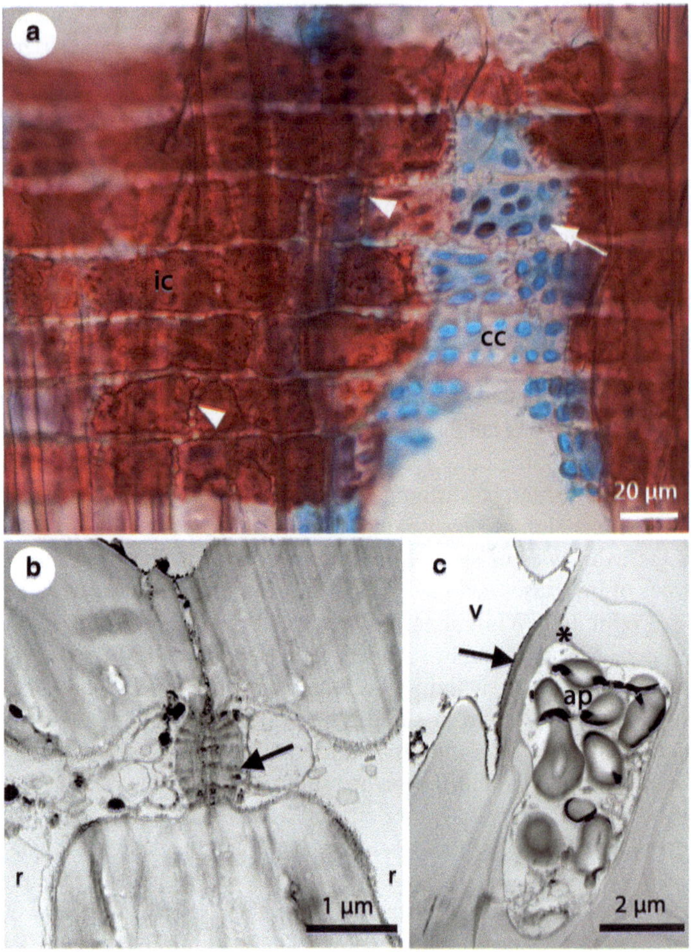

Fig. 8.5 Parenchyma–conduit and parenchyma–parenchyma cell connections in *Quercus robur*. (**a**) A radial section showing a vessel-ray interface observed with a light microscope. Numerous pit connections are apparent at the interface between the ray and vessel (*arrow*) and between the different files of ray cells (arrowheads). A clear distinction between contact (cc) and isolation cells (ic) within the ray is obvious in the picture. (**b**) A slightly bordered pit between two ray cells (r) observed with a transmission electron microscope. The pit membrane is penetrated by plasmodesmata (*arrow*). Various cytoplasmic bodies and vesicles are abundant in the pit channel, suggesting intense transport activity across the pit membrane. (**c**) Cross-section through an axial parenchyma cell (ap) connected to an adjacent vessel (v) via a half-bordered pit. The pit membrane is without plasmodesmata. The amorphous (or protective) layer between the pit membrane and the plasmalemma is thin but still apparent (*asterisk*). From the vessel lumen side, the pit membrane is covered by an electron dense plug, also known as the "black cap" (*arrow*)

perforated by numerous plasmodesmata, such that a symplastic continuum exists within the ray (Fig. 8.5a, b). This continuum can be visualized with symplastic fluorescence tracers (Sokolowska and Zagórska-Marek 2012); however, very little is known about the rate, selectivity, and regulation of this transport pathway. The rate and direction of the bulk symplastic transport of carbohydrates are likely driven by the concentration gradient, similarly to the movement of sucrose during symplastic phloem loading. Based on the dynamics of starch deposition, Sauter and Kloth (1986) calculated a carbohydrate flow rate of 800 pmol cm^{-1} s^{-1} across the tangential ray walls of *Populus × canadensis* and concluded that much of this flux must have occurred via plasmodesmata.

Moreover, the presence of highly ordered microfilaments and microtubules running parallel to the longer (i.e., radial) axis of the ray cells provides possibility for active directional transport (Chaffey and Barlow 2001). Myosin, belonging to the family of ATP-dependent molecular motors, and the polysaccharide callose are localized at the plasmodesmal faces within the pit membranes. Both of these compounds are known to influence the permeability of plasmodesmata (Reichelt et al. 1999; Zavaliev et al. 2011; White and Barton 2011) and could hence facilitate an active regulation of the ray-to-ray cell conductance. Nevertheless, even if plasmodesmata are present and unblocked, the passage of molecules through pits will be associated with a certain resistance. Thus, more efficient radial conduction is expected in rays composed of procumbent rather than square or upright cells, because the number of cell-to-cell connections that need to be crossed is smaller in case of procumbent cells (Carlquist 1975).

Sugars can also move out of the parenchyma cells and enter the conduit lumen. The exchange of carbohydrates between parenchyma cells and conduits is facilitated by conduit–parenchyma pits, which exhibit a different structure than the simple or slightly bordered parenchyma–parenchyma pit pairs (Fig. 8.5a, c). When observed with a transmission electron microscope, the conduit–parenchyma pit membranes appear compact, rather electron dense and free of plasmodesmata (Fig. 8.5c). In addition, a specialized cell wall layer is deposited underneath the pit membrane, lining the entire conduit–parenchyma interface between the plasmalemma and the lignified wall. This so-called amorphous or protective layer may enlarge the actual area available for the exchange of substances (Barnett et al. 1993); however, other functions such as providing a buffer against xylem pressure oscillations were also proposed (Van Bel and Van der Schoot 1988). Both the pit membrane and the amorphous layer are rich in pectins (Wisniewski and Davis 1995; Plavcová and Hacke 2011). The amorphous layer contains also arabinogalactan-rich glycoproteins (AGPs) (Wisniewski and Davis 1995). These extracellular proteins prevent a tight alignment of pectin molecules (Lamport et al. 2006) and hence may increase the porosity and permeability of the amorphous layer. Moreover, AGPs are known to interact with the plasma membrane and act as receptors (Seifert and Roberts 2007), which points to interesting possibilities for a more active role of the amorphous layer in sensing and signaling. Another feature of conduit–parenchyma pits is the formation of an additional pectinaceous plug during winter months observed in several temperate trees (Wisniewski and Davis 1995; Wisniewski et al. 1991a). Because of its high electron density this plug is sometimes referred to as the "black cap." The exact function of the black cap is not known but it might hinder the growth of ice crystals or prevent uncontrolled loss of water and other substances from parenchyma cells during winter dormancy.

Our knowledge of molecular mechanisms involved in the sugar movement between parenchyma cells and conduits is rather limited, with most information coming from a few temperate deciduous trees. In these trees, two opposing sugar fluxes have been identified, namely the sugar efflux from and the sugar influx to the parenchyma cells (Fig. 8.4). The balance between these two fluxes drives the sugar composition of the xylem sap. High sugar concentrations, indicative of high efflux and/or low influx rates, are often found in xylem sap during winter and early spring. For instance, the spring sap concentration of sugar maple (*Acer saccharum*) reaches typically values of 2–3 % (Taylor 1956), while a concentration of about 0.6 % was measured in *Acer platanoides* (Schill et al. 1996) and *Populus × canadensis* (Sauter 1988). In contrast, the summer concentrations are close to 0.1 %.

The efflux of soluble sugars out of parenchyma cells (Fig. 8.4) occurs passively along a concentration gradient (Sauter 1982; Améglio et al. 2004; Münch 1930). Therefore, high efflux rates are expected when the concentration of soluble sugars in parenchyma cells is high. In agreement, high sugar efflux is observed during winter when most of the starch stored in parenchyma cells is hydrolyzed. In walnut, the sap sugar concentration was indeed highest during winter, with sucrose representing the most abundant xylem sap saccharide (Améglio et al. 2002, 2004). The dynamics of sap sugars are different in poplar. In this species, a rapid increase in sap sugar levels was observed during bud break, reaching levels more than three times higher than those measured in winter. Interestingly, hexoses comprised the major fraction of xylem sap sugars during this time, suggesting that sucrose might be hydrolyzed in the apoplast by acid invertase (Sauter 1988). The rapid increase in sap sugar levels indicates the sugar efflux is not just a mere leakage but rather an actively regulated process. The sugar efflux rates are sensitive to inhibitors, suggesting that the efflux is facilitated by membrane channels (Sauter 1982; Améglio et al. 2004). Thus, the modulation of efflux rates can be achieved by changing the abundance and activity of these hitherto uncharacterized proteins.

If sap flow occurs, sugars released into the conduit lumen can be carried via the low-resistance apoplastic pathway toward the canopy. This additional amount of carbon can be valuable to support flushing buds in spring (Bonhomme et al. 2010). However, as vascular connections are often not fully developed during the initial phase of bud reactivation (Ashworth 1982), sugars need to be reabsorbed by parenchyma cells and move to the bud tissue via extraxylary pathways.

In contrast to sugar efflux, the uptake of sugars from the xylem sap by parenchyma cells is an active process facilitated by proton-sugar symporters (Fig. 8.4). Transcript and protein levels of several of these putative transporters have been studied in walnut (Decourteix et al. 2006, 2008). While the sucrose transporter JrSUT1 was strongly up-regulated in xylem parenchyma cells during bud break, two hexose transporters, JrHT1 and JrHT2, were abundant during the period of intense radial growth. This suggests that the sugar uptake is selective and likely tailored to suit specific physiological needs. The symport of sugars is powered by the electrochemical gradient generated by ATP-dependent proton pumps also known as H^+-ATPases (Alves et al. 2007). High expression of H^+-ATPase coincides with a high activity of respiratory enzymes, indicating that the sugar retrieval is energetically demanding but also remarkably efficient. For instance, in willow (*Salix*), the rates of sugar influx have been estimated to be more than five times higher than the rates of sugar efflux (Sauter 1983).

In this section, we described the cellular processes that underlie the dynamics of NSC in sapwood (Fig. 8.4), starting with the conversion of starch into soluble sugars, continuing with the in situ use of sugars, and finishing with their radial and axial transport into more distant sink tissues. We showed that NSC dynamics are driven by the activity of key sugar-modifying enzymes and transport systems, acting within the anatomical and physiological boundaries provided by wood parenchyma cells. In the next section, we will briefly discuss potential implications of these processes for whole plant physiology.

4 The Role of Sapwood NSC at the Whole Plant Level

Carbohydrate storage is important for a tree's ability to withstand periods of unfavorable environmental conditions and to reactivate its growth when favorable conditions are reestablished. Interestingly, the NSC reserves are rarely depleted in trees, leading to the suggestion that tree growth and survival is not limited by carbon supply (Körner 2003). Alternatively, it has recently been proposed that trees actively maintain high NSC concentration at the expense of growth in order to sustain plant functioning under environmental stress (Sala et al. 2012a; Wiley and Helliker 2012). We believe that the wide array of tightly regulated physiological processes taking place in ray and axial parenchyma cells fits well into this picture. While starch accumulation in wood parenchyma at the end of the growing season can be viewed as manifestation of a long-term storage function, the complex dynamics of soluble sugars can be perceived as a suite of active physiological processes—some are related to maintenance respiration and growth while others are mostly involved in stress mitigation (Fig. 8.6). For the sake of simplicity, we will outline these functions as consecutive events progressing over seasons typical for a temperate climate. However, we recognize that not all of these functions are relevant all the time. Instead, different functions can be more important in different tree species or under particular circumstances, resulting in different requirements on structural and physiological properties of wood parenchyma cells.

In winter, two important physiological roles of soluble sugars can be identified, namely the protection of parenchyma cells from freeze injury and reversal of freeze-induced embolism. Subzero temperatures can damage or even kill wood parenchyma cells. Therefore, two strategies for coping with freezing temperatures evolved in these cells—they either tolerate extracellular ice formation or avoid freezing by deep supercooling (Sakai et al. 1987; Kuroda et al. 2003; Burke et al. 1976). In the case of freeze tolerance, an increased concentration of soluble sugars resulting in higher osmotic potential of the cytoplasm helps to prevent cellular dehydration driven by extracellular freezing (Yuanyuan et al. 2009; Cavender-Bares 2005). In case of freeze avoidance, soluble sugars may help to inhibit the formation

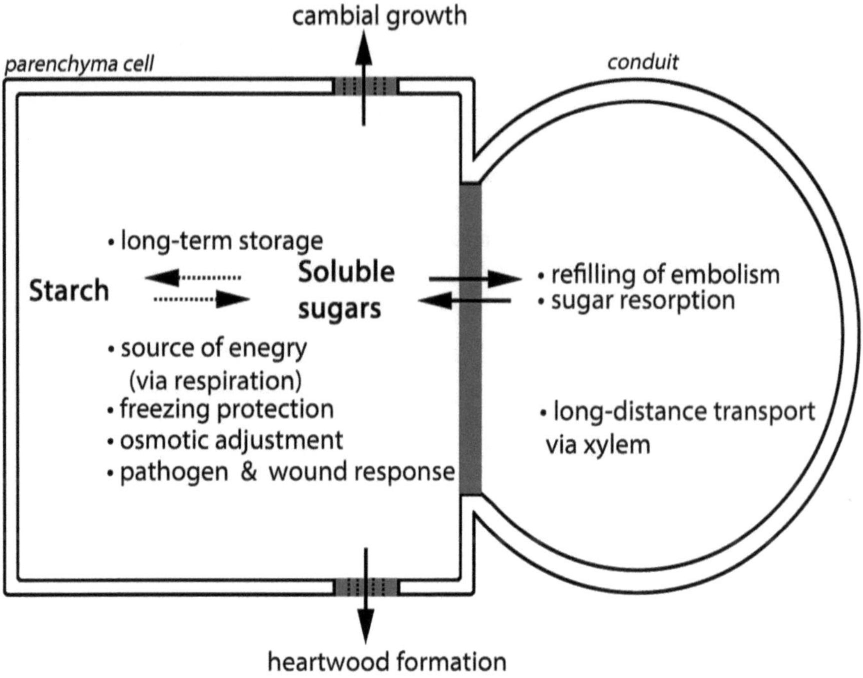

Fig. 8.6 Schematic representation of various functions that starch and soluble sugars fulfill in sapwood. Starch represents the primary form of nonstructural carbohydrates used for the long-term storage. Starch can be converted into soluble sugars that fulfill more active physiological roles. Soluble sugars are used for respiration. They are important for defense against pathogens by providing energy and material for the synthesis of defense chemicals. Increased concentration of soluble sugars in wood parenchyma cells may prevent freeze and desiccation damage. Soluble sugars can move radially via the symplastic continuum in ray parenchyma cells toward the cambium or toward the center of the stem to supply carbon and energy for the formation of new xylem or for the synthesis of heartwood extractives. Soluble sugars are also secreted into the apoplast where they can drive refilling of xylem embolism. Sugars released to the xylem sap can be carried upstream and retrieved closer to the plant apex, thereby supplying carbon to flushing buds

of ice crystals and to stabilize the plasmatic membrane. Furthermore, the integrity of the amorphous layer and its pectin composition are important for the ability of parenchyma cells to undergo supercooling (Wisniewski and Davis 1989; Wisniewski et al. 1991b).

Repeated freeze–thaw cycles are known to induce embolism of xylem conduits even under modest tensions. Ring-porous trees cope with this phenomena by producing new conduits in spring, whereas many diffuse porous species are capable of refilling embolized conduits (Hacke and Sauter 1996). Refilling can be driven by positive root or stem pressure or a combination of the two. While the accumulation of inorganic nutrients in the root apoplast seemed to underlie the development of root pressure in *Juglans*, soluble sugars released by the parenchyma cells were critical for the generation of positive stem pressure in both *Juglans* (Améglio et al. 2002, 2004;

Ewers et al. 2001) and *Acer* (Sauter et al. 1973; Hacke and Sauter 1996). Thus, in climates where freezing occurs, wood parenchyma can be important for the restoration of vascular integrity at the start of a new growing season. On the other hand, a high amount of parenchyma represents a challenge because freeze damage to the living tissue needs to be prevented.

In spring, the main function of soluble sugars is to support new growth in order to quickly reestablish photosynthetic production. Expanding buds and the active cambial zone represent the strongest sinks during this period. Both of these tissues receive carbohydrates stored in sapwood (Hill et al. 1995; Bonhomme et al. 2010; Decourteix et al. 2008). While little is known about the partitioning of reserves between these two tissues, it can be expected that it is closely related to the offset between the cambium and bud phenology. The cambium likely represents a more important sink for stored carbohydrates in ring-porous than diffuse-porous species because a large proportion of the radial stem growth occurs before the onset of photosynthetic activity in ring-porous trees (Barbaroux and Bréda 2002; Panchen et al. 2014). In species that bloom before the leaf-out, the opening flower buds draw strongly on stored reserves as indicated by a pronounced decline in branch wood NSC levels (Hoch et al. 2003). Higher allocation of NSC reserves into the cambium should put more requirements on the radial transport mechanisms via the ray symplast, while the axial transport pathway, which involves sugar exchange between wood parenchyma and xylem apoplast, should be accentuated in case of higher needs for sugar translocation into the buds.

In summer, when the canopy is fully developed and photosynthetically active, soluble sugars found in the sapwood could help to prevent and repair damage caused by environmental stress. In the more traditional sense, the importance of sapwood NSC reserves should be seen in the possibility to regrow leaves in case of severe defoliation caused by environmental stress. From the less traditional point of view, the high NSC pool may be needed for a continuous maintenance of hydraulic integrity that is constantly being perturbed. Drought and the attack of pathogens arguably represent the two most important environmental challenges frequently encountered by trees.

The importance of carbohydrates in the repair of drought-induced embolism has been widely recognized. Despite some recent concerns calling the routine occurrence of refilling under tension into question (Wheeler et al. 2013; Sperry 2013), the active release of both sugars and water into the conduit lumen by xylem parenchyma cells is believed to be at the heart of the putative mechanism that may facilitate rapid reversal of drought-induced embolism (Salleo et al. 1996; Tyree et al. 1999; Hacke and Sperry 2003; Secchi et al. 2011; Secchi and Zwieniecki 2011; Brodersen et al. 2010). Moreover, abundant wood parenchyma, as found in many tropical trees, can help to delay the onset of cavitation by providing high water storage capacity (Borchert and Pockman 2005). It is not known if excessive water loss from parenchyma cells during drought can compromise their physiological functions, although desiccation-induced damage to the protoplasm has been documented in wood parenchyma cells during cold stress (Ristic and Ashworth 1994). If the maintenance of turgor pressure is important, for instance for biomechanical reasons (Chapotin et al. 2006), increased concentration of soluble sugars could provide means for reducing the capacitive discharge from wood parenchyma cells.

The importance of ray and axial parenchyma for wound and pathogen responses in wood is also well documented. Most importantly, parenchyma cells produce tyloses and gums that plug old or damaged xylem conduits, thereby preventing uncontrolled spread of pathogens within the xylem pipeline (Bonsen and Kucera 1990; Nicole et al. 1992). The production of these vascular occlusions involves active secretory processes (Rioux et al. 1998) and hormonal signaling (McElrone et al. 2010), and thus is likely associated with high demands for energy that can be drawn from NSC reserves. On the other hand, a higher proportion of thin-walled parenchyma cells that are rich in carbohydrates can make wood more attractive for nutrient-seeking pathogens and herbivores (Schwarze 2007; Martín et al. 2009). This could result in a faster progression of infection once the pathogen succeeds in overcoming the initial defense mechanisms.

In fall, the wood parenchyma NSC stores should be replenished and available to support the tree's physiological functions in winter and during the next growing season. However, some of the NSC can still be consumed for heartwood formation, which is known to occur predominantly during the period of early dormancy (Taylor et al. 2002). As suggested in a recent review (Spicer 2005), heartwood formation should be viewed as an active developmental program during which a conductive but vulnerable sapwood is transformed in a nonconductive but durable heartwood. This process, initiated within wood parenchyma cells, involves a suite of biochemical reactions that use, at least in part, energy and carbon from carbohydrates stored in sapwood (Hauch and Magel 1998; Magel et al. 1994).

In this section, we summarized the most important ways of how sapwood NSC are used in growth, development, and stress mitigation (Fig. 8.6) and showed the tight links to the well-known functions of ray and axial parenchyma cells. However, it is important to note that reserves other than NSC are also stored in wood parenchyma cells, with nitrogen and phosphorous representing the most important ones (Hoch et al. 2003; Langheinrich and Tischner 1991; Sauter and van Cleve 1991). Thus, it is likely that the tree performance is, at least in some occasions, more strongly limited by the availability of these nutrients than by the availability of carbon (Millard and Grelet 2010; Sala et al. 2012b). Nevertheless, we believe that our analysis of NSC dynamics provides a useful conceptual basis that can be applied to better understand the dynamics of other nutrients as well.

5 Future Perspectives

Research on xylem has a great tradition in integrating structure and function and great advances in understanding the plant water transport have been made by linking the anatomy of xylem conduits to functional hydraulic traits (Hacke et al. 2001, 2006; Jansen et al. 2009). We can envision similar progress in elucidating the functional role of parenchyma cells in carbohydrate storage and dynamics, paved by uncovering the great diversity in ray and axial parenchyma structure and their spatial distribution. Such research would greatly benefit from integrating approaches traditionally used in studies on xylem hydraulics (e.g., perfusion experiments, analysis of pit structure) with methods used to examine phloem physiology (e.g., the application of symplastic and apoplastic tracers, radioactive labelling, molecular methods). Moreover, ecological data on sapwood NSC concentration and composition will help to upscale the processes and imply their importance for whole plant functioning.

Most research to date has been made on temperate species. However, wood structure exhibits great diversity and ray and axial parenchyma seems to be more abundant and exhibit more elaborated patterns in tropical trees. Similarly, carbohydrate metabolism in sapwood seems to be more complex, dynamic, and shifted further from the role in long-term storage in the tropics. Therefore, studies conducted on tropical trees might provide further valuable insights.

Acknowledgements L.P. was supported by a postdoctoral fellowship from the Alexander von Humboldt Foundation and research funding from Ulm University and the Ulm University Society (Ulmer Universitätsgesellschaft). J.S. acknowledges the German Research Foundation (DFG) for financial support. We gratefully acknowledge the support and facilities provided by the Botanical Garden and the Electron Microcopy Unit of Ulm University. We thank Hugh Morris for fruitful discussion and useful comments on an earlier version of this manuscript.

References

Alves G, Sauter JJ, Julien J-L, Fleurat-Lessard P, Améglio T, Guillot A et al (2001) Plasma membrane H^+-ATPase, succinate and isocitrate dehydrogenases activities of vessel-associated cells in walnut trees. J Plant Physiol 158:1263–1271

Alves G, Decourteix M, Fleurat-Lessard P, Sakr S, Bonhomme M, Améglio T et al (2007) Spatial activity and expression of plasma membrane H^+-ATPase in stem xylem of walnut during dormancy and growth resumption. Tree Physiol 27:1471–1480

Améglio T, Bodet C, Lacointe A, Cochard H (2002) Winter embolism, mechanisms of xylem hydraulic conductivity recovery and springtime growth patterns in walnut and peach trees. Tree Physiol 22:1211–1220

Améglio T, Decourteix M, Alves G, Valentin V, Sakr S, Julien J-L et al (2004) Temperature effects on xylem sap osmolarity in walnut trees: evidence for a vitalistic model of winter embolism repair. Tree Physiol 24:785–793

Ashworth EN (1982) Properties of peach flower buds which facilitate supercooling. Plant Physiol 70:1475–1479

Ashworth E, Stirm V, Volenec J (1993) Seasonal variations in soluble sugars and starch within woody stems of *Cornus sericea* L. Tree Physiol 13:379–388

Barbaroux C, Bréda N (2002) Contrasting distribution and seasonal dynamics of carbohydrate reserves in stem wood of adult ring-porous sessile oak and diffuse-porous beech trees. Tree Physiol 22:1201–1210

Barghoorn ES (1941) The ontogenetic development and phylogenetic specialization of rays in the xylem of dicotyledons-III. The elimination of rays. Bull Torrey Bot Club 68:317–325

Barnett J, Cooper P, Bonner LJ (1993) The protective layer as an extension of the apoplast. IAWA J 14:163–171

Bhat K, Bhat K, Dhamodaran T, et al. (1985) Wood and bark properties of branches of selected tree species growing in Kerala. KFRI research report, Kerala Forest Research Institute

Bonhomme M, Peuch M, Améglio T, Rageau R, Guilliot A, Decourteix M et al (2010) Carbohydrate uptake from xylem vessels and its distribution among stem tissues and buds in walnut (*Juglans regia* L.). Tree Physiol 30:89–102

Bonsen KJ, Kucera L (1990) Vessel occlusions in plants: morphological, functional and evolutionary aspects. IAWA Bull 11:393–399

Borchert R, Pockman WT (2005) Water storage capacitance and xylem tension in isolated branches of temperate and tropical trees. Tree Physiol 25:457–466

Braun H (1984) The significance of the accessory tissues of the hydrosystem for osmotic water shifting as the second principle of water ascent, with some thoughts concerning the evolution of trees. IAWA Bull 5:275–294

Brodersen CR, McElrone AJ, Choat B, Matthews MA, Shackel KA (2010) The dynamics of embolism repair in xylem: in vivo visualizations using high-resolution computed tomography. Plant Physiol 154:1088–1095

Burke M, Gusta L, Quamme H, Weiser C, Li P (1976) Freezing and injury in plants. Annu Rev Plant Physiol 27:507–528

Carbone MS, Czimczik CI, Keenan TF, Murakami PF, Pederson N, Schaberg PG et al (2013) Age, allocation and availability of nonstructural carbon in mature red maple trees. New Phytol 200:1145–1155

Carlquist S (1970) Wood anatomy of insular species of *Plantago* and the problem of raylessness. Bull Torrey Bot Club 97(6):353–361

Carlquist S (1975) Wood anatomy of Onagraceae, with notes on alternative modes of photosynthate movement in dicotyledon woods. Ann Mo Bot Gard 62:386–424

Carlquist S (2001) Comparative wood anatomy: systematic, ecological, and evolutionary aspects of dicotyledon wood. Springer, Berlin

Cavender-Bares J (2005) Impacts of freezing on long distance transport in woody plants. In: Holbrook NM, Zwieniecki MA (eds) Vascular transport in plants. Elsevier, San Diego, pp 401–424

Chaffey N, Barlow P (2001) The cytoskeleton facilitates a three-dimensional symplasmic continuum in the long-lived ray and axial parenchyma cells of angiosperm trees. Planta 213:811–823

Chapotin SM, Razanameharizaka JH, Holbrook NM (2006) A biomechanical perspective on the role of large stem volume and high water content in baobab trees (*Adansonia* spp.; Bombacaceae). Am J Bot 93:1251–1264

Chave J, Coomes D, Jansen S, Lewis SL, Swenson NG, Zanne AE (2009) Towards a worldwide wood economics spectrum. Ecol Lett 12:351–366

Cocoletzi E, Angeles G, Sosa V, Patron A (2013) The chloroplasts and unlignified parenchyma of two tropical pioneer forest tree species (Urticaceae). Bot Sci 91:251–260

Decourteix M, Alves G, Brunel N, Améglio T, Guilliot A, Lemoine R et al (2006) JrSUT1, a putative xylem sucrose transporter, could mediate sucrose influx into xylem parenchyma cells and be up-regulated by freeze-thaw cycles over the autumn-winter period in walnut tree (*Juglans regia* L.). Plant Cell Environ 29:36–47

Decourteix M, Alves G, Bonhomme M, Peuch M, Baaziz KB, Brunel N et al (2008) Sucrose (JrSUT1) and hexose (JrHT1 and JrHT2) transporters in walnut xylem parenchyma cells: their potential role in early events of growth resumption. Tree Physiol 28:215–224

Domec J-C, Gartner B (2002) Age-and position-related changes in hydraulic versus mechanical dysfunction of xylem: inferring the design criteria for Douglas-fir wood structure. Tree Physiol 22:91–104

Essiamah S, Eschrich W (1985) Changes of starch content in the storage tissues of deciduous trees during winter and spring. IAWA Bull 6:97–106

Evert RF (2006) Esau's Plant anatomy: meristems, cells, and tissues of the plant body: their structure, function, and development, 3rd edn. Wiley, Hoboken

Ewers FW, Améglio T, Cochard H, Beaujard F, Martignac M, Vandame M et al (2001) Seasonal variation in xylem pressure of walnut trees: root and stem pressures. Tree Physiol 21:1123–1132

Fahn A, Leshem B (1963) Wood fibres with living protoplasts. New Phytol 62:91–98

Giese J-O, Herbers K, Hoffmann M, Klösgen RB, Sonnewald U (2005) Isolation and functional characterization of a novel plastidic hexokinase from *Nicotiana tabacum*. FEBS Lett 579:827–831

Hacke U, Sauter J (1996) Xylem dysfunction during winter and recovery of hydraulic conductivity in diffuse-porous and ring-porous trees. Oecologia 105:435–439

Hacke U, Sperry J (2003) Limits to xylem refilling under negative pressure in *Laurus nobilis* and *Acer negundo*. Plant Cell Environ 26:303–311

Hacke UG, Sperry JS, Pockman WT, Davis SD, McCulloh KA (2001) Trends in wood density and structure are linked to prevention of xylem implosion by negative pressure. Oecologia 126:457–461

Hacke UG, Sperry JS, Wheeler JK, Castro L (2006) Scaling of angiosperm xylem structure with safety and efficiency. Tree Physiol 26:689–701

Hauch S, Magel E (1998) Extractable activities and protein content of sucrose-phosphate synthase, sucrose synthase and neutral invertase in trunk tissues of *Robinia pseudoacacia* L. are related to cambial wood production and heartwood formation. Planta 207:266–274

Hearn DJ (2009) Descriptive anatomy and evolutionary patterns of anatomical diversification in *Adenia* (Passifloraceae). Aliso J Syst Evol Bot 27:13–38

Hill S, Waterhouse J, Field E, Switsur V, Ap RT (1995) Rapid recycling of triose phosphates in oak stem tissue. Plant Cell Environ 18:931–936

Hoch G, Richter A, Körner C (2003) Non-structural carbon compounds in temperate forest trees. Plant Cell Environ 26:1067–1081

Jang J-C, León P, Zhou L, Sheen J (1997) Hexokinase as a sugar sensor in higher plants. Plant Cell 9:5–19

Jansen S, Choat B, Pletsers A (2009) Morphological variation of intervessel pit membranes and implications to xylem function in angiosperms. Am J Bot 96:409–419

Johnson DM, McCulloh KA, Woodruff DR, Meinzer FC (2012) Hydraulic safety margins and embolism reversal in stems and leaves: why are conifers and angiosperms so different? Plant Sci 195:48–53

Koch P (1985) Utilization of hardwoods growing on southern pine sites. U.S. Dept. of Agriculture, Forest Service, Washington, DC

Körner C (2003) Carbon limitation in trees. J Ecol 91:4–17

Kozlowski T (1992) Carbohydrate sources and sinks in woody plants. Bot Rev 58:107–222

Kramer PJ, Kozlowski TT (1979) Physiology of woody plants. Academic, New York

Kribs DA (1937) Salient lines of structural specialization in the wood parenchyma of dicotyledons. Bull Torrey Bot Club 64:177–187

Kuroda K, Kasuga J, Arakawa K, Fujikawa S (2003) Xylem ray parenchyma cells in boreal hardwood species respond to subfreezing temperatures by deep supercooling that is accompanied by incomplete desiccation. Plant Physiol 131:736–744

Lamport DT, Kieliszewski MJ, Showalter AM (2006) Salt stress upregulates periplasmic arabinogalactan proteins: using salt stress to analyse AGP function. New Phytol 169:479–492

Langenfeld-Heyser R (1989) CO_2 fixation in stem slices of *Picea abies* (L.) Karst: microautoradiographic studies. Trees 3:24–32

Langheinrich U, Tischner R (1991) Vegetative storage proteins in poplar induction and characterization of a 32-and a 36-kilodalton polypeptide. Plant Physiol 97:1017–1025

Larcher W, Lütz C, Nagele M, Bodner M (1988) Photosynthetic functioning and ultrastructure of chloroplasts in stem tissues of *Fagus sylvatica*. J Plant Physiol 132:731–737

Lens F, Jansen S, Robbrecht E, Smets E (2000) Wood anatomy of the Vanguerieae (Ixoroideae-Rubiaceae), with special emphasis on some geofrutices. IAWA J 21:443–455

Loescher WH, McCamant T, Keller JD (1990) Carbohydrate reserves, translocation, and storage in woody plant roots. Hortscience 25:274–281

Magel E, Jay-Allemand C, Ziegler H (1994) Formation of heartwood substances in the stemwood of *Robinia pseudoacacia* L. II. Distribution of nonstructural carbohydrates and wood extractives across the trunk. Trees 8:165–171

Martín JA, Solla A, Esteban LG, de Palacios P, Gil L (2009) Bordered pit and ray morphology involvement in elm resistance to *Ophiostoma novo-ulmi*. Can J For Res 39:420–429
Mauseth J, Plemons-Rodriguez B (1997) Presence of paratracheal water storage tissue does not alter vessel characters in cactus wood. Am J Bot 84:815
McCulloh KA, Johnson DM, Meinzer FC, Voelker SL, Lachenbruch B, Domec J-C (2012) Hydraulic architecture of two species differing in wood density: opposing strategies in co-occurring tropical pioneer trees. Plant Cell Environ 35:116–125
McDowell N, Pockman WT, Allen CD, Breshears DD, Cobb N, Kolb T et al (2008) Mechanisms of plant survival and mortality during drought: why do some plants survive while others succumb to drought? New Phytol 178:719–739
McElrone AJ, Grant JA, Kluepfel DA (2010) The role of tyloses in crown hydraulic failure of mature walnut trees afflicted by apoplexy disorder. Tree Physiol 30:761–772
Millard P, Grelet G (2010) Nitrogen storage and remobilization by trees: ecophysiological relevance in a changing world. Tree Physiol 30:1083–1095
Münch E (1930) Die Stoffbewegungen in der Pflanze. Gustav Fischer, Jena
Newell EA, Mulkey SS, Wright JS (2002) Seasonal patterns of carbohydrate storage in four tropical tree species. Oecologia 131:333–342
Nicole M, Geiger J, Nandris D (1992) Defense of angiosperm roots against fungal invasion. In: Timell TE (ed) Defense mechanisms of woody plants against fungi. Springer, Berlin, pp 181–206
Palacio S, Maestro M, Montserrat-Martí G (2007) Seasonal dynamics of non-structural carbohydrates in two species of Mediterranean sub-shrubs with different leaf phenology. Environ Exp Bot 59:34–42
Palacio S, Hoch G, Sala A, Körner C, Millard P (2014) Does carbon storage limit tree growth? New Phytol 201:1096–1100
Panchen ZA, Primack RB, Nordt B, Ellwood ER, Stevens A-D, Renner SS et al (2014) Leaf out times of temperate woody plants are related to phylogeny, deciduousness, growth habit and wood anatomy. New Phytol 203(4):1208–1219
Pate JS, Froend RH, Bowen BJ, Hansen A, Kuo J (1990) Seedling growth and storage characteristics of seeder and resprouter species of Mediterranean-type ecosystems of SW Australia. Ann Bot 65:585–601
Pfanz H, Aschan G, Langenfeld-Heyser R, Wittmann C, Loose M (2002) Ecology and ecophysiology of tree stems: corticular and wood photosynthesis. Naturwissenschaften 89:147–162
Plavcová L, Hacke UG (2011) Heterogeneous distribution of pectin epitopes and calcium in different pit types of four angiosperm species. New Phytol 192:885–897
Pratt R, Jacobsen A, Ewers F, Davis S (2007) Relationships among xylem transport, biomechanics and storage in stems and roots of nine Rhamnaceae species of the California chaparral. New Phytol 174:787–798
Reichelt S, Knight AE, Hodge TP, Baluska F, Samaj J, Volkmann D et al (1999) Characterization of the unconventional myosin VIII in plant cells and its localization at the post-cytokinetic cell wall. Plant J 19:555–567
Richardson AD, Carbone MS, Keenan TF, Czimczik CI, Hollinger DY, Murakami P et al (2013) Seasonal dynamics and age of stemwood nonstructural carbohydrates in temperate forest trees. New Phytol 197:850–861
Rioux D, Nicole M, Simard M, Ouellette G (1998) Immunocytochemical evidence that secretion of pectin occurs during gel (gum) and tylosis formation in trees. Phytopathology 88:494–505
Ristic Z, Ashworth EN (1994) Response of xylem ray parenchyma cells of red osier dogwood (*Cornus sericea* L.) to freezing stress. Plant Physiol 104:737–746
Ruelle J, Clair B, Beauchêne J, Prévost M-F, Fournier M et al (2006) Tension wood and opposite wood in 21 tropical rain forest species. 2. Comparison of some anatomical and ultrastructural criteria. IAWA J 27:341–376
Sakai A, Larcher W et al (1987) Frost survival of plants. Responses and adaptation to freezing stress. Springer, Berlin
Sala A, Hoch G (2009) Height-related growth declines in ponderosa pine are not due to carbon limitation. Plant Cell Environ 32:22–30

Sala A, Woodruff DR, Meinzer FC (2012a) Carbon dynamics in trees: feast or famine? Tree Physiol 32:764–775

Sala A, Hopping K, McIntire EJ, Delzon S, Crone EE (2012b) Masting in whitebark pine (*Pinus albicaulis*) depletes stored nutrients. New Phytol 196:189–199

Salleo S, Lo Gullo M, De Paoli D, Zippo M (1996) Xylem recovery from cavitation-induced embolism in young plants of *Laurus nobilis*: a possible mechanism. New Phytol 132:47–56

Sauter JJ (1982) Efflux and reabsorption of sugars in the xylem I. Seasonal changes in sucrose efflux in *Salix*. Z Pflanzenphysiol 106:325–336

Sauter JJ (1983) Efflux and reabsorption of sugars in the xylem II. Seasonal changes in sucrose uptake in *Salix*. Z Pflanzenphysiol 111:429–440

Sauter JJ (1988) Seasonal changes in the efflux of sugars from parenchyma cells into the apoplast in poplar stems (*Populus* × *canadensis* "robusta"). Trees 2:242–249

Sauter JJ, Kloth S (1986) Plasmodesmatal frequency and radial translocation rates in ray cells of poplar (*Populus* × *canadensis* Moench "robusta"). Planta 168:377–380

Sauter JJ, van Cleve B (1991) Biochemical, immunochemical, and ultrastructural studies of protein storage in poplar (*Populus* × *canadensis* "robusta") wood. Planta 183:92–100

Sauter JJ, van Cleve B (1994) Storage, mobilization and interrelations of starch, sugars, protein and fat in the ray storage tissue of poplar trees. Trees 8:297–304

Sauter JJ, Wellenkamp S (1998) Seasonal changes in content of starch, protein and sugars in the twig wood of *Salix caprea* L. Holzforschung 52:255–262

Sauter JJ, Iten W, Zimmermann MH (1973) Studies on the release of sugar into the vessels of sugar maple (*Acer saccharum*). Can J Bot 51:1–8

Schill V, Hartung W, Orthen B, Weisenseel MH (1996) The xylem sap of maple (*Acer platanoides*) trees—sap obtained by a novel method shows changes with season and height. J Exp Bot 47:123–133

Schoonmaker AL (2013) Resource allocation, water relations and crown architecture examined at the tree and stand-level in northern conifers. PhD thesis, University of Alberta, Edmonton

Schrader S, Sauter JJ (2002) Seasonal changes of sucrose-phosphate synthase and sucrose synthase activities in poplar wood (*Populus* × *canadensis* Moench 'robusta') and their possible role in carbohydrate metabolism. J Plant Physiol 159:833–843

Schwarze FW (2007) Wood decay under the microscope. Fungal Biol Rev 21:133–170

Secchi F, Zwieniecki MA (2011) Sensing embolism in xylem vessels: the role of sucrose as a trigger for refilling. Plant Cell Environ 34:514–524

Secchi F, Gilbert ME, Zwieniecki MA (2011) Transcriptome response to embolism formation in stems of *Populus trichocarpa* provides insight into signaling and the biology of refilling. Plant Physiol 157:1419–1429

Seifert GJ, Roberts K (2007) The biology of arabinogalactan proteins. Annu Rev Plant Biol 58:137–161

Sokolowska K, Zagórska-Marek B (2012) Symplasmic, long-distance transport in xylem and cambial regions in branches of *Acer pseudoplatanus* (Aceraceae) and *Populus tremula* × *P. tremuloides* (Salicaceae). Am J Bot 99:1745–1755

Sperry J (2013) Cutting-edge research or cutting-edge artefact? An overdue control experiment complicates the xylem refilling story. Plant Cell Environ 36:1916–1918

Spicer R (2005) Senescence in secondary xylem: heartwood formation as an active developmental program. In: Holbrook NM, Zwieniecki MA (eds) Vascular transport in plants. Elsevier, San Diego, pp 457–475

Spicer R, Holbrook NM (2005) Within-stem oxygen concentration and sap flow in four temperate tree species: does long-lived xylem parenchyma experience hypoxia? Plant Cell Environ 28:192–201

Spicer R, Holbrook NM (2007a) Effects of carbon dioxide and oxygen on sapwood respiration in five temperate tree species. J Exp Bot 58:1313–1320

Spicer R, Holbrook NM (2007b) Parenchyma cell respiration and survival in secondary xylem: does metabolic activity decline with cell age? Plant Cell Environ 30:934–943

Sturm A (1999) Invertases. Primary structures, functions, and roles in plant development and sucrose partitioning. Plant Physiol 121:1–8

Taylor FH (1956) Variation in sugar content of maple sap. Univ Vermont State Agric College Bull 587:1–39

Taylor AM, Gartner BL, Morrell JJ (2002) Heartwood formation and natural durability—a review. Wood Fiber Sci 34:587–611

Tyree MT, Salleo S, Nardini A, Gullo MAL, Mosca R (1999) Refilling of embolized vessels in young stems of laurel. Do we need a new paradigm? Plant Physiol 120:11–22

Van Bel AJ (1990) Xylem-phloem exchange via the rays: the undervalued route of transport. J Exp Bot 41:631–644

Van Bel AJ, Van der Schoot C (1988) Primary function of the protective layer in contact cells. Buffer against oscillations in hydrostatic pressure in the vessels. IAWA Bull 9:285–288

Verdaguer D, Ojeda F (2002) Root starch storage and allocation patterns in seeder and resprouter seedlings of two Cape Erica (Ericaceae) species. Am J Bot 89:1189–1196

Von Frey-Wyssling A, Aeberli H (1942) Der Anteil von Fasern, Gefäßen und Parenchym verschiedener Holzarten in Dreiecksdarstellung. Holz als Roh-und Werkstoff 5:265–268

Wagenführ R (2007) Holzatlas. VEB Fachbuchverlag, Munich

Wargo PM (1976) Variation of starch content among and within roots of red and white oak trees. For Sci 22:468–471

Wheeler E, Baas P, Rodgers S (2007) Variations in dicot wood anatomy: a global analysis based on the Inside Wood database. IAWA J 28:229

Wheeler JK, Huggett BA, Tofte AN, Rockwell FE, Holbrook NM (2013) Cutting xylem under tension or supersaturated with gas can generate PLC and the appearance of rapid recovery from embolism. Plant Cell Environ 36:1938–1949

White RG, Barton DA (2011) The cytoskeleton in plasmodesmata: a role in intercellular transport? J Exp Bot 62:5249–5266

Wiley E, Helliker B (2012) A re-evaluation of carbon storage in trees lends greater support for carbon limitation to growth. New Phytol 195:285–289

Wisniewski M, Davis G (1989) Evidence for the involvement of a specific cell wall layer in regulation of deep supercooling of xylem parenchyma. Plant Physiol 91:151–156

Wisniewski M, Davis G (1995) Immunogold localization of pectins and glycoproteins in tissues of peach with reference to deep supercooling. Trees 9:253–260

Wisniewski M, Davis G, Schafter K (1991a) Mediation of deep supercooling of peach and dogwood by enzymatic modifications in cell-wall structure. Planta 184:254–260

Wisniewski M, Davis G, Arora R (1991b) Effect of macerase, oxalic acid, and EGTA on deep supercooling and pit membrane structure of xylem parenchyma of peach. Plant Physiol 96:1354–1359

Witt W, Buchholz A, Sauter JJ (1995) Binding of endoamylase to native starch grains from poplar wood. J Exp Bot 46:1761–1769

Woodruff DR, Meinzer FC (2011) Water stress, shoot growth and storage of non-structural carbohydrates along a tree height gradient in a tall conifer. Plant Cell Environ 34:1920–1930

Würth MK, Pelaez-Riedl S, Wright SJ, Körner C (2005) Non-structural carbohydrate pools in a tropical forest. Oecologia 143:11–24

Yamada Y, Awano T, Fujita M, Takabe K (2011) Living wood fibers act as large-capacity "single-use" starch storage in black locust (*Robinia pseudoacacia*). Trees 25:607–616

Yuanyuan M, Yali Z, Jiang L, Hongbo S (2009) Roles of plant soluble sugars and their responses to plant cold stress. Afr J Biotechnol 8:2004–2010

Zavaliev R, Ueki S, Epel BL, Citovsky V (2011) Biology of callose (β-1,3-glucan) turnover at plasmodesmata. Protoplasma 248:117–130

Zeeman SC, Kossmann J, Smith AM (2010) Starch: its metabolism, evolution, and biotechnological modification in plants. Annu Rev Plant Biol 61:209–234

Zieminska K, Butler DW, Gleason SM, Wright IJ, Westoby M (2013) Fibre wall and lumen fractions drive wood density variation across 24 Australian angiosperms. AoB Plants 5:plt046

Chapter 9
Wood Anatomy and Plant Hydraulics in a Changing Climate

William R.L. Anderegg and Frederick C. Meinzer

1 Introduction

Due to their hydraulic system that allows them to transport water from the soil to leaves, woody plants have become incredibly successful in terrestrial ecosystems since their evolution ~400 million years ago (Hartmann 2011). This vascular system lets trees conduct water from the soil up to more than 100 m (Koch et al. 2004), allowing trees to compete for light and absorb several petagrams of carbon from the atmosphere via photosynthesis every year (Le Quéré et al. 2009). Thus, plant hydraulics form the "backbone" of most terrestrial ecosystems, facilitating net primary production and carbon sequestration by the biosphere (Brodribb 2009). The carbon sequestration of global forests alone is estimated at roughly 2.5 Pg carbon, equivalent to 25 % of anthropogenic carbon emissions in 2010 (Pan et al. 2011). Similarly, vascular transport plays a major role in the global hydrological water recycling that drives upwards of 80 % of evapotranspiration over land, influencing global circulation and precipitation patterns (Jasechko et al. 2013).

Hydraulic architecture comprises part of an integrated set of traits and life history trade-offs that allow woody plants to colonize diverse environments, compete, and coexist. Wood anatomy plays a central role in plant hydraulic strategies due to the inherent trade-offs associated with partitioning of wood volume between water transport and structural support functions and a fixed pool of carbon and energy that can be allocated across growth, fecundity, tissue maintenance, and tissue repair (Chave et al. 2009; Domec et al. 2008). Maximizing fitness is thought to involve

W.R.L. Anderegg (✉)
Princeton Environmental Institute, Princeton University, Princeton, NJ 08540, USA
e-mail: anderegg@princeton.edu

F.C. Meinzer
USDA Forest Service, Pacific Northwest Research Station,
3200 SW Jefferson Way, Corvallis, OR, USA

U. Hacke (ed.), *Functional and Ecological Xylem Anatomy*,
DOI 10.1007/978-3-319-15783-2_9

maximum carbon uptake for growth and reproduction, which requires water transport to the canopy, while avoiding the damaging or lethal risks of abiotic stressors such as water limitation and mechanical damage (Chave et al. 2009; Cowan 1978). Wood anatomy interacts with other components in the hydraulic continuum, including the root–rhizosphere interface and the water–air interface in stomatal pores on leaves (Barnard et al. 2011). Furthermore, the allocation to different tissues, including sapwood, fine roots, leaves, as well as rooting distribution and depth, are other major components of individuals' and species' hydraulic strategies.

Wood anatomy and plant hydraulics will be central in understanding species' responses to and ability to cope with rapid environmental change, including anthropogenic climate change (McDowell et al. 2008). While plant hydraulics have evolved to changes in climate over evolutionary timescales (Pittermann et al. 2012), the recent and future rates of climate change are likely to place considerable stress on vascular plants. Global temperatures have increased roughly 0.8 °C since preindustrial times (IPCC 2013), but the rate of temperature increase over 1980–2005 was faster than that of any documented rapid warming period in the geologic past, including warming periods that triggered mass extinctions (Diffenbaugh and Field 2013). The projected magnitude of warming approaches those of the geologic record, but with 10–50-fold faster rates (Diffenbaugh and Field 2013). Due to rising temperatures, an "acceleration" of the hydrological cycle is also expected, leading to wet areas generally getting wetter and dry areas generally getting drier (Betts et al. 2007). Similarly, projections indicate that longer dry spells and increased evaporative demand due to higher temperatures will increase drought stress in many regions of the world, particularly in tropical South America, southern Europe, Australia, western United States, and subtropic Africa (Dai 2013) (Fig. 9.1). Rising atmospheric carbon dioxide concentrations, however, may help trees ameliorate drought stress due to increasing water use efficiency (Keenan et al. 2013), and consequently the balance of these trends is largely unknown (Bonan 2008; Friend et al. 2013).

Though no global datasets currently exist to assess temporal trends, widespread tree mortality triggered by drought and heat stress has been observed on every vegetated continent in recent decades (Allen et al. 2010; Anderegg et al. 2013b) (Fig. 9.2). Some of these events have included regional-scale massive die-offs, such as the mortality of *Pinus edulis* and *Populus tremuloides*, in the western United States following abnormally hot droughts in 2000–2003 (Anderegg et al. 2012b; Breshears et al. 2005; Worrall et al. 2008). Where temporal data are available, plot networks monitored from the 1960s to present indicate that background mortality rates in western North America have increased substantially over that period, likely due to climate stress (Peng et al. 2011; van Mantgem et al. 2009). Due in large part to major uncertainties surrounding how trees die from drought stress, current predictive ability of which species are most vulnerable, when, where, and to what types of drought is very limited (Anderegg et al. 2012a; McDowell et al. 2011).

Plant hydraulics and wood anatomy are likely important traits in assessing species' vulnerability to increasing severity of climatic stresses, such as drought, high temperatures, and consequent high evaporative demand. The biophysical properties

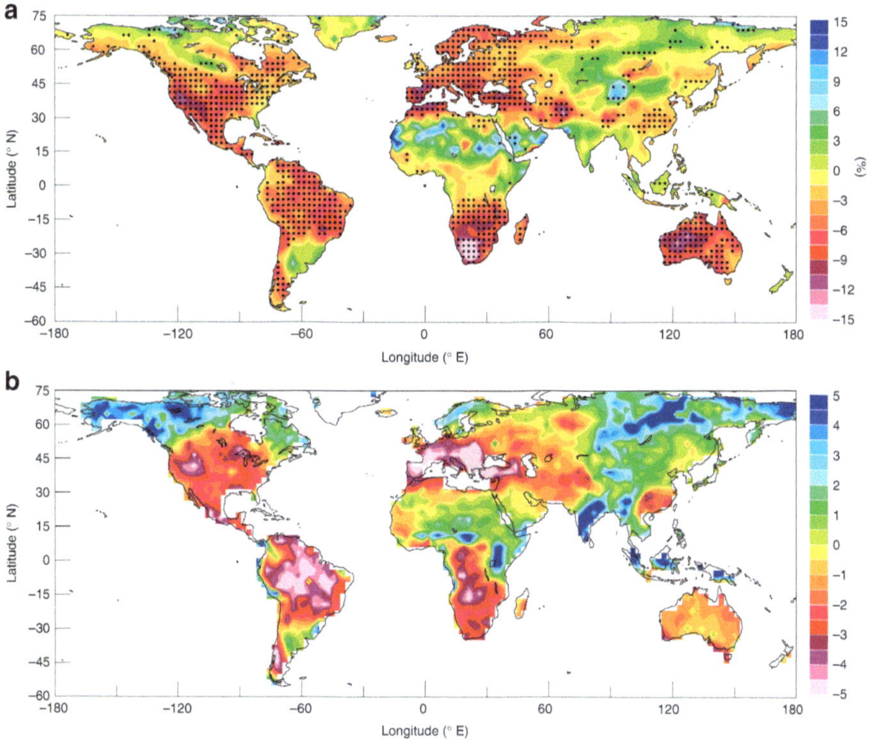

Fig. 9.1 (**a**) Percentage changes from 1980–1999 to 2080–2099 in the multimodel ensemble mean soil-moisture content in the top 10 cm layer (broadly similar for the whole soil layer) simulated by 11 CMIP5 models under the RCP4. Five emissions scenario. Stippling indicates at least 82 % (9 out of 11) of the models agree on the sign of change. (**b**) Mean self-calibrated Palmer Drought Severity Index using the Penman-Monteith formulation (sc_PDSI_pm) averaged over 2090–2099 computed using the 14-model ensemble mean climate (including surface air temperature, precipitation, wind speed, specific humidity, and net radiation) from the CMIP5 simulations under the RCP4.5 scenario. A sc_PDSI_pm value of −3.0 or below indicates severe to extreme droughts for the present climate. Adapted from Dai (2013)

of wood and the structure and configuration of the xylem conduits contained therein are key determinants of the efficiency of whole-plant water transport, resistance to hydraulic dysfunction, and recovery of water transport capacity following periods of hydraulic dysfunction (Tyree and Zimmermann 2002). The structural features that govern the efficiency and safety of water transport at the individual conduit level are discussed in detail elsewhere in this volume. Briefly, the hydraulic conductivity of coniferous tracheids is largely determined by the size and frequency of their bordered pits, the pit membrane pore size, as well as tracheid diameter (Domec et al. 2006; Hacke et al. 2004; Pittermann et al. 2005, 2010), whereas conductivity of angiosperm vessels is most strongly related to their diameter and length (Sperry et al. 2006; Zanne et al. 2010). Conduit hydraulic safety or resistance to embolism is most strongly related to pit membrane pore size in vessels and to bordered pit

Fig. 9.2 Images of climate-induced forest die-off from around the world, adapted from Anderegg et al. (2013b). Photo credits, clock-wise from *top left*: Spain—Rafael Navarro-Cerrillo, China—Youquing Luo, New Mexico—Craig D. Allen, Argentina—Thomas Kitzberger

characteristics, particularly torus overlap at the pit aperture and margo pore size near the torus edge in tracheids (Delzon et al. 2010; Domec et al. 2006, 2008; Hacke and Jansen 2009; Hacke et al. 2004). A third biophysical property of wood related to its structure is hydraulic capacitance, the amount of water released per unit decline in water potential or increase in xylem tension. Capacitive discharge of water into the transpiration stream during the day and recharge of capacitance overnight can play an important role in buffering transpiration-induced fluctuations in xylem tension that could lead to catastrophic levels of embolism in the absence of transient buffering (Hölttä et al. 2009; Meinzer et al. 2003, 2008a, 2009). As might be expected, capacitance is inversely related to wood density (McCulloh et al. 2014; Scholz et al. 2008), which reflects the relative volumes of solid material and pore space available for water storage in wood. Not surprisingly, all of the preceding wood structural features and biophysical properties exhibit strong axial variation from woody roots to terminal branches associated with corresponding axial trends of increasing xylem tension (Domec et al. 2008; Koch et al. 2004; Woodruff et al. 2004). In tall conifers, wood features such as tracheid diameter, specific conductivity, and P_{50} vary dramatically across two orders of magnitude of stem diameter from the trunk base to terminal branches, which can correspond to a height range and path length of >50 m (Fig. 9.3). Thus, the degree of plasticity of wood structure and function

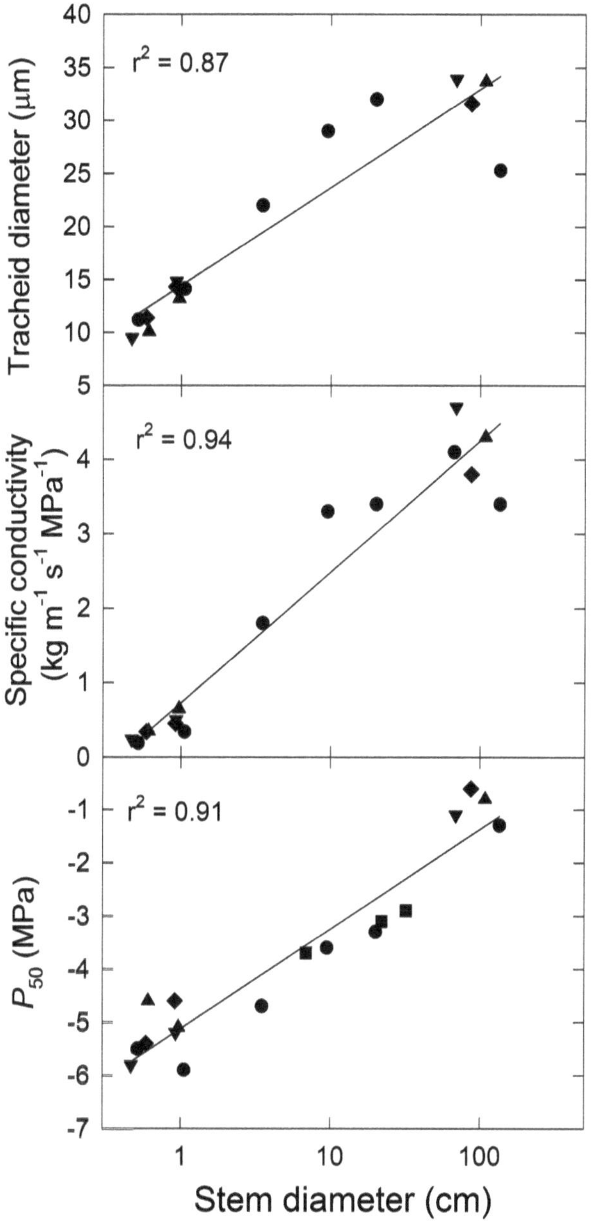

Fig. 9.3 Axial variation in stem anatomical and hydraulic properties in several tall coniferous species. Axial positions range from the trunk base to terminal branches >50 m above the ground. Symbols: (*inverted filled triangle*) *Abies grandis*; (*filled triangle*) *Thuja plicata*; (*filled square*) *Pinus ponderosa*; (*filled diamond*) *Tsuga heterophylla*; (*filled circle*) *Pseudotsuga menziesii*. Note log scale on *x*-axis. Data from Domec and Gartner (2001, 2003); Domec et al. (2006), McCulloh et al. (2014)

within individual plants may serve as a suitable proxy for a species' ability to maintain the integrity of xylem water transport in the face of climate-induced alterations in baseline levels and extremes of xylem tension.

In this chapter, we first provide an overview of the trade-offs present in wood anatomy and physiology by woody plants. We then discuss how rising carbon dioxide concentrations, increasing temperature, and more severe droughts may alter wood anatomy and plant hydraulics and conversely how these plant traits can help predict cross-species vulnerability to climatic changes. We further examine the breadth and potential for anatomical plasticity and its relationship with measured hydraulic properties. We conclude with prominent future research directions and major gaps in the understanding of plant anatomy, physiology, and demography in a world of rapid environmental change.

2 Trade-Offs in Wood Anatomy and Physiology

Species exhibit different operating ranges along continua of wood properties that determine higher order plant hydraulic traits and physiological behavior that conserves hydraulic function. Typically, the trajectory of one wood structural feature or biophysical property dictates the trajectories of other related biophysical, hydraulic and physiological traits and behaviors, resulting in a series of trade-offs. These hydraulic trade-offs are components of overall species strategies for acquisition of other resources in addition to water (Reich 2014). The classic hydraulic trade-off is one of xylem safety versus efficiency, wherein more conductive xylem is less resistant to drought-induced embolism (Meinzer et al. 2010; Sperry et al. 2008; Tyree et al. 1994). As explained above, capacitance can play a central role in avoidance of tension-induced embolism under nonextreme diurnal conditions. However, there appears to be a trade-off of capacitance against resistance to drought-induced embolism across a broad range of woody species (Fig. 9.4). The trade-off is nonlinear and characterized by a threshold lower limit of capacitance below which resistance to embolism increases sharply as capacitance declines. The relationship between capacitance and P_{50} may reflect the rapidly diminishing transient buffering effect of capacitance on xylem tension under conditions of progressively intensifying drought that prevent overnight recharge of tissue water storage compartments. This potential limitation could be partially mitigated if embolism were more readily reversible in species with high capacitance, but this remains to be elucidated.

Wood density is a fundamental biophysical property that can often serve as a robust proxy for an array of simple and complex hydraulic traits over a range of scale from tissue to whole plants (Chave et al. 2009; Meinzer et al. 2008a; Pratt et al. 2007; Zanne et al. 2010). In tropical trees, traits such as trunk-to-branch tapering of vessels, branch leaf-specific conductivity, and whole-plant leaf-specific conductance have been shown to be strongly related to wood density (Fig. 9.5). These and other hydraulic traits often scale uniformly with wood density across species and functional types such as pioneer (low density) and late successional

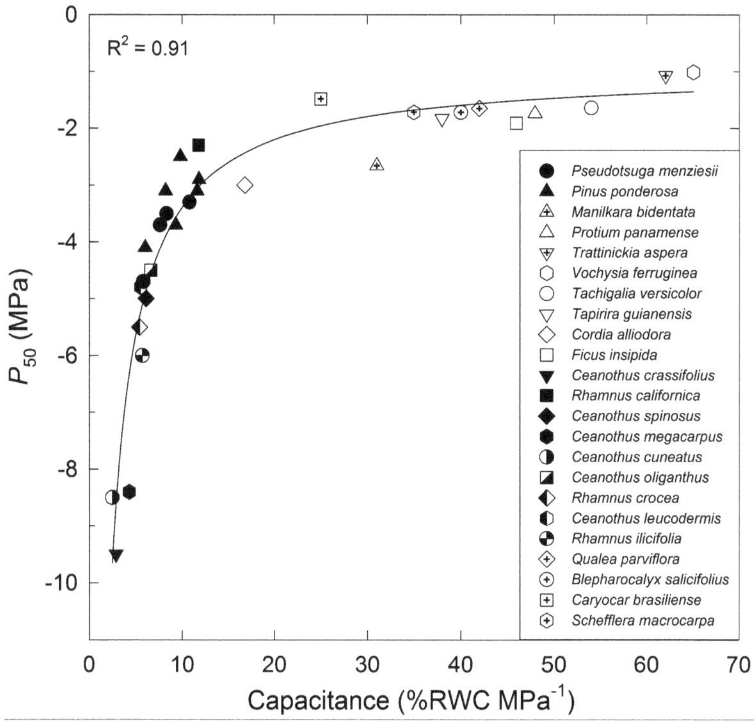

Fig. 9.4 Relationship between sapwood capacitance and the xylem pressure causing 50 % loss of stem conductivity (P_{50}) for several conifer and angiosperm tree species from different ecosystems. Figure from Scholz et al. (2011)

(high density) species (Fig. 9.5). Additional relationships between hydraulic traits and wood density include a positive correlation with resistance to embolism (Bucci et al. 2013; Hacke et al. 2001a; Ogasa et al. 2013; Pratt et al. 2007), a negative correlation with sapwood capacitance (Meinzer et al. 2008b; Scholz et al. 2011), and a negative correlation with minimum leaf water potential and leaf water potential at turgor loss (Meinzer 2003; Meinzer et al. 2008b).

Hydraulic safety margins are an example of higher order physiological regulation associated with basic wood anatomical and biophysical properties. A hydraulic safety margin can be defined as the difference between a given point along a plant organ's xylem vulnerability curve (e.g., P_{50}) and the organ's normal daily minimum xylem pressure determined by stomatal regulation of transpiration (Brodribb et al. 2003; Bucci et al. 2013; Jones and Sutherland 1991; Meinzer et al. 2009; Sparks and Black 1999; Tyree and Sperry 1988). Multispecies plots of branch P_{50} against minimum branch water potential (an estimate of xylem pressure) reveal that safety margins increase as minimum water potential becomes more negative (Pockman and Sperry 2000) and that the tracheid-bearing conifers tend to sustain more negative branch water potentials and maintain larger hydraulic safety margins than the

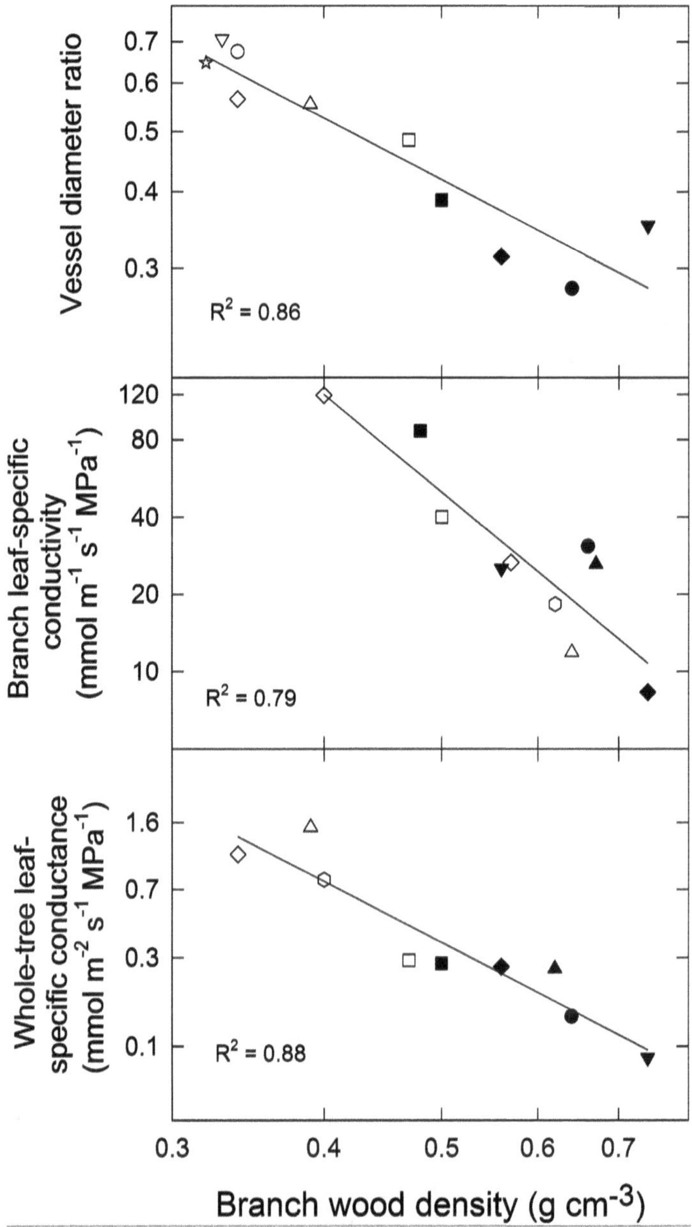

Fig. 9.5 Relationships between branch to trunk vessel diameter ratio (axial tapering), branch leaf-specific conductivity, and whole-tree leaf specific conductance and branch wood density for several Panamanian forest canopy tree species. Each symbol represents a different species. *Open symbols* correspond to early successional (pioneer) species and *closed symbols* correspond to late successional species. Note log scales on all axes. Data from Meinzer et al. (2008a) and McCulloh et al. (2011)

vessel-bearing angiosperms (Choat et al. 2012; Meinzer et al. 2009). Moreover, when safety margins are referenced to the air entry threshold, the inflection point on the xylem vulnerability curve at which loss of conductivity begins to increase steeply with decreasing xylem pressure, many angiosperms exhibit slightly negative safety margins implying that they may normally experience low baseline levels of embolism (Meinzer et al. 2009). These patterns have led to the suggestion that angiosperm trees operate perilously close to their limits for catastrophic hydraulic failure and are at a high risk of mortality during drought events (Choat et al. 2012). However, accumulating evidence points to a robust capacity of many angiosperms to regain stem hydraulic capacity following large drought-induced losses (Martorell et al. 2014; Ogasa et al. 2013; Urli et al. 2013). In contrast, conifers are reputed to have a low ability to recover from massive embolism, consistent with their maintenance of larger safety margins (Brodribb et al. 2010; Brodribb and Cochard 2009). The tendency for conifers to operate at larger hydraulic safety margins than angiosperms is also correlated with lower volume fractions of xylem parenchyma in conifer than in angiosperm sapwood (Johnson et al. 2012). These differences in xylem parenchyma volume are also correlated with higher nonstructural carbohydrate concentrations in angiosperm wood, which has led to the hypothesis that sugars may play a more important role in embolism reversal (Bucci et al. 2003; Salleo et al. 2009) in angiosperms than conifers, reducing the risk of angiosperms operating at smaller hydraulic safety margins (Johnson et al. 2012).

Despite strong relationships between wood anatomy and a series of hydraulic traits, using these traits to reliably predict differences in species performance and survival under current and anticipated climate scenarios remains somewhat elusive. It seems clear that individual traits such as stem P_{50} are not sufficient because surveys of variation in P_{50} within different vegetation types (e.g., Maherali et al. 2004) and among co-occurring species (e.g., Jacobsen et al. 2007; Pratt et al. 2012) indicate that multiple plant hydraulic strategies can be successful in a given environment. Thus, in addition to the hydraulic traits and trade-offs that can be characterized based on wood and conduit structure, plant architectural and behavioral attributes such as rooting depth, degree of iso- or anisohydry, and phenology will likely need to be taken into account.

3 Wood Anatomy, Physiology, and Global Change Drivers

Since preindustrial times (ca. 1750), atmospheric CO_2 concentrations have risen from 280 parts per million to around 400 parts per million in 2013. Rising CO_2 concentrations has generally increased water use efficiency of forests (Keenan et al. 2013), allowing them to take up more carbon per unit water, and possibly buffer water stress as well (Ponce Campos et al. 2013). Elevated CO_2 concentrations have been documented to affect xylem anatomy, leading to generally larger conduit sizes in ring-porous angiosperm species and some gymnosperms, but little changes in conduit sizes in diffuse-porous angiosperm species (Domec et al. 2010; Phillips et al. 2011; Way 2013). Because increases in conduit size largely increase hydraulic conductivity, this can lead to higher vulnerability to water stress, but the interactive effects of CO_2 concentrations with concomitant rises in temperature and drought regimes may offset or counter these patterns (Kilpeläinen et al. 2007; Maherali and DeLucia 2000a).

Similarly, rising temperatures will also affect xylem anatomy and physiology. In a 6-year growth chamber experiment in *Pinus sylvestris*, elevated temperatures differentially affected xylem anatomy more than elevated CO_2 concentrations, leading to increases in tracheid width, length, and coarseness, and the effects were most pronounced in early wood (Kilpeläinen et al. 2007). In *Eucalyptus camaldulensis*, however, growth at higher temperatures led to higher wood density, lower hydraulic conductance, and a shift toward smaller vessel diameters (Thomas et al. 2004). Temperature can affect hydraulic vulnerability to cavitation, both through effects on conduit size and xylem anatomy and through effects on xylem fluid viscosity, although most of the evidence for this comes from studies of contrasting existing environments (e.g., Maherali and DeLucia 2000b).

Naturally, water availability places the largest constraints on xylem anatomy and physiology. Vulnerability to cavitation shows remarkable adaptive radiation across environments, differing widely as a function of water limitation across species (Maherali et al. 2004). Similarly, over evolutionary timescales, dry periods have been shown to drive the adaptation of cavitation-resistant xylem across the hydraulically diverse Cupressaceae family at multiple points in the past 30 million years (Pittermann et al. 2012).

The tight coupling between xylem anatomy, function, and the environment indicates that hydraulic characteristics and traits can provide useful insights into which species, biomes, and locations could be most vulnerable to drought (Nardini et al. 2013). Lethal failure of the plant hydraulic system has been observed in a number of species in response to drought (Anderegg et al. 2012b; Brodribb and Cochard 2009; Hoffmann et al. 2011; Mitchell et al. 2013; Nardini et al. 2013; Urli et al. 2013). Preliminary experiments indicate that mortality risk appears to increase substantially above 50 % loss conductivity in gymnosperms and closer to 80 % loss in angiosperms (Brodribb and Cochard 2009; Urli et al. 2013) (see also Pratt, this volume).

Hydraulic impairment due to embolism and accumulated damage will likely interact with other mechanisms of mortality, including the interdependent plant carbon status (McDowell et al. 2011), but nonetheless appears to be a prominent explanatory variable in explaining cross-species patterns of drought-induced mortality. Hoffmann et al. (2011) found that the percent loss hydraulic conductivity, explained 55 % of the variation in dieback across 22 angiosperm species in the southeastern United States. Nardini et al. (2013) found that the water potential at which 50 % of hydraulic conductivity is lost (P_{50}) captured 90 % of the variation in mortality rates across six species in Italy. Similarly, integrated wood anatomy traits, especially wood density, are correlated with mortality rates after drought in tropical ecosystems globally as well (Kraft et al. 2010; Phillips et al. 2010). Because many drought-induced mortality events are multiyear and last beyond the inciting drought, the extent and limits of regrowth of new xylem and repair of embolism are thought to be quite important as well (Anderegg et al. 2012a; Brodribb et al. 2010), perhaps explaining the differential mortality sensitivities between gymnosperms and angiosperms.

A large array of nonhydraulic characteristics will influence vulnerability to drought as well, including coordination of leaf water balance and stomatal regulation (Mitchell et al. 2013), allocation across tissues such as root-to-leaf-area ratios (West et al. 2008), drought characteristics such as intensity and seasonality (Anderegg et al. 2013a), and soil and topoedaphic characteristics such as characteristic moisture curve (Koepke et al. 2010). Rooting depth is another key trait that can determine water availability and mortality during drought (Paddock et al. 2013). Nevertheless, wood anatomy and hydraulic traits present a promising avenue for evaluating, modeling, and predicting the vulnerability of tree species with future climate changes.

4 Plasticity in Wood Anatomy and Hydraulics over Time and Space

Wood anatomy and plant hydraulics are known to vary in time and in space within the same species. Nevertheless, the magnitude of intraspecific variation in contrast to interspecific variation of hydraulics is less well known (e.g., Lamy et al. 2013) (see also Hacke, this volume). A large number of studies have quantified anatomical and physiological differences in wood in species in contrasting environments, especially along moisture gradients (e.g., Alder et al. 1996; Maherali and DeLucia 2000b). A recent meta-analysis synthesized these studies and found that intraspecific variation of a key hydraulic trait (P_{50}) was ecologically relevant, equivalent to 33 % of the variation across species within the same genus and 20 % of the variation within a plant functional type (Fig. 9.6) (Anderegg 2014). Furthermore, intraspecific variation seemed to be higher in angiosperms than in gymnosperms (Anderegg 2014), which fits with understanding of key anatomical differences between the groups (Johnson et al. 2012). Finally, this key trait was poorly captured in plant functional types used to model vegetation response to drought under climate scenarios (Anderegg 2014).

Temporal variation of plant hydraulics is less well known. Tree ring records indicate that many wood anatomical components are sensitive to climate parameters during the years in which they are formed, which suggests that these anatomical differences could affect hydraulics over time (Fonti et al. 2010; Olano et al. 2013). For example, wood density has been shown to increase in oak trees in years following a severe drought (Corcuera et al. 2004) and an analysis of contemporary and paleoclimate effects on vessels in *Quercus macrocarpa* tree rings showed a significant positive relationship between spring temperatures and earlywood vessel diameter and conductivity (Voelker et al. 2012). The number of functional xylem rings will

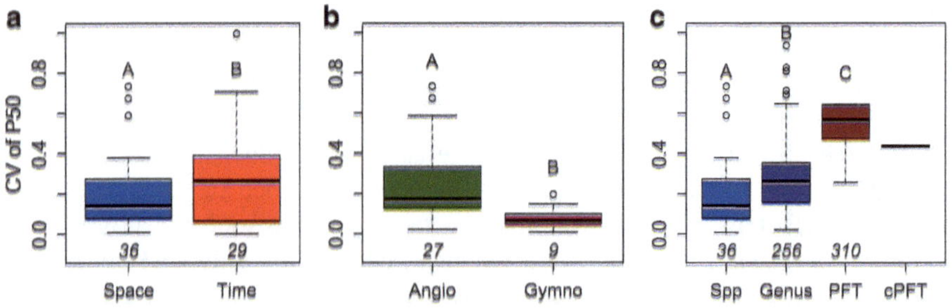

Fig. 9.6 Coefficient of variation of the water potential at which a plant stem reaches percent loss of hydraulic conductivity (P_{50}) across (**a**) space versus time within a given angiosperm species, (**b**) spatially across angiosperms versus gymnosperms within a given species, and (**c**) comparing within-species spatial variation, within-genus variation, within-plant functional type (PFT) variation (tree PFTs only), and across PFT variation (here "cPFT" for cross-PFT). Numbers below bars indicate the sample size of the number of species included. Letters denote statistical significance at $p < 0.05$. Adapted from Anderegg (2014)

also determine the degree of potential plasticity of hydraulic architecture in response to changes in climate. For example, in the oaks described above, the species' water transport relies strongly upon xylem formed in the same year, giving rise to high plasticity in response to climate (Voelker et al. 2012). Anatomical studies have indicated that individual rings differ in their hydraulic properties and vulnerability to embolism (Melcher et al. 2003). And at a whole-plant level, accumulated damage through a mechanism like cavitation fatigue (Hacke et al. 2001b) or acclimation following a severe stress can also likely occur. Nonetheless, the magnitude and ubiquity of these processes is largely unknown.

Spatial and temporal variation in wood anatomy and hydraulic function can arise from two processes—genetic variation or phenotypic plasticity. The balance of these two holds immense importance for the ability of woody species to adapt to shifting climate regimes (Nicotra et al. 2010). If intrapopulation genetic variability is limited and interpopulation variability is large and plasticity is low, leading to local adaptation, then many regions of a species' range would be vulnerable to a change in climate (Franks et al. 2013). On the other hand, if plasticity is high, many individual plants may have scope for adapting to climate shifts in situ, allowing for greater resilience (Nicotra et al. 2010). A compilation of common garden experiments found that local adaptation and clinal variation in genetics and plant traits are quite common in woody plants (Alberto et al. 2013), but the ranges and prevalence of phenotypic plasticity are less well known. Nevertheless, as mentioned above, wood structure and its hydraulic properties can vary dramatically with axial position in tall trees, consistent with substantial plasticity during ontogeny that allows them to cope with increasing xylem tension driven by gravity and cumulative frictional resistances to flow as they increase in height (Fig. 9.3). In Douglas-fir trees ranging in height from 6 to 85 m, P_{50} ranged from about −2 MPa in trunks to −8 MPa in uppermost branches and tracheid pit aperture conductance declined exponentially with increasing resistance to embolism. Extrapolation of the height-related trend in pit aperture conductance suggested that pit conductance would fall to zero (i.e., zero water transport) at a height of 99–123 m in trunks, consistent with recorded maximum heights for this species (Domec et al. 2008). Thus, in tall tree species, height-related trends in wood anatomy and hydraulics may serve as an adequate proxy for responses to increasing aridity.

5 Future Directions

The emergence of global databases of wood anatomy and plant hydraulic traits, combined with the urgent need to understand and predict woody plant responses to global environmental change, presents a number of exciting and promising research directions. A major challenge revolves around connecting wood anatomy to demographic outcomes and fitness in a changing environment (e.g., Pratt et al. 2014). For example, how meaningful are hydraulic safety margins and are they indicative of true thresholds in vulnerability to drought? Similarly, after what hydraulic thresholds do plants risk mortality and what components and degrees of hydraulic dysfunction can be repaired and survived? Cross-species experiments in glasshouses that can carefully control and monitor water stress and whole-tree fluxes (Urli et al. 2013) and experiments that monitor recovery after a given stress (Ogasa et al. 2013) are greatly needed. To date, we lack a detailed and mechanistic understanding of embolism refilling, and how much it actually occurs in nature, particularly given recent conflicting interpretations of the potential impact of sampling artifacts on the apparent magnitude of daily embolism and refilling in some studies (Wheeler et al. 2013). This uncertainty is a major obstacle to understanding levels of hydraulic risk and lethal thresholds in response to drought and temperature stress.

A second promising research avenue involves the investigation of spatial and temporal variation in wood anatomy and plant hydraulics, and how much of this variation is genetically determined versus phenotypic plasticity (Anderegg 2014). In particular, there are relatively few common garden experiments that have measured anatomical or hydraulic traits (but see Lamy et al. 2013; Schreiber et al. 2011; Wortemann et al. 2011), which are useful in determining the relative influence of adaptation versus plasticity, and also relatively few studies that quantify trait variation of multiple coordinated traits such as gas exchange, morphology, and hydraulic traits across a species range (but see Martínez-Vilalta et al. 2009). Additionally, the temporal scales and potential for turnover and plasticity in anatomy and hydraulic traits over multiple years is not well known. Together, these types of studies can help illuminate how pervasive is local adaptation, when and where in a species' range it could be vulnerable to drought or temperature stress, and how much plasticity can buffer a rapidly changing environment due to climate change.

Ultimately, climate change will have a profound effect on woody plants globally and the ecosystem functions and services they provide. The rates of current and future climate change meet or exceed the fastest rates observed in previous mass extinction events in the paleoclimate record. Better understanding of the mechanisms through which climate affects plant hydraulics and, in turn, how plant characteristics mediate physiological and demographic responses to a change environment will facilitate prediction of climate change impacts on ecosystems and society.

Acknowledgements We thank Uwe Hacke for feedback on the manuscript. W.R.L.A. was supported by a National Oceanic and Atmospheric Administration Climate and Global Change Postdoctoral Fellowship, administered by the University Corporation for Atmospheric Research in Boulder, Colorado.

References

Alberto FJ, Aitken SN, Alía R, González-Martínez SC, Hänninen H, Kremer A, Lefèvre F, Lenormand T, Yeaman S, Whetten R, Savolainen O (2013) Potential for evolutionary responses to climate change—evidence from tree populations. Glob Chang Biol 19:1645–1661

Alder N, Sperry J, Pockman W (1996) Root and stem xylem embolism, stomatal conductance, and leaf turgor in Acer grandidentatum populations along a soil moisture gradient. Oecologia 105:293–301

Allen CD, Macalady AK, Chenchouni H, Bachelet D, McDowell N, Vennetier M, Kitzberger T, Rigling A, Breshears DD, Hogg EH, Gonzalez P, Fensham R, Zhang Z, Castro J, Demidova N, Lim JH, Allard G, Running SW, Semerci A, Cobb N (2010) A global overview of drought and heat-induced tree mortality reveals emerging climate change risks for forests. For Ecol Manage 259:660–684

Anderegg WR (2014) Spatial and temporal variation in plant hydraulic traits and their relevance for climate change impacts on vegetation. New Phytol 205(3):1008–1014

Anderegg WRL, Berry JA, Field CB (2012a) Linking definitions, mechanisms, and modeling of drought-induced tree death. Trends Plant Sci 17:693–700

Anderegg WRL, Berry JA, Smith DD, Sperry JS, Anderegg LDL, Field CB (2012b) The roles of hydraulic and carbon stress in a widespread climate-induced forest die-off. Proc Natl Acad Sci U S A 109:233–237

Anderegg LDL, Anderegg WRL, Berry JA (2013a) Not all droughts are created equal: translating meteorological drought into woody plant mortality. Tree Physiol 33:672–683

Anderegg WRL, Kane JM, Anderegg LDL (2013b) Consequences of widespread tree mortality triggered by drought and temperature stress. Nat Clim Change 3:30–36

Barnard DM, Meinzer FC, Lachenbruch B, McCulloh KA, Johnson DM, Woodruff DR (2011) Climate-related trends in sapwood biophysical properties in two conifers: avoidance of hydraulic dysfunction through coordinated adjustments in xylem efficiency, safety and capacitance. Plant Cell Environ 34:643–654

Betts RA, Boucher O, Collins M, Cox PM, Falloon PD, Gedney N, Hemming DL, Huntingford C, Jones CD, Sexton DM (2007) Projected increase in continental runoff due to plant responses to increasing carbon dioxide. Nature 448:1037–1041

Bonan GB (2008) Forests and climate change: forcings, feedbacks, and the climate benefits of forests. Science 320:1444–1449

Breshears DD, Cobb NS, Rich PM, Price KP, Allen CD, Balice RG, Romme WH, Kastens JH, Floyd ML, Belnap J, Anderson JJ, Myers OB, Meyer CW (2005) Regional vegetation die-off in response to global-change-type drought. Proc Natl Acad Sci U S A 102:15144–15148

Brodribb TJ (2009) Xylem hydraulic physiology: the functional backbone of terrestrial plant productivity. Plant Sci 177:245–251

Brodribb TJ, Cochard H (2009) Hydraulic failure defines the recovery and point of death in water-stressed conifers. Plant Physiol 149:575–584

Brodribb TJ, Holbrook NM, Edwards EJ, GutiÉRrez MV (2003) Relations between stomatal closure, leaf turgor and xylem vulnerability in eight tropical dry forest trees. Plant Cell Environ 26:443–450

Brodribb TJ, Bowman DJMS, Nichols S, Delzon S, Burlett R (2010) Xylem function and growth rate interact to determine recovery rates after exposure to extreme water deficit. New Phytol 188:533–542

Bucci S, Scholz F, Goldstein G, Meinzer F, Sternberg L (2003) Dynamic changes in hydraulic conductivity in petioles of two savanna tree species: factors and mechanisms contributing to the refilling of embolized vessels. Plant Cell Environ 26:1633–1645

Bucci SJ, Scholz FG, Peschiutta ML, Arias NS, Meinzer FC, Goldstein G (2013) The stem xylem of Patagonian shrubs operates far from the point of catastrophic dysfunction and is additionally protected from drought-induced embolism by leaves and roots. Plant Cell Environ 36:2163–2174

Chave J, Coomes D, Jansen S, Lewis SL, Swenson NG, Zanne AE (2009) Towards a worldwide wood economics spectrum. Ecol Lett 12:351–366

Choat B, Jansen S, Brodribb TJ, Cochard H, Delzon S, Bhaskar R, Bucci SJ, Feild TS, Gleason SM, Hacke UG, Jacobsen AL, Lens F, Maherali H, Martinez-Vilalta J, Mayr S, Mencuccini M, Mitchell PJ, Nardini A, Pittermann J, Pratt RB, Sperry JS, Westoby M, Wright IJ, Zanne AE (2012) Global convergence in the vulnerability of forests to drought. Nature 491:752–755

Corcuera L, Camarero J, Gil-Pelegrín E (2004) Effects of a severe drought on Quercus ilex radial growth and xylem anatomy. Trees 18:83–92

Cowan IR (1978) Stomatal behaviour and environment. In: Preston RD, Woolhouse HW (eds) Advances in botanical research. Academic, New York, pp 117–228

Dai A (2013) Increasing drought under global warming in observations and models. Nat Clim Change 3:52–58

Delzon S, Douthe C, Sala A, Cochard H (2010) Mechanism of water-stress induced cavitation in conifers: bordered pit structure and function support the hypothesis of seal capillary-seeding. Plant Cell Environ 33:2101–2111

Diffenbaugh NS, Field CB (2013) Changes in ecologically critical terrestrial climate conditions. Science 341:486–492

Domec J-C, Gartner BL (2001) Cavitation and water storage capacity in bole xylem segments of mature and young Douglas-fir trees. Trees 15:204–214

Domec J-C, Gartner BL (2003) Relationship between growth rates and xylem hydraulic characteristics in young, mature and old-growth ponderosa pine trees. Plant Cell Environ 26:471–483

Domec J-C, Lachenbruch B, Meinzer FC (2006) Bordered pit structure and function determine spatial patterns of air-seeding thresholds in xylem of Douglas-fir (Pseudotsuga menziesii; Pinaceae) trees. Am J Bot 93:1588–1600

Domec J-C, Lachenbruch B, Meinzer FC, Woodruff DR, Warren JM, McCulloh KA (2008) Maximum height in a conifer is associated with conflicting requirements for xylem design. Proc Natl Acad Sci 105:12069–12074

Domec J-C, Schäfer K, Oren R, Kim HS, McCarthy HR (2010) Variable conductivity and embolism in roots and branches of four contrasting tree species and their impacts on whole-plant hydraulic performance under future atmospheric CO_2 concentration. Tree Physiol 30:1001–1015

Fonti P, von Arx G, García-González I, Eilmann B, Sass-Klaassen U, Gärtner H, Eckstein D (2010) Studying global change through investigation of the plastic responses of xylem anatomy in tree rings. New Phytol 185:42–53

Franks SJ, Weber JJ, Aitken SN (2014) Evolutionary and plastic responses to climate change in terrestrial plant populations. Evol Appl 7(1):123–139

Friend AD, Lucht W, Rademacher TT, Keribin R, Betts R, Cadule P, Ciais P, Clark DB, Dankers R, Falloon PD, Ito A, Kahana R, Kleidon A, Lomas MR, Nishina K, Ostberg S, Pavlick R, Peylin P, Schaphoff S, Vuichard N, Warszawski L, Wiltshire A, Woodward FI (2013) Carbon residence time dominates uncertainty in terrestrial vegetation responses to future climate and atmospheric CO_2. Proc Natl Acad Sci 111(9):3280–3285

Hacke UG, Jansen S (2009) Embolism resistance of three boreal conifer species varies with pit structure. New Phytol 182:675–686

Hacke UG, Sperry JS, Pockman WT, Davis SD, McCulloh KA (2001a) Trends in wood density and structure are linked to prevention of xylem implosion by negative pressure. Oecologia 126:457–461

Hacke UG, Stiller V, Sperry JS, Pittermann J, McCulloh KA (2001b) Cavitation fatigue. Embolism and refilling cycles can weaken the cavitation resistance of xylem. Plant Physiol 125:779–786

Hacke UG, Sperry JS, Pittermann J (2004) Analysis of circular bordered pit function II. Gymnosperm tracheids with torus-margo pit membranes. Am J Bot 91:386–400

Hartmann H (2011) Will a 385 million year-struggle for light become a struggle for water and for carbon?—how trees may cope with more frequent climate change-type drought events. Glob Chang Biol 17:642–655

Hoffmann WA, Marchin RM, Abit P, Lau OL (2011) Hydraulic failure and tree dieback are associated with high wood density in a temperate forest under extreme drought. Glob Chang Biol 17:2731–2742

Hölttä T, Cochard H, Nikinmaa E, Mencuccini M (2009) Capacitive effect of cavitation in xylem conduits: results from a dynamic model. Plant Cell Environ 32:10–21

IPCC (2013) Summary for policymakers. Climate change 2013: the physical science basis. Cambridge University Press, Cambridge

Jacobsen AL, Pratt RB, Davis SD, Ewers FW (2007) Cavitation resistance and seasonal hydraulics differ among three arid Californian plant communities. Plant Cell Environ 30:1599–1609

Jasechko S, Sharp ZD, Gibson JJ, Birks SJ, Yi Y, Fawcett PJ (2013) Terrestrial water fluxes dominated by transpiration. Nature 496:347–350

Johnson DM, McCulloh KA, Woodruff DR, Meinzer FC (2012) Hydraulic safety margins and embolism reversal in stems and leaves: why are conifers and angiosperms so different? Plant Sci 195:48–53

Jones H, Sutherland R (1991) Stomatal control of xylem embolism. Plant Cell Environ 14:607–612

Keenan TF, Hollinger DY, Bohrer G, Dragoni D, Munger JW, Schmid HP, Richardson AD (2013) Increase in forest water-use efficiency as atmospheric carbon dioxide concentrations rise. Nature 499:324–327

Kilpeläinen A, Gerendiain AZ, Luostarinen K, Peltola H, Kellomäki S (2007) Elevated temperature and CO_2 concentration effects on xylem anatomy of Scots pine. Tree Physiol 27:1329–1338

Koch GW, Sillett SC, Jennings GM, Davis SD (2004) The limits to tree height. Nature 428:851–854

Koepke DF, Kolb TE, Adams HD (2010) Variation in woody plant mortality and dieback from severe drought among soils, plant groups, and species within a northern Arizona ecotone. Oecologia 163:1079–1090

Kraft NJB, Metz MR, Condit RS, Chave J (2010) The relationship between wood density and mortality in a global tropical forest data set. New Phytol 188:1124–1136

Lamy JB, Delzon S, Bouche PS, Alia R, Vendramin GG, Cochard H, Plomion C (2013) Limited genetic variability and phenotypic plasticity detected for cavitation resistance in a Mediterranean pine. New Phytol 201(3):874–886

Le Quéré C, Raupach MR, Canadell JG, Marland G, Bopp L, Ciais P, Conway TJ, Doney SC, Feely RA, Foster P (2009) Trends in the sources and sinks of carbon dioxide. Nat Geosci 2:831–836

Maherali H, DeLucia EH (2000a) Interactive effects of elevated CO_2 and temperature on water transport inponderosa pine. Am J Bot 87:243–249

Maherali H, DeLucia EH (2000b) Xylem conductivity and vulnerability to cavitation of ponderosa pine growing in contrasting climates. Tree Physiol 20:859–867

Maherali H, Pockman WT, Jackson RB (2004) Adaptive variation in the vulnerability of woody plants to xylem cavitation. Ecology 85:2184–2199

Martínez-Vilalta J, Cochard H, Mencuccini M, Sterck F, Herrero A, Korhonen J, Llorens P, Nikinmaa E, Nolè A, Poyatos R (2009) Hydraulic adjustment of Scots pine across Europe. New Phytol 184:353–364

Martorell S, Diaz-Espejo A, Medrano H, Ball MC, Choat B (2014) Rapid hydraulic recovery in Eucalyptus pauciflora after drought: linkages between stem hydraulics and leaf gas exchange. Plant Cell Environ 37:617–626

McCulloh KA, Johnson DM, Meinzer FC, Woodruff DR (2014) The dynamic pipeline: hydraulic capacitance and xylem hydraulic safety in four tall conifer species. Plant Cell Environ 37(5):1171–1183

McCulloh KA, Meinzer FC, Sperry JS, Lachenbruch B, Voelker SL, Woodruff DR, Domec J-C (2011) Comparative hydraulic architecture of tropical trees representing a range of successional stages and wood density. Oecologia 167:27–37

McDowell N, Pockman WT, Allen CD, Breshears DD, Cobb N, Kolb T, Plaut J, Sperry J, West A, Williams DG, Yepez EA (2008) Mechanisms of plant survival and mortality during drought: why do some plants survive while others succumb to drought? New Phytol 178:719–739

McDowell NG, Beerling DJ, Breshears DD, Fisher RA, Raffa KF, Stitt M (2011) The interdependence of mechanisms underlying climate-driven vegetation mortality. Trends Ecol Evol 26:523–532

Meinzer FC (2003) Functional convergence in plant responses to the environment. Oecologia 134:1–11

Meinzer FC, James SA, Goldstein G, Woodruff D (2003) Whole-tree water transport scales with sapwood capacitance in tropical forest canopy trees. Plant Cell Environ 26:1147–1155

Meinzer FC, Campanello PI, Domec J-C, Gatti MG, Goldstein G, Villalobos-Vega R, Woodruff DR (2008a) Constraints on physiological function associated with branch architecture and wood density in tropical forest trees. Tree Physiol 28:1609–1617

Meinzer FC, Woodruff DR, Domec J-C, Goldstein G, Campanello PI, Gatti MG, Villalobos-Vega R (2008b) Coordination of leaf and stem water transport properties in tropical forest trees. Oecologia 156:31–41

Meinzer FC, Johnson DM, Lachenbruch B, McCulloh KA, Woodruff DR (2009) Xylem hydraulic safety margins in woody plants: coordination of stomatal control of xylem tension with hydraulic capacitance. Funct Ecol 23:922–930

Meinzer FC, McCulloh KA, Lachenbruch B, Woodruff DR, Johnson DM (2010) The blind men and the elephant: the impact of context and scale in evaluating conflicts between plant hydraulic safety and efficiency. Oecologia 164:287–296

Melcher PJ, Zwieniecki MA, Holbrook NM (2003) Vulnerability of xylem vessels to cavitation in sugar maple. Scaling from individual vessels to whole branches. Plant Physiol 131:1775–1780

Mitchell PJ, O'Grady AP, Tissue DT, White DA, Ottenschlaeger ML, Pinkard EA (2013) Drought response strategies define the relative contributions of hydraulic dysfunction and carbohydrate depletion during tree mortality. New Phytol 197:862–872

Nardini A, Battistuzzo M, Savi T (2013) Shoot desiccation and hydraulic failure in temperate woody angiosperms during an extreme summer drought. New Phytol 200(2):322–329

Nicotra AB, Atkin OK, Bonser SP, Davidson AM, Finnegan E, Mathesius U, Poot P, Purugganan MD, Richards C, Valladares F (2010) Plant phenotypic plasticity in a changing climate. Trends Plant Sci 15:684–692

Ogasa M, Miki NH, Murakami Y, Yoshikawa K (2013) Recovery performance in xylem hydraulic conductivity is correlated with cavitation resistance for temperate deciduous tree species. Tree Physiol 33:335–344

Olano JM, Arzac A, García-Cervig6n AI, von Arx G, Rozas V (2013) New star on the stage: amount of ray parenchyma in tree rings shows a link to climate. New Phytol 198(2):486–495

Paddock W, Davis SD, Pratt RB, Jacobsen AL, Tobin MF, Lopez-Portillo J, Ewers F (2013) Factors determining mortality of adult chaparral shrubs in an extreme drought year in California. Aliso 31:49–57

Pan Y, Birdsey RA, Fang J, Houghton R, Kauppi PE, Kurz WA, Phillips OL, Shvidenko A, Lewis SL, Canadell JG, Ciais P, Jackson RB, Pacala SW, McGuire AD, Piao S, Rautiainen A, Sitch S, Hayes D (2011) A large and persistent carbon sink in the world's forests. Science 333:988–993

Peng C, Ma Z, Lei X, Zhu Q, Chen H, Wang W, Liu S, Li W, Fang X, Zhou X (2011) A drought-induced pervasive increase in tree mortality across Canada's boreal forests. Nat Clim Change 1:467–471

Phillips OL, van der Heijden G, Lewis SL, Lopez-Gonzalez G, Aragao LEOC, Lloyd J, Malhi Y, Monteagudo A, Almeida S, Alvarez Davila E, Amaral I, Andelman S, Andrade A, Arroyo L, Aymard G, Baker TR, Blanc L, Bonal D, Alves de Oliveira AC, Chao K-J, Davila Cardozo N, da Costa L, Feldpausch TR, Fisher JB, Fyllas NM, Freitas MA, Galbraith D, Gloor E, Higuchi N, Honorio E, Jimenez E, Keeling H, Killeen TJ, Lovett JC, Meir P, Mendoza C, Morel A, Nunez Vargas P, Patino S, Peh KSH, Pena Cruz A, Prieto A, Quesada CA, Ramirez F, Ramirez H, Rudas A, Salamao R, Schwarz M, Silva J, Silveira M, Slik JWF, Sonke B, Thomas AS, Stropp J, Taplin JRD, Vasquez R, Vilanova E (2010) Drought-mortality relationships for tropical forests. New Phytol 187:631–646

Phillips NG, Attard RD, Ghannoum O, Lewis JD, Logan BA, Tissue DT (2011) Impact of variable [CO_2] and temperature on water transport structure–function relationships in Eucalyptus. Tree Physiol 31:945–952

Pittermann J, Sperry JS, Hacke UG, Wheeler JK, Sikkema EH (2005) Torus-margo pits help conifers compete with angiosperms. Science 310:1924–1924

Pittermann J, Choat B, Jansen S, Stuart SA, Lynn L, Dawson TE (2010) The relationships between xylem safety and hydraulic efficiency in the Cupressaceae: the evolution of pit membrane form and function. Plant Physiol 153:1919–1931

Pittermann J, Stuart SA, Dawson TE, Moreau A (2012) Cenozoic climate change shaped the evolutionary ecophysiology of the Cupressaceae conifers. Proc Natl Acad Sci 109:9647–9652

Pockman WT, Sperry JS (2000) Vulnerability to xylem cavitation and the distribution of Sonoran Desert vegetation. Am J Bot 87:1287–1299

Ponce Campos GE, Moran MS, Huete A, Zhang Y, Bresloff C, Huxman TE, Eamus D, Bosch DD, Buda AR, Gunter SA, Scalley TH, Kitchen SG, McClaran MP, McNab WH, Montoya DS, Morgan JA, Peters DPC, Sadler EJ, Seyfried MS, Starks PJ (2013) Ecosystem resilience despite large-scale altered hydroclimatic conditions. Nature 87:144–146

Pratt R, Jacobsen A, Ewers F, Davis S (2007) Relationships among xylem transport, biomechanics and storage in stems and roots of nine Rhamnaceae species of the California chaparral. New Phytol 174:787–798

Pratt RB, Jacobsen AL, Jacobs SM, Esler KJ (2012) Xylem transport safety and efficiency differ among fynbos shrub life history types and between two sites differing in mean rainfall. Int J Plant Sci 173:474–483

Pratt RB, Jacobsen AL, Ramirez AR, Helms AM, Traugh CA, Tobin MF, Heffner MS, Davis SD (2014) Mortality of resprouting chaparral shrubs after a fire and during a record drought: physiological mechanisms and demographic consequences. Glob Chang Biol 20:893–907

Reich PB (2014) The world-wide 'fast–slow' plant economics spectrum: a traits manifesto. J Ecol 102:275–301

Salleo S, Trifilò P, Esposito S, Nardini A, Gullo MAL (2009) Starch-to-sugar conversion in wood parenchyma of field-growing Laurus nobilis plants: a component of the signal pathway for embolism repair? Funct Plant Biol 36:815–825

Scholz FG, Bucci SJ, Goldstein G, Moreira MZ, Meinzer FC, Domec JC, Villalobos-Vega R, Franco AC, Miralles-Wilhelm F (2008) Biophysical and life-history determinants of hydraulic lift in Neotropical savanna trees. Funct Ecol 22:773–786

Scholz FG, Phillips NG, Bucci SJ, Meinzer FC, Goldstein G (2011) Hydraulic capacitance: biophysics and functional significance of internal water sources in relation to tree size. In: Meinzer FCC, Lachenbruch B, Dawson TEE (eds) Size- and age-related changes in tree structure and function. Springer, Dordrecht, pp 341–361

Schreiber SG, Hacke UG, Hamann A, Thomas BR (2011) Genetic variation of hydraulic and wood anatomical traits in hybrid poplar and trembling aspen. New Phytol 190:150–160
Sparks JP, Black RA (1999) Regulation of water loss in populations of Populus trichocarpa: the role of stomatal control in preventing xylem cavitation. Tree Physiol 19:453–459
Sperry JS, Hacke UG, Pittermann J (2006) Size and function in conifer tracheids and angiosperm vessels. Am J Bot 93:1490–1500
Sperry JS, Meinzer FC, McCulloh KA (2008) Safety and efficiency conflicts in hydraulic architecture: scaling from tissues to trees. Plant Cell Environ 31:632–645
Thomas DS, Montagu KD, Conroy JP (2004) Changes in wood density of Eucalyptus camaldulensis due to temperature—the physiological link between water viscosity and wood anatomy. For Ecol Manage 193(1):157–165
Tyree MT, Sperry JS (1988) Do woody plants operate near the point of catastrophic xylem dysfunction caused by dynamic water stress? Answers from a model. Plant Physiol 88:574–580
Tyree MT, Davis SD, Cochard H (1994) Biophysical perspectives of xylem evolution: is there a trade-off of hydraulic efficiency for vulnerability to dysfunction? IAWA J 15:335–360
Tyree MT, Zimmermann MH (2002) Xylem structure and the ascent of sap. Springer, Berlin
Urli M, Porté AJ, Cochard H, Guengant Y, Burlett R, Delzon S (2013) Xylem embolism threshold for catastrophic hydraulic failure in angiosperm trees. Tree Physiol 33:672–683
van Mantgem PJ, Stephenson NL, Byrne JC, Daniels LD, Franklin JF, Fule PZ, Harmon ME, Larson AJ, Smith JM, Taylor AH, Veblen TT (2009) Widespread increase of tree mortality rates in the Western United States. Science 323:521–524
Voelker SL, Noirot-Cosson P-E, Stambaugh MC, McMurry ER, Meinzer FC, Lachenbruch B, Guyette RP (2012) Spring temperature responses of oaks are synchronous with North Atlantic conditions during the last deglaciation. Ecol Monogr 82:169–187
Way DA (2013) Will rising CO_2 and temperatures exacerbate the vulnerability of trees to drought? Tree Physiol 33:775–778
West AG, Hultine KR, Sperry JS, Bush SE, Ehleringer JR (2008) Transpiration and hydraulic strategies in a Piñon-Juniper woodland. Ecol Appl 18:911–927
Wheeler JK, Huggett BA, Tofte AN, Rockwell FE, Holbrook NM (2013) Cutting xylem under tension or supersaturated with gas can generate PLC and the appearance of rapid recovery from embolism. Plant Cell Environ 36:1938–1949
Woodruff D, Bond B, Meinzer F (2004) Does turgor limit growth in tall trees? Plant Cell Environ 27:229–236
Worrall JJ, Egeland L, Eager T, Mask RA, Johnson EW, Kemp PA, Shepperd WD (2008) Rapid mortality of Populus tremuloides in southwestern Colorado, USA. For Ecol Manage 255:686–696
Wortemann R, Herbette S, Barigah TS, Fumanal B, Alia R, Ducousso A, Gomory D, Roeckel-Drevet P, Cochard H (2011) Genotypic variability and phenotypic plasticity of cavitation resistance in Fagus sylvatica L. across Europe. Tree Physiol 31:1175–1182
Zanne AE, Westoby M, Falster DS, Ackerly DD, Loarie SR, Arnold SE, Coomes DA (2010) Angiosperm wood structure: global patterns in vessel anatomy and their relation to wood density and potential conductivity. Am J Bot 97:207–215

Chapter 10
The Anatomical Determinants of Leaf Hydraulic Function

Lawren Sack, Christine Scoffoni, Daniel M. Johnson, Thomas N. Buckley, and Timothy J. Brodribb

1 Introduction

Leaves are enormously diverse in size and shape, and especially in their internal anatomy, including their venation architecture (Figs. 10.1, 10.2, and 10.3) (Esau 1977; Ellis et al. 2009). Across species, venation systems vary in the branching and arrangements of leaf major veins, i.e., the first-order vein(s) entering the leaf from the petiole, and the second- and third-order veins branching off—and the minor veins, i.e., the 3–5 additional orders of smaller veins embedded within the lamina and forming a continuous mesh with the major veins (Figs. 10.1 and 10.2). Additionally, leaf veins vary strongly in their internal anatomy—i.e., that of the parenchyma, xylem, and phloem inside—and in their external anatomy—including the parenchymatous and sometimes sclerenchymous bundle sheath (Fig. 10.3). Leaves are additionally enormously variable in the anatomy of the lamina outside the veins—i.e., the arrangement and sizes of the spongy and palisade mesophyll tissues, and sometimes water storage and accessory transport tissues (Fig. 10.3). Of all the plant lineages, angiosperms evolved the most exceptional diversity across species in all of these vein and outside-vein anatomical traits, as for the rest of their

L. Sack (✉) • C. Scoffoni
Department of Ecology and Evolutionary Biology, University of California, Life Sciences 3218, 621 Charles E. Young S, Los Angeles, CA 90095, USA
e-mail: lawrensack@ucla.edu

D.M. Johnson
Department of Forest, Rangeland and Fire Sciences, University of Idaho, Moscow, ID 83844, USA

T.N. Buckley
IA Watson Grains Research Centre, The University of Sydney, Narrabri, NSW, Australia

T.J. Brodribb
School of Plant Science, University of Tasmania, Hobart, TAS 7001, Australia

U. Hacke (ed.), *Functional and Ecological Xylem Anatomy*,
DOI 10.1007/978-3-319-15783-2_10

Fig. 10.1 Diversity of leaf major venation. Chemically cleared and stained whole leaves of *Arabidopsis thaliana* Col-0 (Brassicaceae, *left*) and *Viburnum molle* (Adoxaceae, *right*). Photo credit: *A. thaliana* (M. Caringella)

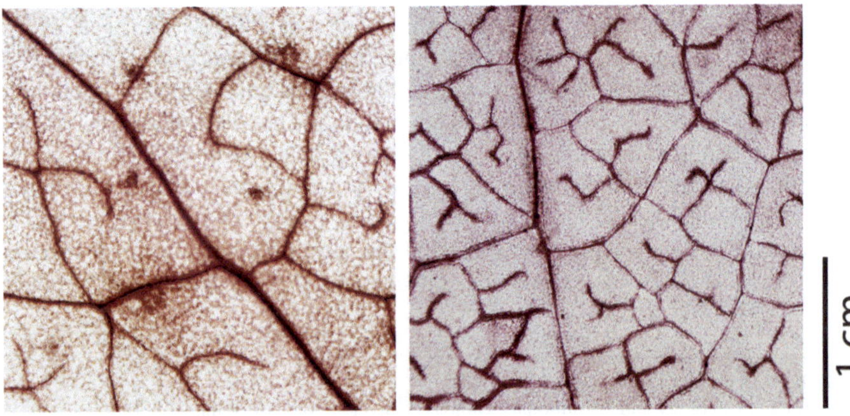

Fig. 10.2 Diversity of leaf minor venation. Chemically cleared and stained micrographs of *Arabidopsis thaliana* Col-0 (Brassicaceae, *left*) and *Viburnum molle* (Adoxaceae, *right*). Photo credit: *A. thaliana* (M. Caringella)

morphology and physiology (Augusto et al. 2014). The variation across species in vein and lamina anatomy carries extensive information about the physiological function, development, evolution, ecology, and paleohistory of leaves (Haberlandt 1914; Roth-Nebelsick et al. 2001; Brodribb et al. 2010; Sack and Scoffoni 2013). Further, much of the functional importance of the leaf's anatomy relates to its role

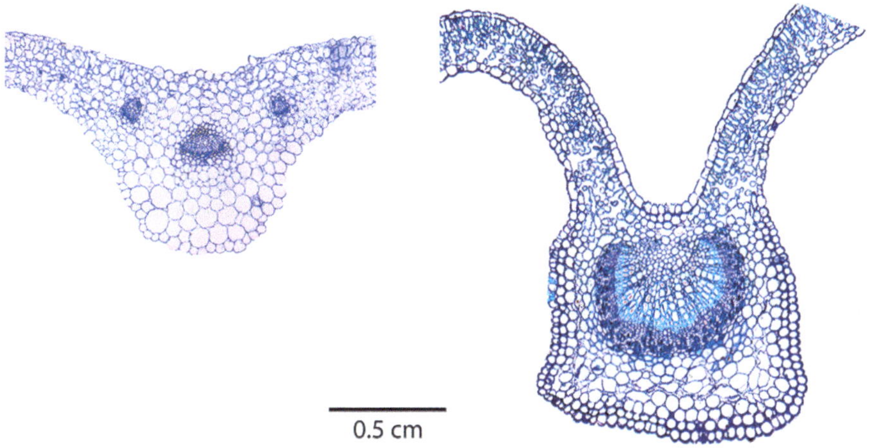

Fig. 10.3 Diversity of leaf midrib and lamina anatomy. Transverse cross-sections of *Arabidopsis thaliana* Col-0 (Brassicaceae, *left*) and *Viburnum molle* (Adoxaceae, *right*). Photo credit: *A. thaliana* (M. Caringella)

within the plant water transport system. This review focuses on the contribution of both the vein system and the outside-vein system to the leaf hydraulic conductance (K_{leaf}) and to K_{leaf} dynamics with leaf water status.

We focus on hydraulic function because it is a backbone of whole plant physiological performance (Tyree and Zimmermann 2002; Brodribb 2009). When the stomata open for CO_2 assimilation, the leaf's moist internal airspaces are exposed to the dry outside air, resulting in transpirational water loss. Consequently, if the hydraulic system did not replace sufficient water, the mesophyll would desiccate and stomata would close (Sack and Holbrook 2006). The plant hydraulic system thus imposes a major constraint on the ability of the stomata to remain open to allow photosynthesis. Indeed, leaves generally evolve and develop with hydraulic and stomatal traits matched such that transpiration and photosynthesis can proceed with sufficient water supply (Sack et al. 2003; Dunbar-Co et al. 2009; Brodribb and Jordan 2011; Carins Murphy et al. 2012). The hydraulic system can be analyzed using concepts from electronics or plumbing: the plant is considered as a series of resistors in series or parallel (also including capacitors in fully elaborated models), where the efficiency of the system can be summarized with a single ratio: the flow rate divided by the pressure driving force. Thus, the plant hydraulic conductance (K_{plant}) is determined as the transpiration rate divided by the water potential gradient from soil to leaf, and K_{leaf} as the transpiration rate divided by the water potential gradient from petiole to evaporation sites within the leaf; in these calculations, the water potential of the evaporating sites is often approximated for practical purposes as the bulk leaf water potential. Studies of plant communities around the world have revealed strong correlations of the hydraulic conductances of organs or whole plants with gas exchange and photosynthetic rates, and with species' ecological specializations. Indeed, a number of studies have shown close correlations of K_{leaf} with stomatal conductance, transpiration rate, and maximum photosynthetic rates across diverse species, consistent with the leaf being a major bottleneck in the whole plant hydraulic system with consequent influence on photosynthetic gas exchange (Sack et al. 2003; Franks 2006; Sack and Holbrook 2006; Brodribb et al. 2007).

The K_{leaf} varies by over 65-fold across species (Sack and Holbrook 2006). This variation is consistent with the nature of K_{leaf} as a single value summarizing a complex micro-hydrological system, influenced by both the leaf xylem and outside-xylem compartments. The xylem compartment relates to the leaf venation, and the outside-xylem compartment to the flow of water across the xylem parenchyma and bundle sheath, and then through and/or around mesophyll cells to the site of water evaporation within the leaf (Sack and Scoffoni 2013). Early work had assumed that K_{leaf} would be mainly limited by the outside-xylem compartment, because it involves flow across membranes, whereas flow through the venation is via xylem conduits that are dead and hollow at maturity. However, this understanding has changed. There is now general recognition of the role of aquaporins in membrane flow, increasing the permeability of membranes by orders of magnitude relative to that of a simple phospholipid bilayer, and appreciation that water flows through the leaf xylem in very narrow vessels and tracheids. Most importantly, measurements have been made of the low hydraulic conductance in vein xylem (K_x), which is typically of the same order as that outside the xylem, i.e., across the bundle sheath and mesophyll (K_{ox}) (Zwieniecki et al. 2002; Sack et al. 2004). Current understanding is that there is substantial hydraulic resistance both inside and outside the xylem and thus the anatomy of both the vein system and the outside-vein lamina can influence K_{leaf}, and scale up to influencing leaf gas exchange and whole plant performance (Sack and Holbrook 2006).

Hydraulic conductance is not a constant. Indeed, declines of hydraulic conductance may occur throughout the plant during strong transpiration or especially during soil and atmospheric drought. Hydraulic decline is often stronger in leaves than stems due to more negative water potentials, lower resistance to embolism, and/or the collapse of mesophyll tissues (Hao et al. 2008; Brodribb 2009; Johnson et al. 2011, 2012a).

Our aim is to review the correlations in the literature of K_{leaf} and K_{leaf} vulnerability to dehydration with cell, tissue, and whole leaf structure and composition (Tables 10.1, 10.2, and 10.3). Our overall thesis is that K_{leaf} and its vulnerability depend on multiple traits in concert, at different levels of tissue and organ construction.

2 Venation Traits

A number of vein traits are causal drivers or correlates across species of K_{leaf} for hydrated leaves (Table 10.1). The best supported anatomical correlate of K_{leaf} is the minor vein length per leaf area (minor VLA), or the total vein length per area (VLA, also known as "vein density"), which are themselves tightly correlated because

Table 10.1 Vein traits that contribute to leaf hydraulic conductance in hydrated leaves at full irradiance (K_{max}) and/or influence leaf hydraulic vulnerability

	K_{max}	K_{leaf} vulnerability
Greater xylem conduit numbers or sizes	Contributes to higher vein conductivity within and across species (Sack and Frole 2006; Maherali et al. 2008; Dunbar-Co et al. 2009; Taneda and Terashima 2012). In turn higher vein conductivity, especially in lower-order veins, can contribute to higher K_{leaf} as shown in computer modeling (McKown et al. 2010), and across diverse angiosperm species, and within species (Nardini et al. 2005; Sack and Frole 2006)	Narrow xylem conduits in midribs and greater xylem conduit cell wall thickness/lumen breadth ratios in minor veins may contribute to resistance to hydraulic decline within crowns, or across species within a genus or across diverse species adapted to a gradient of moisture (Cochard et al. 2004; Johnson et al. 2009; Blackman et al. 2010; Nardini et al. 2012; Jordan et al. 2013)
Major vein length per area	High major VLA contributes to higher K_{leaf} as shown in computer modeling and across closely related species, and varieties of a given species (McKown et al. 2010; Sommerville et al. 2012; Nardini et al. 2014)	High major VLA provides tolerance of disruption to the hydraulic system caused by damage or drought as shown in computer modeling and across diverse angiosperm species, and within angiosperm genera (Sack et al. 2008; McKown et al. 2010; Scoffoni et al. 2011; Nardini et al. 2012, 2014)
Minor vein length per area	High minor VLA can contribute to higher K_{leaf} as shown in computer modeling (McKown et al. 2010), across tropical tree species (Sack and Frole 2006) and species of diverse lineages (Brodribb et al. 2007), and correlates with higher stomatal conductance (Boyce et al. 2009; Feild et al. 2011), and higher light-saturated photosynthetic rates per leaf area and mass across diverse species (Brodribb et al. 2010; Feild et al. 2011; Walls 2011; Sack et al. 2013)	
Free ending veins per unit area	Numerous FEVs per unit area can correlate with higher K_{leaf} across diverse species (Scoffoni et al. 2011)	
Vein topology		Increased looping provides optimal transport given fluctuating flow or damage according to mathematical and computer models (Corson 2010; Katifori et al. 2010)

Table 10.2 Bundle sheath and bundle sheath extension traits that contribute to leaf hydraulic conductance in hydrated leaves at full irradiance (K_{max}) and/or influence leaf hydraulic vulnerability

	K_{max}	K_{leaf} vulnerability
Bundle sheath cell size	Correlates with K_{leaf} across ontogenetically different leaves of *Ginkgo biloba* (Leigh et al. 2011)	Hypothesized to act in cavitation repair in the xylem in C_3 and C_4 grasses (Griffiths et al. 2013)
Bundle sheath extensions length and volume per area	Correlates with K_{leaf} across ontogenetically different leaves of *Ginkgo biloba* (Leigh et al. 2011) and across *Acacia* species (Sommerville et al. 2012)	
	A tomato mutant with reduced BSEs had lower K_{leaf} (Zsögön et al. 2015)	
	Correlates with the response of K_{leaf} to irradiance across diverse species (Scoffoni et al. 2008)	
	Correlates with the water delivery to the epidermis which can damp responses of K_{leaf} and gas exchange to vapor pressure deficit (Buckley et al. 2011)	
Bundle sheath permeability	ABA signaling and aquaporin activation/deactivation in the bundle sheath control K_{leaf} (Lee et al. 2008, 2009; Ache et al. 2010; Shatil-Cohen et al. 2011; Flexas et al. 2013; Griffiths et al. 2013; Pantin et al. 2013; Prado et al. 2013; Prado and Maurel 2013; Sack and Scoffoni 2013; Secchi and Zwieniecki 2013; Chaumont and Tyerman 2014)	Aquaporins in BS and xylem parenchyma may play a role in xylem refilling in leaves and stems (Laur and Hacke 2014a, b)

minor veins make up the bulk of leaf veins length (Sack and Frole 2006; Brodribb et al. 2007). Across vascular plants, early branching clades such as ferns exhibit low VLA and K_{leaf} compared with later groups such as angiosperms (Boyce et al. 2009). Even among angiosperms there has been a tendency for VLA and K_{leaf} to increase over time (Brodribb and Feild 2010). These strong trends found across large clades are often weaker among specific groups such as the eudicots, suggesting that other factors are also important in determining K_{leaf} (McKown et al. 2010; Scoffoni et al. 2011; Sack et al. 2013; Sack and Scoffoni 2013). Notably, many traits influence K_{leaf}, as explored throughout this chapter, and thus in certain species sets K_{leaf} may be weakly related or uncorrelated with any given trait in certain lineages or species sets due to a greater variation in other key traits. Additional vein traits that influence the maximum value of K_{leaf} (K_{max}, which occurs at high water potential and high irradiance) include larger xylem conduit numbers and sizes, higher major vein length per area, vein topology with greater reticulation, and a greater number of free ending veins per area (Table 10.1).

Leaf hydraulic vulnerability also depends on vein traits (Table 10.1). Leaf dehydration may lead to embolism in the vein xylem, leading to decline of K_x in combination with declines in K_{ox} that could arise due to tissue shrinkage (see following section). The question of whether embolism occurs in all orders of leaf veins

Table 10.3 Leaf lamina mesophyll anatomy traits that contribute to leaf hydraulic conductance in hydrated leaves at full irradiance (K_{max}) and/or influence leaf hydraulic vulnerability

	K_{max}	K_{leaf} vulnerability
Spongy mesophyll: palisade mesophyll thickness ratio (or spongy mesophyll + epidermis: palisade mesophyll thickness ratio)	Hypothesized to contribute to K_{leaf}; negatively correlated with vein length per area and positively correlated with K_{leaf} across diverse angiosperms (Wylie 1946; Sack and Frole 2006)	
Leaf mesophyll thickness	Correlated with K_{leaf} across species (Aasamaa et al. 2001; Sack et al. 2003; Sack and Frole 2006) and for sun vs. shade leaves within canopies (Sack et al. 2003; Brodribb and Jordan 2011)	Tissue shrinkage may contribute to K_{leaf} decline, and thus leaves with higher elastic modulus might be more resistant to hydraulic decline (Charra-Vaskou et al. 2012; Scoffoni et al. 2014)
Accessory transport elements	Accessory transport elements may contribute to K_{leaf} in cycads, conifers, and angiosperms (Esau 1977; Brodribb et al. 2007, 2010; Sack and Scoffoni 2013)	Accessory transport elements might act as water storage to buffer cell water potentials from transiently high transpiration rates (Takeda 1913) Collapse of transfusion tracheids in conifers is associated with K_{leaf} decline in dehydrating leaves (Brodribb and Holbrook 2005; Zhang et al. 2014)

remains open, but both cavitation and collapse appear to be greater in conduits with larger lumens, and thus the major vein xylem may be more likely to embolize than collapse because it contains the widest vessels in the leaf, and because it is more mechanically reinforced (Blackman et al. 2010; Sack and Scoffoni 2013). Studies reporting embolism-induced reductions in K_{leaf} have found support for cavitation events occurring in leaf petioles or midribs using a range of hydraulic measurements and visualization approaches, and for collapse of tracheids in some pine species (Nardini et al. 2001; Bucci et al. 2003; Cochard et al. 2004; Johnson et al. 2009, 2012b; Charra-Vaskou et al. 2012). Similarly, K_{leaf} vulnerability may be lower for species with narrower xylem conduits in the midrib and/or minor veins, if these embolize or collapse at a lower leaf water status. Two studies of angiosperms found that species having smaller leaf midrib conduits were more resistant to embolism (Johnson et al. 2009; Nardini et al. 2012). Further, across four species of pines, the two species with smaller diameter leaf vein tracheids had much more negative conduit collapse pressures than did the two species with larger tracheid diameters (Cochard et al. 2004). Additionally, along a height gradient in Douglas fir, the hydraulic vulnerability of leafy shoots was lower for shoots with needles with narrower xylem conduits, fewer tracheids per cross-sectional leaf area, and fewer pits per tracheid (Woodruff et al. 2008). In dicotyledons, narrower conduits in minor veins may be protected during collapse under strong xylem tensions arising during drought. Thus, leaf minor vein conduit cell wall thickness to lumen breadth ratio ($[t/b]^3$) was positively related to leaf hydraulic vulnerability for 20 angiosperm species (Blackman et al. 2010), and positively correlated with moisture availability in 67 species from the Proteaceae (Jordan et al. 2013).

Higher leaf major vein length per area may also confer resistance to hydraulic dysfunction. Indeed, small leaves from drier habitats tend to have higher leaf major VLA, which would provide a greater level of drought tolerance. Thus, in studies of ten species from diverse angiosperm families (Scoffoni et al. 2011), and six angiosperm species (three *Acer* sand three *Quercus* species) (Nardini et al. 2012) and four *Coffea arabica* varieties, a greater resistance to leaf hydraulic dysfunction correlated with higher major VLA (Nardini et al. 2014).

Additionally, the topology of the vein system (i.e., arrangement of veins) may confer resistance to hydraulic decline. Greater vein connectivity should render the system relatively tolerant to the effects of embolism if it can route water around blockages through highly conductive veins (Sack et al. 2008; Corson 2010; Katifori et al. 2010; Scoffoni et al. 2011).

3 Bundle Sheath and Bundle Sheath Extension Traits

The tissues immediately surrounding the veins provide a hydraulic interface with the mesophyll and can strongly influence K_{leaf} (Table 10.2). The bundle sheath (BS) is a cylinder of parenchyma cells that surround the vascular tissues in the leaf vein (Esau 1977) and as such is the tissue that water must move through between xylem and the mesophyll. Classically, researchers expected the BS and bundle sheath extensions (BSE), a tissue composed of parenchyma and sometimes sclerenchyma that in many species extends between the veins and epidermis, to influence water transport in the leaf (Wylie 1952; Canny 1990a). Indeed, recent research shows that these features influence K_{leaf} and its dynamics in relation to light and water supply (Scoffoni et al. 2008; Buckley et al. 2011). In some species, the BS has suberized layers in its anticlinal walls, analogous to the Casparian strip in root endodermis, which could greatly reduce the conductivity of the BS by preventing extracellular water transport (Lersten 1997). A Casparian strip in the BS is consistent with accumulation of tracers at the proximal margins of BS cells (Canny 1990b), and with genes being expressed during development of the leaf BS in common with the root endodermis (Slewinski et al. 2012), but the distribution of a BS Casparian strip across species is largely unknown. Indeed, BS cells might act on controlling the import and export of water just as they are used for loading phloem cells (Ache et al. 2010; Nardini et al. 2010), and thus prevent water loss and solute leakage from the veins (O'Brien and Carr 1970; Canny 1990a; Mertz and Brutnell 2014). This is consistent with growing evidence that the BS is a central "control point" for integrating whole leaf function (Flexas et al. 2013; Griffiths et al. 2013; Sack and Scoffoni 2013).

If flow across the BS is predominantly across the plasma membrane, then aquaporin function probably determines its contribution to K_{leaf} (Shatil-Cohen et al. 2011; Moshelion et al. 2015). Consequently, recent work has implied that the BS is a major influence on the outside component of K_{leaf}, i.e., K_{ox}, and thus on K_{leaf} itself (Buckley et al., unpublished data). Indeed, *Arabidopsis* and *Populus* mutants in which aquaporins are disrupted in the BS and vein parenchyma show diminished K_{leaf} (Prado et al. 2013; Prado and Maurel 2013; Secchi and Zwieniecki 2013; Chaumont and Tyerman 2014). Aquaporins present in BS cells, phloem cells, and transfusion parenchyma may facilitate radial water flow in *Picea glauca* needles (Laur and Hacke 2014a). Additionally, in *Arabidopsis*, abscisic acid (ABA) influences K_{leaf} via the BS (Shatil-Cohen et al. 2011; Pantin et al. 2013). In maize, a reduction of turgor in the midrib BS reduced K_{leaf} (Kim and Steudle 2007). The BS may also influence the distribution of water from veins across the lamina; BS cell permeability was higher in minor than major veins in tobacco (Lee et al. 2009).

The BS and vein internal parenchyma may also control K_{leaf} dynamics in response to the leaf's external environment and internal water status (Table 10.2). Thus, in *Arabidopsis*, whole rosette hydraulic conductance declined under high irradiance, and was correlated with the permeability of BS and vein parenchyma cells, but not with that of mesophyll cells (Prado et al. 2013). Notably, however, the BS cell permeability would have a complex relationship with K_{leaf} for given species, given that it may increase with aquaporin activity but decline with reduced turgor. Thus, K_{leaf} increased under higher irradiance for fig leaf gourd and tobacco, though BS cell permeability declined, as the cells lost turgor with the higher transpiration, due to potassium efflux and reduced aquaporin activity (Lee et al. 2008, 2009).

Bundle sheath cells and vein parenchyma cells have been hypothesized to play a major role in xylem cavitation repair, through the same process as root exudation (Nardini et al. 2008; Laur and Hacke 2014b). A correlation of BS size with low precipitation and aridity index across both C_3 and C_4 grasses supported the inference of a possible role for cavitation repair, and indeed, that selection on BS cell size could enable C_4 Kranz anatomy evolution (Griffiths et al. 2013). Foliar uptake and aquaporins in BS might even play a role in stem xylem refilling in *Picea glauca* (Laur and Hacke 2014a).

BSEs are more frequent in sun-adapted species, and in exposed leaves within canopies (Kenzo et al. 2007; Sack and Scoffoni 2013), consistent with a role in facilitating greater transpiration and photosynthetic rates under high irradiance. Indeed, BSEs apparently enhance K_{leaf}, by providing a larger surface area for water movement between xylem and the epidermis or mesophyll and/or for water to evaporate into the intercellular spaces (Sheriff and Meidner 1974; Zwieniecki et al. 2007; Ye et al. 2008). After the water is delivered to the epidermis via the BSEs, conduction may be more efficient from the epidermis to the mesophyll than through the mesophyll (Wylie 1943; Sheriff and Meidner 1974). Thus, dye experiments with lead showed water is transported to the epidermis via the apoplast of BSEs in *Tradescantia virginiana* (Byott and Sheriff 1976). *Acacia* phyllodes with more BSEs had higher K_{leaf} (Sommerville et al. 2012), and in *Ginkgo biloba*, long shoot leaves have higher K_{leaf} than short shoot leaves, corresponding with their having larger BSEs

(Leigh et al. 2011). Further, a tomato mutant with reduced BSEs had lower K_{leaf} (Zsögön et al. 2015). Additionally, BS and/or BSEs may influence the dynamics of K_{leaf} to light and VPD. Thus, species with BSEs tend to more frequently show K_{leaf} responses to irradiance (Scoffoni et al. 2008). Species with BSEs may maintain an especially close hydraulic connection between epidermis and vascular tissues. In these species, stomatal closure may be more closely connected with bulk leaf water potential, whereas, in leaves without BSEs where epidermis is thus isolated from the veins, stomatal dynamics may be expected to be less influenced by bulk leaf turgor loss. Consequently, species with BSEs also tend to have faster responses of stomatal conductance to changes in water supply and demand, but more so for supply (Buckley et al. 2011), which is consistent with enhancement of K_{leaf} by BSEs.

4 Mesophyll Traits

The tissues outside the vein xylem are an important component of the leaf hydraulic system and thus can strongly influence K_{leaf} (Table 10.3). Once water passes the BS, it flows as a liquid and/or as vapor through or around mesophyll and epidermal cells to the terminal evaporation sites. While the terminal sites are still not known, classical observations and recent modeling shows that water may evaporate deep within the leaf or near the stomata and travel as vapor through the leaf, and this may depend on the leaf anatomy and also on the leaf's environment (Sheriff 1977; Boyer 1985; Rockwell et al. 2014; Buckley 2015; Scoffoni 2015). Modeling suggests that the bulk of water transport in leaves is in the liquid phase, but that leaves with low tissue density may have significant vapor phase transport of water under conditions of high irradiance when heat is absorbed by chlorophyllous mesophyll layers (Rockwell et al. 2014; Buckley 2015). The degree that mesophyll anatomy will influence K_{ox} depends on its resistance relative to that of the BS and in fact the sites of evaporation may be dictated by the resistances of different pathways through the tissues (Buckley 2015).

Whether the anatomy of the mesophyll is a large influence on K_{ox}, and therefore K_{leaf} is a subject for current debate. The thickness of the mesophyll layer between the veins and the stomata ("vein–epidermal distance"; VED) has been proposed to negatively influence K_{leaf} (Brodribb et al. 2007), as a potential determinant of the flow path between veins and the sites of evaporation (referred to as the "mesophyll-distance" for flow outside the xylem, D_m). A physical model supported that idea, and the corollary that VLA and VED should be negatively correlated across species, as expected if they were colimiting to K_{leaf} (Noblin et al. 2008; Zwieniecki and Boyce 2014). However, that idea presumes that the bulk of water outside the xylem is transported across the mesophyll to the substomatal cavity before evaporation, but others have suggested that much water may evaporate near the BS. Which scenario is most accurate may vary with species' leaf anatomy, leaf water status, and ambient environment (Sack et al. 2013). Indeed, several studies have reported positive rather than negative correlations of K_{leaf} with leaf thickness and/or with spongy/palisade ratio (Aasamaa et al. 2001; Sack et al. 2003; Sack and Frole 2006), and leaves adapted or acclimated to high irradiance generally tend to be thicker and yet tend to have a higher K_{leaf} than accounted for by simply a higher VLA (Brodribb and Jordan 2011). These facts would suggest that the VED might impose little constraint on K_{leaf} per se, or indeed, that outside xylem flow efficiency may in some cases be increased by parallel horizontal layers. Indeed, the classical expectation was that water transport through the spongy mesophyll and epidermis is effective (Wylie 1939, 1946). There is a general trend across species for low VLA to correlate with a high ratio of the thickness of spongy mesophyll (and/or the thickness of spongy-mesophyll-plus-epidermis) to the thickness of palisade mesophyll, and this was interpreted as showing that a greater K_{ox} would be conferred by a thicker spongy mesophyll to compensate for larger flow distances outside the vein in leaves with lower VLA (Wylie 1946). A recent theoretical analysis also found that mesophyll cell porosity and connectivity were key parameters influencing K_{leaf} (Buckley 2015). Under high irradiance, when vapor phase transport driven by vertical temperature gradients is important, the greater tissue porosity of spongy mesophyll may contribute even more strongly to a high K_{leaf}. Because temperature is predicted to peak within the palisade mesophyll, temperature gradients are probably greater across the lower half of the leaf, thus increasing the spongy mesophyll's potential to contribute to the vapor phase component of K_{leaf}. The role of outside-xylem anatomy in determining K_{leaf} remains an important focus for future modeling and experimental work.

Mesophyll anatomy may also have an influence on K_{ox} vulnerability (Table 10.3). The degree of mesophyll shrinkage during leaf dehydration correlates with hydraulic declines (Charra-Vaskou et al. 2012; Scoffoni et al. 2014). The resulting declines in K_{ox} could be due to reduced cell surface area for evaporation, severed cell–cell connections and thus fewer water flow pathways, and/or to lower aquaporin activity. Potentially, a higher modulus of elasticity, conferred by a thicker or denser cell wall relative to the size of the cell lumens, may thus confer tolerance to hydraulic decline by reducing cell volumetric shrinkage (Scoffoni et al. 2014).

Many species have "accessory transport or storage" tissues within their BS or lamina mesophyll tissues (Brodribb et al. 2010). Such additional transport cells in leaves can be observed as sclerified cells that can be isolated in the mesophyll or connect to the leaf veins branching out into areoles. Indeed, transfusion tracheids are present in all gymnosperm leaves (Hu and Yao 1981) and have long been thought to serve as either sites for water storage (Takeda 1913) or for transport of water and solutes between the mesophyll and the vascular tissue inside leaf veins (Esau 1977). Species with such tissues appear to have higher K_{leaf} than would be expected simply from their VLA (Brodribb et al. 2007), indicating a potential shortening of flow pathways, or additional evaporative surface (Tomlinson and Fisher 2005; Brodribb et al. 2007, 2010; Sack and Scoffoni 2013; Zhang et al. 2014). Transfusion tracheids collapse during leaf dehydration in *Podocarpus* (Brodribb and Holbrook 2005) and in *Taxus* (Zhang et al. 2014), associated with dehydration-induced declines in K_{leaf}, and may recover with rehydration (Zhang et al. 2014). However, a study of two *Pinus* species (Johnson et al. 2009) found that transfusion tracheids did not collapse during dehydration but did empty at water potentials less negative than would cause loss of K_{leaf}. The role of transfusion tracheids in leaf hydraulic vulnerability is still unclear.

5 Relationship with Photosynthetic Anatomy and Coordinated Development

Anatomical traits may be correlated with K_{leaf} across species for several distinct reasons (Sack et al. 2013; Sack and Scoffoni 2013). Correlations may arise due to a mechanistic causality, to a common developmental mechanism, to coselection during evolution across environments, and/or to a common plastic trajectory during growth in given environments (Brodribb et al. 2013; John et al. 2013; Sack et al. 2013). The specific type of correlation among traits may vary across species sets, and elucidating the basis for these correlations is essential both to understand their functional significance and also their generality and predictiveness.

Many of the correlations described above and in Tables 10.1, 10.2, and 10.3 are mechanistic, i.e., they arise because given traits contribute directly to the flow efficiency through the leaf. However, coselection of traits that are not directly linked in hydraulic function, for overall optimality in adaptation to environment is equally common. The leaf hydraulic conductance is not an isolated system, but rather fully integrated within the leaf gas exchange system. Thus, additional correlations are frequently observed between hydraulic traits and traits related to gas exchange at the leaf and plant scales. Thus, for example, stomatal density and pore area are also often correlated across species with maximum K_{leaf}, which arises at least in part due to coordination between VLA and stomatal density (Aasamaa et al. 2001; Sack et al. 2003, 2005; Dunbar-Co et al. 2009; Feild et al. 2011; Carins Murphy et al. 2012, 2014; Zhang et al. 2012). While stomatal density does not directly influence VLA or K_{leaf}, selection for rapid gas exchange should act to increase all three traits. Further, much of the anatomy relevant to K_{leaf} also influences mesophyll control of CO_2 assimilation. For example, the conductance between the sites of evaporation and the sites of carboxylation, or "mesophyll conductance" (g_m) correlates across species with K_{leaf} (Flexas et al. 2013). This link may arise due to shared dependence of K_{ox} and/or g_m on mesophyll surface area, and on a shared role of aquaporins in both H_2O and CO_2 transport.

Indeed, many anatomical traits show coordinated development, which provides a direct and effective route for selection to optimize a species' function. In particular, some anatomical features that are important for hydraulic function and gas exchange depend on cell or leaf size, and thus they develop in a coordinated way (Brodribb et al. 2013; John et al. 2013). For example, smaller leaves have higher major VLA, which will tend to confer lower K_{leaf} vulnerability, and if the smaller leaves under consideration are formed from smaller cells, as is true in certain cases, then they will also tend to have higher minor VLA and higher stomatal density, which will confer higher K_{leaf} and higher potential gas exchange rates. In such cases, development is a means to coordinate all traits to match function.

6 Conclusions

Many traits within and outside the xylem contribute to leaf hydraulic conductance and its dynamics in response to water stress. For over 100 years, these questions have been recognized as essential for understanding the limits of plant performance. Computer modeling, focused experiments, and new visualization technologies will continue to reveal the anatomical underpinnings of K_{leaf} at higher resolution. However, fundamental biophysical questions that impact strongly on K_{leaf} still remain unanswered by experiment, including the hydraulic conductivity of cell walls to long distance transport and the osmotic water permeabilities of mesophyll and epidermal cells. Models are especially needed that allow the effects of individual anatomical traits on K_{leaf} to be examined both in isolation from one another and in concert. Experiments that "tweak" individual features are needed on model plants such as Arabidopsis and poplar. Additionally, further comparisions are needed within and among lineages with strong variation in anatomical features. Using all these approaches in combination will result in a full appreciation of the functional consequences of the great variation in leaf hydraulic anatomy, with certain benefits for predictive ecology and for optimal crop design.

References

Aasamaa K, Sober A, Rahi M (2001) Leaf anatomical characteristics associated with shoot hydraulic conductance, stomatal conductance and stomatal sensitivity to changes of leaf water status in temperate deciduous trees. Aust J Plant Physiol 28:765–774

Ache P, Bauer H, Kollist H, Al-Rasheid KAS, Lautner S, Hartung W et al (2010) Stomatal action directly feeds back on leaf turgor: new insights into the regulation of the plant water status from non-invasive pressure probe measurements. Plant J 62:1072–1082

Augusto L, Davies TJ, Delzon S, De Schrijver A (2014) The enigma of the rise of angiosperms: can we untie the knot? Ecol Lett 17:1326–1338

Blackman CJ, Brodribb TJ, Jordan GJ (2010) Leaf hydraulic vulnerability is related to conduit dimensions and drought resistance across a diverse range of woody angiosperms. New Phytol 188:1113–1123

Boyce CK, Brodribb TJ, Feild TS, Zwieniecki MA (2009) Angiosperm leaf vein evolution was physiologically and environmentally transformative. Proc R Soc B Biol Sci 276:1771–1776

Boyer JS (1985) Water transport. Annu Rev Plant Physiol Plant Mol Biol 36:473–516

Brodribb TJ (2009) Xylem hydraulic physiology: the functional backbone of terrestrial plant productivity. Plant Sci 177:245–251

Brodribb TJ, Feild TS (2010) Leaf hydraulic evolution led a surge in leaf photosynthetic capacity during early angiosperm diversification. Ecol Lett 13:175–183

Brodribb TJ, Holbrook NM (2005) Water stress deforms tracheids peripheral to the leaf vein of a tropical conifer. Plant Physiol 137:1139–1146

Brodribb TJ, Jordan GJ (2011) Water supply and demand remain balanced during leaf acclimation of *Nothofagus cunninghamii* trees. New Phytol 192:437–448

Brodribb TJ, Feild TS, Jordan GJ (2007) Leaf maximum photosynthetic rate and venation are linked by hydraulics. Plant Physiol 144:1890–1898

Brodribb TJ, Feild TS, Sack L (2010) Viewing leaf structure and evolution from a hydraulic perspective. Funct Plant Biol 37:488–498

Brodribb TJ, Jordan GJ, Carpenter RJ (2013) Unified changes in cell size permit coordinated leaf evolution. New Phytol 199:559–570

Bucci SJ, Scholz FG, Goldstein G, Meinzer FC, Sternberg LDL (2003) Dynamic changes in hydraulic conductivity in petioles of two savanna tree species: factors and mechanisms contributing to the refilling of embolized vessels. Plant Cell Environ 26:1633–1645

Buckley TN (2015) The contributions of apoplastic, symplastic and gas phase pathways for water transport outside the bundle sheath in leaves. Plant Cell Environ 38(1):7–22

Buckley TN, Sack L, Gilbert ME (2011) The role of bundle sheath extensions and life form in stomatal responses to leaf water status. Plant Physiol 156:962–973

Byott GS, Sheriff DW (1976) Water movement into and through *Tradescantia virginiana* (L.) leaves 2. Liquid flow pathways and evaporative sites. J Exp Bot 27:634–639

Canny MJ (1990a) What becomes of the transpiration stream? New Phytol 114:341–368

Canny MJ (1990b) Fine veins of dicotyledon leaves as sites for enrichment of solutes of the xylem sap. New Phytol 115:511–516

Carins Murphy MR, Jordan GJ, Brodribb TJ (2012) Differential leaf expansion can enable hydraulic acclimation to sun and shade. Plant Cell Environ 35:1407–1418

Carins Murphy MR, Jordan GJ, Brodribb TJ (2014) Acclimation to humidity modifies the link between leaf size and the density of veins and stomata. Plant Cell Environ 37:124–131

Charra-Vaskou K, Badel E, Burlett R, Cochard H, Delzon S, Mayr S (2012) Hydraulic efficiency and safety of vascular and non-vascular components in Pinus pinaster leaves. Tree Physiol 32:1161–1170

Chaumont F, Tyerman SD (2014) Aquaporins: highly regulated channels controlling plant water relations. Plant Physiol 164:1600–1618

Cochard H, Froux F, Mayr S, Coutand C (2004) Xylem wall collapse in water-stressed pine needles. Plant Physiol 134:401–408

Corson F (2010) Fluctuations and redundancy in optimal transport networks. Phys Rev Lett 104:048703

Dunbar-Co S, Sporck MJ, Sack L (2009) Leaf trait diversification and design in seven rare taxa of the Hawaiian *Plantago* radiation. Int J Plant Sci 170:61–75

Ellis B, Daly DC, Hickey LJ, Mitchell JD, Johnson KR, Wilf P et al (2009) Manual of leaf architecture. Cornell University Press, Ithaca

Esau K (1977) Anatomy of seed plants, 2nd edn. Wiley, New York

Feild TS, Upchurch GR Jr, Chatelet DS, Brodribb TJ, Grubbs KC, Samain M-S et al (2011) Fossil evidence for low gas exchange capacities for Early Cretaceous angiosperm leaves. Paleobiology 37:195–213

Flexas J, Scoffoni C, Gago J, Sack L (2013) Leaf mesophyll conductance and leaf hydraulic conductance: an introduction to their measurement and coordination. J Exp Bot 64:3965–3981

Franks PJ (2006) Higher rates of leaf gas exchange are associated with higher leaf hydrodynamic pressure gradients. Plant Cell Environ 29:584–592

Griffiths H, Weller G, Toy L, Dennis RJ (2013) You're so vein: bundle sheath physiology, phylogeny and evolution in C_3 and C_4 plants. Plant Cell Environ 36(2):249–261

Haberlandt G (1914) Physiological plant anatomy, 4th edn. Macmillan, London

Hao GY, Hoffmann WA, Scholz FG, Bucci SJ, Meinzer FC, Franco AC et al (2008) Stem and leaf hydraulics of congeneric tree species from adjacent tropical savanna and forest ecosystems. Oecologia 155:405–415

Hu YS, Yao BJ (1981) Transfusion tissue in gymnosperm leaves. Bot J Linn Soc 83:263–272

John GP, Scoffoni C, Sack L (2013) Allometry of cells and tissues within leaves. Am J Bot 100:1936–1948

Johnson DM, Meinzer FC, Woodruff DR, McCulloh KA (2009) Leaf xylem embolism, detected acoustically and by cryo-SEM, corresponds to decreases in leaf hydraulic conductance in four evergreen species. Plant Cell Environ 32:828–836

Johnson DM, McCulloh KA, Meinzer FC, Woodruff DR, Eissenstat DM (2011) Hydraulic patterns and safety margins, from stem to stomata, in three eastern US tree species. Tree Physiol 31:659–668

Johnson DM, McCulloh KA, Woodruff DR, Meinzer FC (2012a) Hydraulic safety margins and embolism reversal in stems and leaves: why are conifers and angiosperms so different? Plant Sci 195:48–53
Johnson DM, McCulloh KA, Woodruff DR, Meinzer FC (2012b) Evidence for xylem embolism as a primary factor in dehydration-induced declines in leaf hydraulic conductance. Plant Cell Environ 35:760–769
Jordan GJ, Brodribb TJ, Blackman CJ, Weston PH (2013) Climate drives vein anatomy in Proteaceae. Am J Bot 100:1483–1493
Katifori E, Szollosi GJ, Magnasco MO (2010) Damage and fluctuations induce loops in optimal transport networks. Phys Rev Lett 104:048704
Kenzo T, Ichie T, Watanabe Y, Hiromi T (2007) Ecological distribution of homobaric and heterobaric leaves in tree species of Malaysian lowland tropical rainforest. Am J Bot 94:764–775
Kim YX, Steudle E (2007) Light and turgor affect the water permeability (aquaporins) of parenchyma cells in the midrib of leaves of *Zea mays*. J Exp Bot 58:4119–4129
Laur J, Hacke UG (2014a) Exploring *Picea glauca* aquaporins in the context of needle water uptake and xylem refilling. New Phytol 203:388–400
Laur J, Hacke UG (2014b) The role of water channel proteins in facilitating recovery of leaf hydraulic conductance from water stress in *Populus trichocarpa*. Plos One 9:e111751
Lee SH, Zwiazek JJ, Chung GC (2008) Light-induced transpiration alters cell water relations in figleaf gourd (*Cucurbita ficifolia*) seedlings exposed to low root temperatures. Physiol Plant 133:354–362
Lee SH, Chung GC, Zwiazek JJ (2009) Effects of irradiance on cell water relations in leaf bundle sheath cells of wild-type and transgenic tobacco (*Nicotiana tabacum*) plants overexpressing aquaporins. Plant Sci 176:248–255
Leigh A, Zwieniecki MA, Rockwell FE, Boyce CK, Nicotra AB, Holbrook NM (2011) Structural and hydraulic correlates of heterophylly in *Ginkgo biloba*. New Phytol 189:459–470
Lersten NR (1997) Occurrence of endodermis with a casparian strip in stem and leaf. Bot Rev 63:265–272
Maherali H, Sherrard ME, Clifford MH, Latta RG (2008) Leaf hydraulic conductivity and photosynthesis are genetically correlated in an annual grass. New Phytol 180:240–247
McKown AD, Cochard H, Sack L (2010) Decoding leaf hydraulics with a spatially explicit model: principles of venation architecture and implications for its evolution. Am Nat 175:447–460
Mertz RA, Brutnell TP (2014) Bundle sheath suberization in grass leaves: multiple barriers to characterization. J Exp Bot 65:3371–3380
Moshelion M, Halperin O, Wallach R, Oren R, Way DA (in press) Role of aquaporins in determining transpiration and photosynthesis in water-stressed plants: crop water-use efficiency, growth and yield. Plant Cell Environ. http://onlinelibrary.wiley.com/doi/10.1111/pce.12410/abstract
Nardini A, Tyree MT, Salleo S (2001) Xylem cavitation in the leaf of *Prunus laurocerasus* and its impact on leaf hydraulics. Plant Physiol 125:1700–1709
Nardini A, Gortan E, Salleo S (2005) Hydraulic efficiency of the leaf venation system in sun- and shade-adapted species. Funct Plant Biol 32:953–961
Nardini A, Ramani M, Gortan E, Salleo S (2008) Vein recovery from embolism occurs under negative pressure in leaves of sunflower (*Helianthus annuus*). Physiol Plant 133:755–764
Nardini A, Raimondo F, Lo Gullo MA, Salleo S (2010) Leafminers help us understand leaf hydraulic design. Plant Cell Environ 33:1091–1100
Nardini A, Peda G, La Rocca N (2012) Trade-offs between leaf hydraulic capacity and drought vulnerability: morpho-anatomical bases, carbon costs and ecological consequences. New Phytol 196:788–798
Nardini A, Ounapuu-Pikas E, Savi T (2014) When smaller is better: leaf hydraulic conductance and drought vulnerability correlate to leaf size and venation density across four *Coffea arabica* genotypes. Funct Plant Biol 41:972–982
Noblin X, Mahadevan L, Coomaraswamy IA, Weitz DA, Holbrook NM, Zwieniecki MA (2008) Optimal vein density in artificial and real leaves. Proc Natl Acad Sci U S A 105:9140–9144

O'Brien TP, Carr DJ (1970) A suberized layer in the cell walls of the bundle sheath of grasses. Aust J Biol Sci 23:275–287

Pantin F, Monnet F, Jannaud D, Costa JM, Renaud J, Muller B et al (2013) The dual effect of abscisic acid on stomata. New Phytol 197(1):65–72

Prado K, Maurel C (2013) Regulation of leaf hydraulics: from molecular to whole plant levels. Front Plant Sci 4:255

Prado K, Boursiac Y, Tournaire-Roux C, Monneuse J-M, Postaire O, Da Ines O et al (2013) Regulation of *Arabidopsis* leaf hydraulics involves light-dependent phosphorylation of aquaporins in veins. Plant Cell 25:1029–1039

Rockwell FE, Holbrook NM, Stroock AD (2014) The competition between liquid and vapor transport in transpiring leaves. Plant Physiol 164:1741–1758

Roth-Nebelsick A, Uhl D, Mosbrugger V, Kerp H (2001) Evolution and function of leaf venation architecture: a review. Ann Bot 87:553–566

Sack L, Frole K (2006) Leaf structural diversity is related to hydraulic capacity in tropical rain forest trees. Ecology 87:483–491

Sack L, Holbrook NM (2006) Leaf hydraulics. Annu Rev Plant Biol 57:361–381

Sack L, Scoffoni C (2013) Leaf venation: structure, function, development, evolution, ecology and applications in the past, present and future. New Phytol 198:983–1000

Sack L, Cowan PD, Jaikumar N, Holbrook NM (2003) The 'hydrology' of leaves: co-ordination of structure and function in temperate woody species. Plant Cell Environ 26:1343–1356

Sack L, Streeter CM, Holbrook NM (2004) Hydraulic analysis of water flow through leaves of sugar maple and red oak. Plant Physiol 134:1824–1833

Sack L, Tyree MT, Holbrook NM (2005) Leaf hydraulic architecture correlates with regeneration irradiance in tropical rainforest trees. New Phytol 167:403–413

Sack L, Dietrich EM, Streeter CM, Sanchez-Gomez D, Holbrook NM (2008) Leaf palmate venation and vascular redundancy confer tolerance of hydraulic disruption. Proc Natl Acad Sci U S A 105:1567–1572

Sack L, Scoffoni C, John GP, Poorter H, Mason CM, Mendez-Alonzo R et al (2013) How do leaf veins influence the worldwide leaf economic spectrum? Review and synthesis. J Exp Bot 64:4053–4080

Scoffoni C (2015) Modelling the outside-xylem hydraulic conductance: towards a new understanding of leaf water relations. Plant Cell Environ 38(1):4–6

Scoffoni C, Pou A, Aasamaa K, Sack L (2008) The rapid light response of leaf hydraulic conductance: new evidence from two experimental methods. Plant Cell Environ 31:1803–1812

Scoffoni C, Rawls M, McKown A, Cochard H, Sack L (2011) Decline of leaf hydraulic conductance with dehydration: relationship to leaf size and venation architecture. Plant Physiol 156:832–843

Scoffoni C, Vuong C, Diep S, Cochard H, Sack L (2014) Leaf shrinkage with dehydration: coordination with hydraulic vulnerability and drought tolerance. Plant Physiol 164:1772–1788

Secchi F, Zwieniecki MA (2013) The physiological response of *Populus tremula* × *alba* leaves to the down-regulation of *PIP1* aquaporin gene expression under no water stress. Front Plant Sci 4:507

Shatil-Cohen A, Attia Z, Moshelion M (2011) Bundle-sheath cell regulation of xylem-mesophyll water transport via aquaporins under drought stress: a target of xylem-borne ABA? Plant J 67:72–80

Sheriff DW (1977) Evaporation sites and distillation in leaves. Ann Bot 41:1081–1082

Sheriff DW, Meidner H (1974) Water pathways in leaves of Hedera helix L. and Tradescantia virginiana L. J Exp Bot 25:1147–1156

Slewinski TL, Anderson AA, Zhang C, Turgeon R (2012) Scarecrow plays a role in establishing Kranz anatomy in maize leaves. Plant Cell Physiol 53:2030–2037

Sommerville KE, Sack L, Ball MC (2012) Hydraulic conductance of *Acacia* phyllodes (foliage) is driven by primary nerve (vein) conductance and density. Plant Cell Environ 35:158–168

Takeda H (1913) A theory of 'transfusion-tissue'. Ann Bot 27:361–363

Taneda H, Terashima I (2012) Co-ordinated development of the leaf midrib xylem with the lamina in *Nicotiana tabacum*. Ann Bot 110:35–45

Tomlinson PB, Fisher JB (2005) Development of nonlignified fibers in leaves of *Gnetum gnemon* (Gnetales). Am J Bot 92:383–389

Tyree MT, Zimmermann MH (2002) Xylem structure and the ascent of sap. Springer, Berlin

Walls RL (2011) Angiosperm leaf vein patterns are linked to leaf functions in a global scale data set. Am J Bot 98:244–253

Woodruff DR, Meinzer FC, Lachenbruch B (2008) Height-related trends in leaf xylem anatomy and shoot hydraulic characteristics in a tall conifer: safety versus efficiency in water transport. New Phytol 180:90–99

Wylie RB (1939) Relations between tissue organization and vein distribution in dicotyledon leaves. Am J Bot 26:219–225

Wylie RB (1943) The role of the epidermis in foliar organization and its relations to the minor venation. Am J Bot 30:273–280

Wylie RB (1946) Relations between tissue organization and vascularization in leaves of certain tropical and subtropical dicotyledons. Am J Bot 33:721–726

Wylie RB (1952) The bundle sheath extension in leaves of dicotyledons. Am J Bot 39:645–651

Ye Q, Holbrook NM, Zwieniecki MA (2008) Cell-to-cell pathway dominates xylem-epidermis hydraulic connection in *Tradescantia fluminensis* (Vell. Conc.) leaves. Planta 227:1311–1319

Zhang S-B, Guan Z-J, Sun M, Zhang J-J, Cao K-F, Hu H (2012) Evolutionary association of stomatal traits with leaf vein density in *Paphiopedilum*, Orchidaceae. Plos One 7(6):e40080

Zhang Y-J, Rockwell FE, Wheeler JK, Holbrook NM (2014) Reversible deformation of transfusion tracheids in *Taxus baccata* Is associated with a reversible decrease in leaf hydraulic conductance. Plant Physiol 165:1557–1565

Zsögön A, Alves Negrini AC, Peres LE, Nguyen HT, Ball MC (2015) A mutation that eliminates bundle sheath extensions reduces leaf hydraulic conductance, stomatal conductance and assimilation rates in tomato (*Solanum lycopersicum*). New Phytol 205(2):618–626

Zwieniecki MA, Boyce CK (2014) Evolution of a unique anatomical precision in angiosperm leaf venation lifts constraints on vascular plant ecology. Proc R Soc B Biol Sci 281

Zwieniecki MA, Melcher PJ, Boyce CK, Sack L, Holbrook NM (2002) Hydraulic architecture of leaf venation in *Laurus nobilis* L. Plant Cell Environ 25:1445–1450

Zwieniecki MA, Brodribb TJ, Holbrook NM (2007) Hydraulic design of leaves: insights from rehydration kinetics. Plant Cell Environ 30:910–921

Index

U. Hacke (ed.), *Functional and Ecological Xylem Anatomy*,
DOI 10.1007/978-3-319-15783-2

MIX
Papier aus verantwortungsvollen Quellen
Paper from responsible sources
FSC® C105338

If you have any concerns about our products,
you can contact us on
ProductSafety@springernature.com

In case Publisher is established outside the EU,
the EU authorized representative is:
Springer Nature Customer Service Center GmbH
Europaplatz 3, 69115 Heidelberg, Germany

Printed by Libri Plureos GmbH
in Hamburg, Germany